PRINCIPLES OF FOOD PACKAGING
SECOND EDITION

PRINCIPLES OF FOOD PACKAGING

SECOND EDITION

2nd ed.

Stanley Sacharow, B.A., M.A.
Professional Member,
The Packaging Institute
M. Institute of Packaging (U. K.)

Roger C. Griffin, Jr., B.S., M.S.
Professional Member,
Institute of Food Technologists

avi

AVI PUBLISHING COMPANY
Westport, Connecticut

Frontispiece: Market Scene (1603) by Jacob Matham. *Courtesy of* the Metropolitan Museum of Art, the Elisha Wittelsey Collection, the Elisha Wittelsey Fund, 1964.

ISBN-0-87055-347-X

Printed in the United States of America

Preface to the Second Edition

In the more than 10 years since the first edition of "Food Packaging" was published, there have been many significant changes in the industry. Packages have come and gone, costs have skyrocketed, ecology has become a household term, material shortages are an ever present reality, and energy; everybody is talking about energy!

The present edition reflects the state of food packaging in 1980. Although the food packaging industry is different than it was in 1970, it still remains a field of study that learns from its past and applies this knowledge to its future. The composite can, which is used to package frozen orange juice and fabricated potato chips, was used during World War II to package ammunition items. The tear-strip was used for ease of opening packages of cannon fuses in the early battles of the American Civil War.

We have tried in this edition to keep to the objective of the original "to serve as a guide for new or inexperienced workers" . . . "whose duties involve them in the creation, marketing or use of food packages." We have also tried to incorporate the many helpful suggestions sent to us over the last decade.

Our thanks to our wives and children for the encouragement needed to revise this pioneering text and also to Dr. D.K. Tressler, Dr. N.W. Desrosier and Ms. Lisa E. Melilli of the AVI Publishing Company for their assistance in bringing this book into being.

The authors would also like to thank Scott Hunter Sacharow for preparing the index. His diligence and hard work are reflected in the many hours of study needed for this task. We appreciate his efforts.

STANLEY SACHAROW
ROGER C. GRIFFIN, JR.

July, 1980

Preface from First Edition

Ever since man appeared on this Earth, he has fought for survival. One of his greatest battles has been against famine. Over the centuries food packaging has played an ever increasing role in the war against starvation. The final crucial battle was won with the 19th century discoveries in the fields of food processing and microbiology, and the 20th century discoveries in the science of agriculture.

Twentieth century food packaging is less concerned with man's survival and more dedicated, rather, to his comfort and convenience. Today's affluent society and modern technology have combined to bring a myriad of new packages into the marketplace.

During the last 50 years packaging has evolved from an art to a science with all the agonies that ensue in such a transition. During the past several decades the food industry has recognized the packaging function as an important facet of their operations and has created packaging departments staffed with professional packaging scientists and engineers. Colleges and universities have added packaging courses to their curricula and at least two now grant graduate degrees in packaging science.

The pioneer workers in this field were so busy creating that few took time to record their accomplishments. There is no recorded comprehensive history of the development of packaging machinery. The reports on food packaging developments are found scattered in many diverse journals. No text exists which covers the whole field of commercial food packaging in such a way as to serve as a guide for new or inexperienced workers, whether in industry, government, or the academic world, whose duties involve them in the creation, marketing, or use of food packages. It is hoped that this book will serve this purpose. Like all first efforts it is bound to have imperfections. The authors hope each reader will submit criticisms or additional material that can be used to improve subsequent editions and make them more useful to all.

Appreciation is here expressed to all who have helped us complete this book. We expressly acknowledge our gratitude to Reynolds Metals Company whose pioneering efforts in the creation of food packages have resulted in an organization that is both knowledgeable and dedicated; to A.I. Totten, Jr., General Director of Packaging Research, and his staff who have encouraged our efforts; and to our Librarian, Mrs. Marie Davis Walker, who has unflaggingly tracked down elusive references; and to many who have contributed information, photographs, reference material, or just plain enthusiasm.

STANLEY SACHAROW
ROGER C. GRIFFIN, JR.

Richmond, Virginia
September 1, 1969

Foreword

This book is aimed at the person who wants an overall view of the subject which is "basic" in the sense of general knowledge. Through nontechnical language it presents the various facets of the food industry and their packaging requirements. The book is a highly lucid text of general interest. The marketing executive, economist or technician will find a wealth of useful knowledge of the elements and practices of food packaging.

SEYMOUR G. GILBERT, Ph.D.
Professor, Food Science
Rutgers University
New Brunswick, N.J.

Contents

"Our forebears feared that ptomaine poisoning lurked in every tin . . . food lay about in bulk at shop doors and was visited by visitors who were no customers. The grocer displayed it on an open shelf at the mercy of dust and flies, handled and fingered it lovingly, and gave it to his customer in a paper bag which often stuck and had to be separated by atmospheric pressure supplied by his lungs."

"Turn-of-the-Century Food Packaging"
(Anon.)

1

Food Packaging—Its Background

> One of the most fascinating aspects of packaging history is that its pace tends to reflect our life style, priorities and concerns.
>
> *Walter Stern*

The foods we eat have come to us from many lands and many cultures. A great proportion of them predate written history. Food packaging is similarly entwined with earlier civilizations and its earliest beginnings are rooted in man's advancements in the production and processing of foods.

In Paleolithic times, food was consumed where found, and when needed, man used natural containers such as hollow tree trunks, gourds, hollow rocks, shells, leaves and pieces of bark. In later times he learned to fashion better containers from these natural materials. He deliberately hollowed out logs or stones and utilized animal parts such as bladders, skins, horns, bones, sinews and hair. Earliest fabrics evolved from furs. First, fibers were matted into felts; later they were plaited and woven. Weaving of thin vines or wooden withes into shelters led to the making of baskets out of rushes and grasses. Mesolithic man stored food surpluses in this type of basket. Neolithic man fashioned metal containers and discovered pottery. Some of the earliest metal "cups" were drinking horns shaped like natural horns. Even the ancients who carefully wrapped Yuya (King Tutankhamun's maternal grandfather) around 1360 B.C. had specific "packaging" requirements. When this mummy was discovered in the Valley of the Kings in 1905, it was in excellent condition.

The founding of the art of pottery making also provided an infinite number of clay vessels of many shapes and sizes: from tiny amphoras to huge urns. Over 4000 years ago, the people of Moenjo-Daro (one of the world's first cities which was located in the area which is now known as Pakistan) sealed pottery jars to keep moisture and pests from their wheat and barley. Herodotus recorded that in 530 B.C. the Persians supplied conquered Egyptian cities with water and wine in earthenware jars. These were collected later and returned for reuse, an early form of recycling. Glass was known in Sumerian times; the small glass jar or bottle was known in 1500 B.C. Hollowed alabaster was used for small quantities of precious liquids or ointments. In 79 A.D. the Romans were using glass bottles as well as ceramic pots and jars, but they preferred leather bags for large quantities of liquids and solids. The leather bag wineskin was referred to by Jesus of Nazareth (Matthew 9:17) : "Neither do men put new wine into old wineskins; else the skins burst." In Pompeii, the fortunate seaport location allowed the lucrative exportation of its fine wine, fruit and garum, a fermented fish sauce basic to Roman cooking. To store and ship these commodities, large clay amphoras (two-handled jugs) were produced. Many of these remained intact through the ages and can be seen in various museums.

The barrel was invented by Alpine tribes, according to Pliny. Its carefully fitted staves and heads were fashioned from wood, interlocked by mortises and the whole bound together with iron hoops. Wooden chests were made in similar fashion from fitted and pegged boards because nails were handmade and too expensive for such use.

Methods of physical distribution were apparently well advanced in China during the Ming dynasty (1368–1644). In the 15th century, China had a well-established trade in porcelain in south and west Asia and Egypt. Ming historical records relate that merchants from Tartary and Arabia, when leaving Peking for home, would take back with them several cartloads of porcelain, piled up several meters high.

In order to lessen product damage over the vast transport distance, the merchants thought of an interesting method for packing these products. A small amount of sand would be placed in each piece to be shipped, along with some bean and wheat seeds. Scores of the wares were tied together and stored in a moist atmosphere with frequent watering. The wheat and bean soon sprouted and the package was sturdily entwined with vines. The packages were then thrown to the ground as a quality control measure to ensure that only the most sturdily packed were transported. Although shipment entailed an overland journey of 10,000 miles, the merchants could still make a large profit.

From prehistoric times until about 1200 A.D. the status of packaging could be summed up as follows:

Material	*Package Form and Use*
Leather	Wrappings, bags, bottles
Cloth	Wrappings, sacks
Wood	Barrels, boxes, kegs, chests

Material	*Package Form and Use*
Grass, or split wood[1]	Baskets, matting
Stone	Small pots or jars
Earthenware	Pots, jars, urns, ewers, bowls, etc.
Metal	Pots, bowls, cups, etc.
Glass	Jars, bottles, cups, bowls, etc.

ORIGINS OF PACKAGING MATERIALS

Paper

Paper and papyrus were developed originally as writing materials to replace parchment (animal skin) and vellum (skins of newborn calves, kids or lambs). Papyrus was formed from flattened strips of the center pith of the papyrus reed. It was used by the ancient Egyptians and Greeks as early as 500 B.C. The earliest paper was made in China by T'Sai Lun from mulberry bark about 105 A.D. Papers were then made from flax fibers and later from old linen rags. The Arabs learned the art when a Chinese army attacked Samarkand in 751 A.D. and some papermakers were captured. Although some say the Crusaders brought paper home with them from the Holy Land, which is not unlikely, it is certain that the Arabs introduced paper to Sicily and thence to Italy and southern Germany. The Moors introduced paper to Spain in the 12th century, from whence it spread through France, western Germany, the Netherlands, Belgium and England. Earliest records of English papermaking date to about 1310 A.D. The first papermaking in America was in 1690 in Germantown, Pennsylvania. Cardboard was also a Chinese invention of about the 16th century. The first papermaking machinery, in which fibers were laid down on a moving wire cloth, began production in 1799 in England. The use of wood pulp in papermaking was introduced in 1867. Corrugated paper was invented in the mid 1800s and shipping cartons made from faced corrugated paperboard began to replace wooden crates and boxes about the turn of the century. Ovenable paperboard, a child of the 1970s, now appears to be a serious challenger to other more conventional ovenable materials. Known since the late 1960s, the surging market in 1978 for microwave ovens has rekindled interest in it.

Glass

Glass is believed to be an offshoot of pottery. First usages were as glass glazes on beads or amulets. Green glazed stone beads dating to 12,000

[1]In the Canyon de Chelly, Arizona about 100 A.D., the basket was created. Archaeologists called these people "basket makers." The early baskets were used to store corn and squash and to prevent waste. Being tightly woven, they were used to hold water as well. They were also used as stew pots when hot stones were dropped into them.

Courtesy of Reynolds Metals Co.

FIG. 1.1. REPRESENTATION OF ANCIENT EGYPTIANS WRAPPING POULTRY IN PALM LEAVES FOR PROTECTION AGAINST CONTAMINATION

B.C. and a blue pure glass amulet dating to 7000 B.C. were found in Egypt. By 1550 B.C. glassmaking was an important industry in Egypt, and various small bottles and ornaments were produced in many colors—but none were clear. The art of pressing glass in molds to produce bowls and cups dates to 1200 B.C. The Phoenecians were believed to have invented the blowpipe about 300 B.C., which vastly speeded up production; nevertheless even in later Roman times (400 A.D.) fine glassware was more valuable than silver or gold. Clear transparent glass was discovered about the beginning of the Christian Era and in the 3rd century Roman glassmakers were casting window glass on flat stones. During the following 1000 years glassmaking spread over Europe and the art of making stained glass windows for cathedrals was developed. Up until the 18th century, Venetian and other Italian glassmakers were considered the world leaders. By the 18th century Bohemian and English glass began to challenge this reputation. Cut glass was invented in 1600 by a court jeweler in Prague (Bohemia). Still, glass remained expensive until improved techniques of the 18th and 19th century brought the price of bottles to a more reasonable level. In this era, the demand for bottles that packaged whiskey and other potables was always insistant and insatiable. Without it, the glass industry might not have developed as rapidly as it did in the United States following the War of 1812.

Courtesy of Walter Landor and Associates

FIG. 1.2. CERAMIC MARMALADE JAR ILLUSTRATING EARLY METHOD OF PRINTING
This package is still used for this type of product.

Tinplate

Sometime around the year 1200 A.D. the artisans of Bohemia discovered a hot dip process for plating tin onto thin sheets of iron which had been hammered out by hand from rods. This secret was guarded zealously for 400 years, but in 1620 the Duke of Saxony managed to steal it. The Iron Age brought new metals and machinery for continuous rolling and plating. Steel has replaced iron and stronger alloys have made it possible to use thinner gauges of metal. Changes in processing techniques have led to thinner and thinner coatings of tin. Developments of the past decade have produced tinless cans, steel foil and aluminum-clad steel alloys.

Courtesy of Walter Landor and Associates

FIG. 1.3. LITHOGRAPHED TINPLATE CANISTERS
These were very popular for dry products.

Tin, Lead and Other Metals

In ancient times boxes and cups were made from silver and gold but these were too valuable for common usage. We know the Romans used lead in many ways, including for water pipes. The only known evidence in packaging is their application of lead seals to ointment jars. It is likely they found ways to hammer the soft metal into thin foils. Not knowing its poisonous nature they could have wrapped foods in it. Leadfoil was used to line tea boxes in 1826 and in cigarette packaging as late as the 1930s.

Tinfoil is an all embracing term. Most tin alloys were made from tin and either lead, antimony, zinc or copper. Pure tin is difficult to roll. While tin was used for wrapping chocolates in the 1840s, cheese wrap foils used as late as the 1930s were made of tin plus antimony. Tinlead and tin-clad lead foils were used in packaging also.

Aluminum

In 1825 Oersted produced the first aluminum particles. The metal was so difficult to extract from its ore that as late as 1852 it cost $545 per lb. French Emperor Napoleon III used aluminum forks and spoons for special guests, others had to dine off silver or gold. Early Victorian jewelry was often made of aluminum. These are now highly prized collectible items. In 1854, Deville and Bunsen developed more efficient processes, which by 1885 had brought the price down to $11.33 per lb, but this was still too high. In 1886, Heroult and Hall discovered the modern electrolytic process for extracting aluminum from alumina. In 1888, Bayer found a less expensive way to extract alumina from bauxite ores. By 1892 the price had dropped to $0.57 per lb and it kept dropping to a low of $0.14 in 1942.

As soon as the price became low enough, aluminum found its way into many uses. Its properties lent well to easy fabrication. The first commercial aluminum foil was rolled sometime after 1910, and during the early 1920s techniques for rolling and printing were perfected. Continuous casting techniques for foil production appeared in the late 1970s.

Stoneware

In the taverns and kitchens of early America, stoneware was the common container and dispenser. Before the advent of vacuum canning and refrigeration in the 19th century, distinctive gray or brown stoneware, often incised with blue cobalt-oxide or purple manganese decoration, was used to store everything from butter and pickled oysters to cider and Holland gin.

Tradesmen used stoneware jars to hold turpentine, acids and other industrial liquids. Merchants sometimes used stoneware jars inscribed with advertisements for their stock. A public house without its initialed stoneware inkwells and spittoons was a questionable establishment in which to seek shelter.

TABLE 1.1. SOME GREAT HIGHLIGHTS IN PACKAGING[1]

Invention	Date	Where Used
Moistureproof cellophane	1929	E. I. DuPont & Co.—many markets
Flexible vacuum package	1947	Bacon packed by Armour
Form/fill/seal machine	1932	Henry Heide, Inc.—candies
Blister packages	1952	Shaw-Randall Co.—notions
Multipacks	1922	Beverages
Composite can	1961	Sun Oil—motor oil
Aerosol	1948	Low pressure type developed based on Army work
One-way bottle	1940s	Owens-Illinois "Stubby"
Beer can	1935	American Can—G. Kruege Co.
Squeeze bottle	1947	"Stopette"—deodorant
Twist-off closures	1964	Pittsburgh Brewing Co.—"Iron City" beer

[1] This is not an all-inclusive list. Some additional highlights include the pop-top beer can, the "stay-on" beer can tab and various safety closures for child-resistance.

Stoneware first manufactured in 15th century Germany was popular because its salt glazes made for harder and safer containers than porous earthenware vessels with lead-based glazes. New Jersey became the center of the 18th and 19th century trade in this country, which eventually spread south to Philadelphia and north to Boston. Industrialization, graniteware, Mason jars and refrigeration eventually led to a decline in handmade household stoneware.

Plastics

A search for a substitute for ivory billiard balls led to an investigation of cellulose nitrate, which itself was first described in 1845. In 1862, molded articles made from "parkesine" were displayed at the Great Exposition in London. In 1870 Hyatt patented "celluloid," a mixture of cellulose nitrate and camphor. This was the first commercially-successful, manmade plastic and also one of the first examples of a solid plasticizer. Cellulose nitrate was also the first commercially successful plastics film.

Celluloid and other cellulose nitrate products were so flammable that investigators sought other less flammable materials. Cellulose acetate dissolved in acetone was widely used in cements and lacquers. The latter were used to "dope" the fabrics on World War I airplane wings.

The viscose or cellulose xanthate process for making regenerated cellulose fibers and films was developed in 1892. The first commercial applications of viscose films were in Europe between 1912 and 1914. Cellophane film was introduced in the United States in 1924. The addition of coatings of cellulose nitrate added heat-sealability and better resistance to moisture. Up until 1950, cellophane reigned supreme among packaging films.

Casein plastics were introduced in 1899. Casein-formaldehyde buttons and buckles were used extensively during the 1920s. Casein plastics are still used today in the familiar casein-latex emulsion adhesives. Other plastics which were made from polymerization with formaldehyde in the latter part of the 19th century were phenol-, urea- and melamine-formaldehyde. During the 1930s and 1940s, melamine-formaldehyde resins were introduced as wet-strength additives for papers.

Styrene was first distilled from a tree balsam in 1831 and its spontaneous polymerization was noted. In 1866 it was produced from coaltar benzene. The first commercially successful polystyrene was introduced in 1930. Molded polystyrene articles were widely used but found to be brittle. The hazard from shattered toys or food containers led in the 1950s to the blending of polystyrene with synthetic rubbers to gain impact resistance. Polystyrene films and molded polystyrene containers achieved commercial acceptance thereafter. The final evolution occurred when styrene, acrylonitrile and butadiene were copolymerized to produce the ABS plastics. Also in 1950 expanded polystyrene foam was introduced and soon found a market as an insulation and cushioning material. Consumers and store decorators used expanded polystyrene foam for Christmas decorations.

Studies on natural rubber led to the discoveries of rubber coatings and rubber cements. In 1934, rubber hydrochloride films were introduced. Even before this, scientists isolated the monomer of natural rubber and isoprene, and then polymerized it. In 1915, a synthetic methyl rubber (2,3 methyl butadiene) was produced in the laboratory and was followed by a series of conjugated diene rubbers. In 1931 neoprene was discovered. In 1933 butadiene-styrene rubber was introduced—the familiar Buna S of World War II. Its cousin Buna N, butadiene acrylonitrile rubber, was also a war-baby. By the 1950s rubber chemistry had made such great strides that a whole spectrum of elastomers was available. Elastomers have been used widely as adhesive components in packaging.

Vinyl chloride (VC) monomer was first prepared in 1835. The first polyvinyl chloride (PVC) was reported in 1872. By 1912 commercial processes were available for making cable insulation. Polyvinyl alcohol was prepared in 1924 from hydrolyzed polyvinyl acetate. The polyvinyl acetates were found to be too soft and the chlorides too hard, so in 1928 a copolymer of vinyl acetate and vinyl chloride was prepared. The copolymers proved to be very useful. With the addition of plasticizers a series of coated and calendered fabrics were produced. The first shower curtains and raincoats of the early 1940s had strong odors from their migratory plasticizers; however, as technology improved, better polymers and better plasticizers were found. In 1958, heat shrinkable PVC film was introduced commercially—the first strong entry of vinyls into the packaging field. VC-acrylic copolymers have made excellent rigid-molded articles. Vinyl chloride and vinyl ether copolymers have found use in lacquers.

Polyvinylidene chloride (PVDC) was discovered in 1838. The pure monomer was synthesized in 1872 and homopolymers were made in 1916. They were found to have very narrow melting ranges and to be very hard and brittle. In 1936, a copolymer of VC and VDC was prepared, the first of the "Sarans." Emulsion polymerization was developed between 1940 and 1947.

The first polyester was prepared in 1847. Acrylate esters were made in 1873 and methyl acrylate was polymerized in 1880. The first commercial production of polyacrylates was in 1927; of methyl methacrylate in 1933, just in time for service as cockpit covers and gun turrets on World War II aircraft. Alkyd polyesters were discovered in 1901 and eventually found extensive use in paints, lacquers and coatings. Polyethylene-glycol-terephthalate was first prepared in 1941 and marketed in the 1950s as "Mylar" film. Other polyester films have since become available.

One of the most important plastics developments as far as packaging is concerned occurred in 1933 when Fawcett and Gibson of Imperial Chemical Industry, Ltd. (ICI) ran some experiments on ethylene and found a white, waxy solid coated on the wall of the pressurized vessel. This was the first high-pressure reacted polyethylene. The first commercial use of this material was as a film wrap for submarine telephone cable and during World War II as a radar cable wrap. In the early 1940s, Karl Ziegler began research which led to a catalytic method for producing polyethylenes by low pressure reactions. The Ziegler (or Philips) process produced higher density polymers than were achieved by the ICI process. These high density polyethylenes were commercialized in 1956 and very rapidly achieved high volume sales.

In 1961, Dr. Richard W. Rees of DuPont invented "ionomers" by chemically linking metal ions to ethylene polymers containing acid groups. His development, the "surlyn" family of low density polyethylene resins, opened up new areas and markets to flexible packaging converters and users.

In 1954, Natta in Italy used a stereospecific catalyst to produce polypropylene and firmly established the principle that polymer chains can be tailored to fit a preconceived pattern. In recent years, a whole family of copolymers and homopolymers of ethylene, propylene and butylene have been streaming into the marketplace.

Other plastics of interest in packaging include the polyamides (nylons), which were first developed in 1937 as fibers and graduated to packaging films in the late 1950s and early 1960s, and the polycarbonates marketed in 1959.

As time goes on, more and more new plastics are being commercialized. Current trends are not only adding new families such as the ionomers and polyamides, and new family members such as nylon 13 and polybutene, but also second generation blends and combinations, such as PVDC-coated polyester, polyethylene-coated polyamide, rubber modified polystyrene, ethylene-vinyl-acetate-copolymers and the new coextruded films, such as polyethylene-polypropylene-polyethylene. Possibilities appear to be unlimited.

PACKAGE FORMS

The barrel, wooden box, ceramic jar or pot, leather bag and cloth sack are as old as civilization. Some of the other important milestones in the historical development of package forms are the following.

Metal Boxes

With the invention of tinplate in 1200 A.D., the fabrication of soldered metal boxes was made possible. There are, however, no records of such containers until about 1764 when tobacconists in London began selling snuff in metal canisters. In the early 1830s, both matches and cookies were sold in tin boxes. Some boxes were designed with embossed lettering or patterns, others carried printed paper labels. Between 1850 and 1900, processes were developed for printing the metal itself (see discussion of Graphic Arts later in this chapter). The first printed metal boxes in the United States were made in 1866 for cakes of Dr. Lyon's tooth powder. Characterized by excellent print quality, turn-of-the-century tinplate cans are now highly collectible. Many early cans have designs with eight or nine colors in contrast to today's four or five colors. Today a can design with more than five colors is usually prohibitive by cost. Outstanding examples of early tinplate cans with excellent print and design include Huntley and Palmer's range of biscuit cans, the "Log Cabin Maple Syrup" containers and the Art Nouveau British designs featuring Edward VII and his six namesake predecessors.

Metal Cans

The cylindrical tinplate can was devised by Peter Durand in 1810. First cans were soldered by hand leaving a 1½ in. diameter hole in the top.

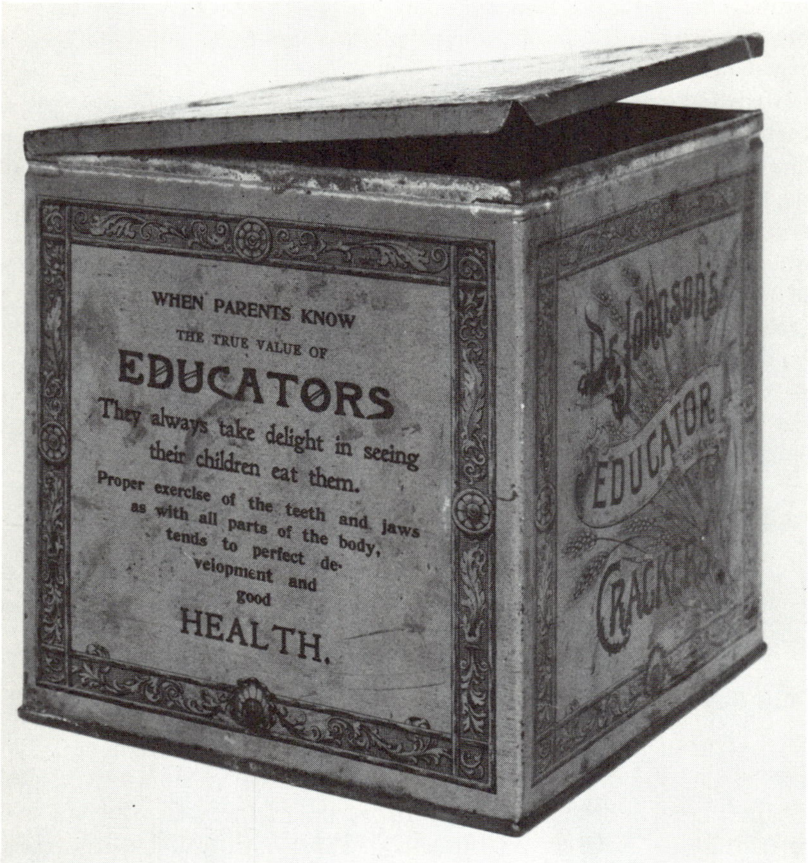

Courtesy of Walter Landor and Associates

FIG. 1.4. EARLY CRACKER BOX
This is an excellent example of an early cracker box made from soldered tinplate.

After the food was forced through the hole (not without damage), the hole was closed by soldering a patch over it. Sometimes a small hole was punched to let air escape during cooking and then this was closed with a drop of solder. Canmakers could turn out 60 cans a day when making them by hand. Various tools and jigs were devised to make the fabrication job easier. In 1868, enamels were put on can interiors to halt corrosion or discoloration of the food. The Max Ams double-seamed method of closure using sealing compound was introduced in 1888, and by 1900 machinery was available to make 2500 cans in 1 hr. With such production rates the soldered can was soon replaced, but the hole and cap can did not disappear until 1922. Some soldered-end cans are still used today (for condensed milk). In 1957 aluminum cans were introduced for lubricating oil and by 1963 were being used for beer. In the past few years cans have been introduced with side seams that are heat-sealed using a hot melt as the sealant.

The first cans were opened by using a hammer and chisel. The key-wind metal tear-strip can was invented in 1866 and was carried to the extreme in the sardine can where nearly the whole top panel could be torn out by the key. The can opener based on the lever principle was first introduced in 1875. All manner of variations on the lever principle including key wind, crank wind and motor driven rotary can openers have been developed over the past 75 years. The pop-top or tear-tab can lid of the 1950s depended upon aluminum for its success. Other vacuum-packed cans are now being fabricated with tear tapes made from plastic or laminated foil and plastic, which provide both the seal and a means for easy opening.

Aerosol cans were invented in 1889 but were first successfully marketed as "bug bombs" in World War II. Today, they are used for a wide variety of products, including foods of suitable consistency. Although the "ozone layer" controversy has dimmed the future of aerosols, new valves and propellants appear to be overcoming public reaction.

Bottles and Jars

Glass bottles and jars date back to antiquity. Glass jars were first fashioned from ropes of molten glass coiled into cylindrical shapes and fused together. The invention of the blowpipe made the manufacture of round containers and bottles much easier. The invention of the split mold permitted irregular shapes and raised decorations to be made. Most bottles of the 17th and 18th centuries utilized these features to identify the bottlemaker or the product contained. The first fully automatic rotary bottlemaking machine was patented in 1889. A modern Owens machine can produce 20,000 bottles in 1 day.

In Victorian and Edwardian England, specially designed "poison bottles" were used by pharmacists. These were made so that anyone handling such a container, either in the dark or without care, would immediately find it strange to the feel of the hand. An added feature was the use of the earliest known safety closure. Bottles were used with wired-in stoppers, with wires attached to a bell alarm system and even with a bell inside the bottle. Dozens of styles were made ranging from wedge-shaped to a coffin-shaped bottle with star-shaped embossing.

Closures have come a long way from whittled wooden plugs and hand-carved corks which were used as far back as 1000 B.C., according to

TABLE 1.2. SOME COMMON CAN SIZES[1]

Can Name	Can Dimensions (millimeters)	Approx. Net Wt.[3] (grams)	Net Contents Liquid Product[4] (milliliters)	Some Products for Which Ordinarily Used
2z Mushroom	508 × 572	92.1	96.1	Mushrooms
5z Baby food	540 × 730	134.6	125.7	Baby foods, chocolate syrup
6z	540 × 889	170.1	155.3	Tomato paste, tomato sauce, juices
6½z	540 × 984	170.1	177.4	Frozen concentrates, juices
Evaporated milk	603 × 635	170.1	177.4	Evaporated milk
Meat spread	635 × 397	85.0	81.3	Meat spreads
	635 × 635	226.8	140.5	Meat products
4z Pimiento	683 × 508	134.6	118.3	Pimientos, chopped olives
	683 × 635	198.4	162.7	Cranberry sauce
4z Mushroom	683 × 699	191.3	192.2	Mushrooms
8z Short	683 × 762	219.7	207.0	Baked beans, tomato sauce, shrimp, specialties
8z Tall	683 × 826	240.9	229.2	Vegetables and fruits, juices, specialties
No. 1 Picnic	683 × 1016	297.6	280.9	Vegetables, some fruit juices, soups, meat, fish, specialties
Beer	683 × 1222	—	354.9	Beer and carbonated beverages
Half-quart beer	683 × 1588	—	473.2	Beer
16z Domed	683 × 1588	—	236.6	Whipped cream, toppings
No. 211 cylinder	683 × 1238	368.5	354.9	Fruit juices, nectars, tomato juice
Evaporated milk	746 × 1000	—	428.8	Evaporated milk
4z Flat pimiento	762 × 381	113.4	110.9	Pimientos
	762 × 889	326.0	332.7	Pork and beans
Chocolate syrup	762 × 1008	453.5	384.4	Chocolate syrup
	762 × 1048	411.0	384.4	Infant formulas
No. 300	762 × 1127	411.0	399.2	Vegetables, some fruits, juices, soups, meat, fish, pet foods, specialties
	762 × 1159	453.5	414.0	Meat products
No. 300 cylinder	762 × 1413	538.6	502.7	Soups, pork and beans, specialties
¼ lb Flat	778 × 349	106.3	103.5	Salmon
No. 1 Tall	778 × 1191	453.5	443.6	Fruits, some vegetables, juices, fish, specialties
No. 303	810 × 1111	453.5	443.6	Most commonly used size for vegetables, fruits, juices, soups, specialties
No. ½	873 × 460	198.4	170.0	Tuna
½ lb Flat	873 × 520	219.7	192.2	Salmon
No. 1 Flat	873 × 556	255.1	236.6	Pineapple
	873 × 635	—	266.2	Sausage, fish flakes, coffee

TABLE 1.2. *(Continued)*

Can Name	Can Dimensions (millimeters)	Approx. Net Wt.[3] (grams)	Net Contents Liquid Product[4] (milliliters)	Some Products for Which Ordinarily Used
No. 2 Squat[5]	873 × 794	226.8	—	Nuts
12z Vacuum	873 × 857	340.1	384.4	Vacuum packed corn
No. 95	873 × 1016	481.9	473.2	Breads, sea foods
No. 2	873 × 1159	566.9	532.3	Vegetables, fruits, juices, soups, specialties
Jumbo	873 × 1429	708.6	680.2	Pork and beans, mushrooms
32z (Quart)	873 × 1937	963.7	946.3	Fruit juices and drinks
No. 1¼	1032 × 627	411.0	369.7	Pineapple
Shortening[5]	1032 × 881	453.5	—	Shortening
No. 2½	1032 × 1191	822.0	768.9	Fruits, some vegetables and juices, meat products
	1032 × 1413	992.1	946.3	Frozen concentrates
	1032 × 1556	1077.1	1086.8	Spaghetti, beans in tomato sauce
	1080 × 905	680.3	650.6	Meat products
No. 3 Cylinder	1080 × 1778	1417.2	1360.4	Fruit and vegetable juices
Vacuum Coffee[5]	1302 × 889	453.5	—	Coffee
Vacuum Coffee[5]	1302 × 1635	907.0	—	Coffee
Shortening[5]	1302 × 1492	1360.6	—	Shortening
No. 10	1572 × 1778	3004.6	2839.0	Institutional size for vegetables, fruits, juices, meat and fish products, soups, specialties
	1572 × 2222	3741.5	3785.3	Soft drink syrups
12z Oblong[5]	984 × 540 × 810	340.1	—	Meat products
Pullman Base[5]	1048 × 921 × 1651	1360.6	—	Meat products
Pullman Base[5]	1048 × 921 × 3175	2721.0	—	Meat products
Pullman Base[5]	1238 × 1175 × 2794	3628.0	—	Meat products
Miniature Base Ham[5]	1460 × 1016 × 683	680.3	—	Ham, pear-shaped
#1 Base Ham[5]	1937 × 1365 × 762	1360.6	—	Ham, pear-shaped
#1 Base Ham[5]	1937 × 1365 × 952	1814.0	—	Ham, pear-shaped
#2 Base Ham[5]	2350 × 1619 × 889	2352.5	—	Ham, pear-shaped
#4 Base Ham[5]	2699 × 1921 × 1206	4620.0	—	Ham, pear-shaped
¼ Drawn	1095 × 778 × 230	92.1	—	Sardines
No. 1 Oval	1635 × 1111 × 381	425.2	—	Sardines, sea foods

[1] Adapted from American Can Co., *The Canned Food Reference Manual.*
[2] Dimensions vary slightly within manufacturing tolerances. Diameter is listed first, followed by height.
[3] The net weights of various foods in the same size can will vary with the density of the product. The weights cited are for foods of average density, except where the container is largely used for one specific class of product.
[4] The volume figures cited are average commercial fills.
[5] Key-opened cans.

From the private collection of Roger C. Griffin, Jr.

FIG. 1.5. HANDBLOWN GLASS BOTTLE
An example of early American blown glass is shown here in a bottle from Wistar glassworks (early 1700s).

Horace. For best seals the plugs were daubed with wax or pitch. Corks remained the chief closure used for bottles until the invention of the screw cap in 1875 and of the crown seal bottle cap in 1892. At first the crown cap liners were plain cork, then linoleum and then in 1945 composition cork which had been used until very recent years. In the mid-1930s, experimentation was conducted on use of rubber latex and plastic liners. In 1941 a polyethylene liner was tried. In 1955, vinyl plastisols had some success and in 1957 cellular vinyl compositions were used. Aluminum facings were introduced in 1960. The crown itself has remained essentially unchanged, however, in the past few years bottle mouths have been modified to permit the crown to be twisted off. Other noncorrugated capseals have been used as far back as 1906. The latest types are made of aluminum and so designed that they are torn off. When they are twisted off, a lower ring tears away to indicate the package has been opened. This latter type is common on whiskey bottles.

For nonpressurized products, e.g., ketchup, bottles have been closed with molded plastic or metal screw-on caps.

Wide-mouth jars have been closed with wax and waxed cloth on paper, and with metal foils. Tighter closures have been made with rubbergasketed metal screw caps or pry-off lids. Some jars today have a heat-sealed membrane barrier initial closure plus a protective screw-on cover. The first provides product freshness while the second provides protection and reclosure.

Metal Tubes

Collapsible tubes were first made from soft metal for artists' paints in 1841 and replaced animal-bladders which had previously served the purpose. They found early use in dispensing glues, medicinal salves and toothpastes, but little use was made of them for food products until the 1960s. Plastic collapsible tubes have come on the market containing sandwich pastes, cake icings, pudding toppings and the like. The first closure was by a (male) metal screw with double head. The (female) molded screw cap was a later innovation probably after the turn of the century. The latest trend in tubes is the laminated tube. Made from a flexible laminate, these tubes are formed on a machine and filled in-line.

Trays, Pans and Other Containers

In this category must be listed packages that are formed into dishes, cups, bowls, pans or trays, such as pie pans and the now familiar TV dinner tray. Molded paper and pulpboard picnic plates, both rectangular and round, with and without compartmentation, have been made for several decades. Pulpboard trays are also used to package meats and produce in supermarkets. Aluminum foil containers were test-marketed in 1948 and soon after were available in many shapes. Convolute and spiral-wound canisters made from paper were marketed toward the latter part of the 19th century and were used extensively, with asphalt providing moisture barrier, as shippers for war materials and artillery shells in World War I. Few changes were made until the 1940s when aluminum foil labels were used to provide superior moisture barriers. Still later, foil was incorporated as an innerliner, thereby making it possible to package liquids. With the introduction of plastics that could be thermoformed, a wide variety of molded plastic boxes, trays, pans and the like became possible. Ovenable paperboard now appears to have become intermeshed with the microwave oven development. It is starting to appear on various supermarket shelves.

Boxes of Wood or Cardboard

In earliest times, if you wanted a box you made it yourself. Later, boxmaking became a trade. In the 1630s, boxes made of thin wood were

Courtesy of Reynolds Metals Co.

FIG. 1.6. FORMED ALUMINUM CONTAINERS WITH CRIMPED LIDS USED FOR FROZEN ENTREES

advertised. Similar boxes, some of wood and some of paperboard, were still being manufactured by hand labor in the 19th century. By this time printed labels were being pasted on the outside. The oldest box surviving today is a wooden box made in 1801.

Commercial boxmaking is supposed to have begun in England in 1817. The oldest surviving cardboard box is believed to have been made prior to 1825. The first boxmakers in the United States setup their trade in 1810 in Philadelphia. Because of the difficulty in making corners, wooden boxes and cardboard boxes of the 19th century were usually round.

By 1855 machinery was being used for cutting and creasing boxboard. The setup box was first used for jewelry, later for pills and drugs, and still later (about 1840) for candies.

When stores began to supply boxes to their customers, folding boxes were introduced to save storage space. The first are believed to have been made in the United States in 1850 to package tacks. In 1870, Robert Gair, who had already made a success from paper bag manufacture, invented the first automatic cutter-creaser. By 1897, there were over 800 patents relating to folding boxes. By the early 1900s cereals, such as Grape-Nuts (1898), Post Toasties (1904) and the original Uneeda biscuits were being packaged in waxed paper-lined unprinted cartons

Courtesy of Container Corp. of America

FIG. 1.7. WOODEN SHIPPING CONTAINERS
Wooden boxes were used extensively as shipping cases. Shapes could be round, tapered round, or rectangular.

with printed overwraps carrying the product brand name and advertising copy.

Wrappers and Labels

Wrappers have been used since primitive days when leaves were employed. In the days of handmade paper, the sheets were so expensive that the papermakers wrapped them with cast-offs to protect against soil and damage. As far back as the 1550s and earlier such wrappers were being printed with the manufacturer's name. Medicinals and tobacco were sold in "sealed" paperwrappers in the 1660s. By the early 1700s one could buy papers containing pins, tobacco, tea and wig powder. With the coming of machine-made paper and lithography, printed labels were introduced. These were applied to boxes, bottles, jars and cans for all products. Soon food products were being marked with brand names so that the consumer familiar with the quality of the contents would be able to identify his/her future repurchase.

Parisian bonbon vendors wrapped their wares in twists of colored paper in 1847. Even as early as 1840 metal foil was used to wrap chocolate

FIG. 1.8. ONE OF THE FIRST FOLDED CARTONS USED FOR BREAKFAST CEREALS

Courtesy of Walter Landor and Associates

"cakes" (probably wafers). French novelist and satirist Anatole France described such a package, which included a printed paper overwrap sealed with wax. This could have been leadfoil, but tinfoil was known and used at that time for both tea and coffee. In the 1850s, Cadbury's drinking chocolate was wrapped in tinfoil, greaseproof paper and a printed paper overwrap. During the latter half of the 19th century, paperwrappers were used for more and more products including butter and margarine.

A candlemaker who enjoyed fishing but didn't like the mess he encountered when carrying the fish home in a newspaper, found that newspaper coated with candlewax solved his problem. In 1877 he founded a company which manufactured wax paper. By 1894 paraffin waxed paper was being used as a carton inner liner for crackers, by 1900 for candy wraps, by 1902 as a breadwrapper and in the same year rolls of waxed paper in a cutter-edged carton were sold to consumers.

Waxed paper proved the most commonly used moisture-barrier material until the introduction of cellophane and aluminum foil in 1912. It

FIG. 1.9. AN EARLY TEA PACKAGE
This is an early type of flexible package for loose tea. Note the metal foil showing at the top.

Courtesy of Walter Landor and Associates

wasn't long before wax paper and foil were combined to produce superior-barrier wrappers. In the late 1950s came the introduction of heat-sealable and shrinkable plastic films such as polyvinylchloride, polyvinylidene chloride, polyethylene and polypropylene. The 1960s and 1970s were the years of second generation polymers—ionomers, copolymers and the more exotic special films—"Aclar," etc.

Bags

The making of bags out of paper dates back at least to the Thirty Years' War (1618–1648). The first commercial English paper bag manufacturer was established in 1844 in Bristol. Printing was done by way of a steam-driven litho press. The first bagmaking machine was developed in 1852 by Francis Wolle in the United States. By 1873, the gusseted bag design had been patented. Still, as late as 1902 some bags were being made by hand. In the 1870s, the glued-paper sack was invented to replace the cotton flour sack, but the multiwalled paper sack for larger quantities did not replace cloth until 1925 when a means of sewing the ends was invented. As previously noted, by 1905 machinery was available for automatically producing in-line printed paper bags. When plastic films came on the scene, machinery was adapted to make bags from them also. The bag is even considered to be a form of contemporary art. In 1976, an exhibition called "Homage to the Bag" was held at the Museum of Contemporary Crafts in New York City. It featured all the many forms of bags—from Andy Warhol's silkscreened shopping bag to an Afghanistan bag used for carrying salt.

GRAPHIC ARTS

Decoration and ornamentation predate history. Ancient artifacts show that primitive man learned to paint, carve and sculpt. For many thousands of years all such work was applied directly to the object that was to be decorated. No one knows when the first experimentation with indirect methods occurred. Whoever first applied a color or ink to one surface and then pressed it against another to transfer the color was the first printer. We have evidence that the first printing on paper from carved wooden blocks took place in China about 868 A.D.[2] Also in China about 1041 A.D. individual wooden blocks for printing of characters were used. The art of printing on paper from wooden blocks undoubtedly existed in Europe prior to Gutenberg's use of movable type in 1454, and probably was predated by block printing on textiles.

The earliest surviving printed package was a wrapper used by a German papermaker, Andreas Bernhart, in the 1550s. During the next 100 years, the use of printed paperwrappers spread to other products, from patent medicines and dentifrices to tobacco and various foods. In the early 1600s few, if any, products were identified by the manufacturer's name.

As various new items were imported to England through overseas trade, as for instance tobacco (1610), tea (1614), coffee (1648) and chocolate (1655), the opportunity for cheating the public by adulteration or by disguising inferior quality was seized upon by some. Honest merchants began to mark their wares in a distinctive manner, first by newspaper advertising, then by printed "trade cards" and finally by printed wrappers or labels.

In the mid 1700s, engraved copper or steel plates were being used in place of wood blocks for printed labels. A transfer method was invented for printing on pottery. Paper was first printed from engraved plates and then the colored side was adhered to the unglazed, unfired clay which absorbed the colored ink. After the paper was washed away, the sub-

[2]It appears that before this Chinese development, Japan's Empress Shotoku in 770 A.D. sanctioned the first printing on paper—1 million prayers to ward off an epidemic of smallpox.

FIG. 1.10. THE SHOPPING BAG
Before modern packaging began to exert its influence with prepackaging of produce the shopper carried food home in the utilitarian paper sack.

Courtesy of Reynolds Metals Co.

sequent firing burned the color in permanently. In 1798, the principles of lithography were discovered by Senefelder in Bavaria. This, together with the advent of cheaper machine-made paper, gave the graphic arts a strong impetus. Lithography was used for transfer printing on pottery in more than one color, and color printing on paper became increasingly popular. Some printers, as for example Currier and Ives, printed one color from a lithoplate and then added the other colors by hand painting. As the 19th century progressed, printers vied with one another in designing new movable type faces and fancier labels.

At the turn of the 18th century, men used a bear-grease compound in their hair as a pomade. These pomades were packaged in decorative pots in order to promote their contents. The quality of the finish, color and printing was better on the early lids than on the later ones. The early printed lids had thin gold borders and gaily colored pictures that followed the theme of the bear. Other lids illustrated a more whimsical humor such as "Bears at School" and "Bears Reading A Newspaper." Although bear grease pots were probably the first to be decorated in this manner, the idea of selling a fairly mundane product on the strength of an attractively produced container spread quickly. These pots were subsequently used for fish pastes, mustard and potted meats. The rising costs of color printing in the 1880s put an end to pot-lid production. However, decorative pot-lids form one of the most interesting subjects in early consumer packaging.

In the 1830s printed color was introduced on match box labels. In 1835, George Baxter produced multicolor work using a copper plate for the base color and wooden blocks for the registered colors.

In 1859, papermakers were selling colored papers and many printers were using the Baxter process. In 1870, Richard Cadbury painted landscapes in oil and then had them reproduced in lithographed labels for the Cadbury chocolate box tops. Powered presses in 1865 were flat bed and required very large lithograph stones. About 600 impressions an hour were standard because of the reciprocal action. The discovery in 1820 that zinc could be substituted for stone made the rotary press possible and increased production to 1100 impressions per hour.

Direct printing on primed tinplate was not successfully done in 1864 as the engraved printing plates or lithograph stones would not make good contact. The transfer method held sway for about 10 years. Here the ink was printed on thin paper which was applied to the tinplate and then the paper was washed off. In 1875, Robert Barclay in England invented offset lithography. He applied slow-drying inks to a glazed cardboard cylinder from which they were transferred to the tinplate. Later the cardboard was replaced with rubber-covered canvas, still later by a rubber roll. In 1903, this principle was converted from flatbed to rotary presses and up to 4000 impressions an hour became possible.

Flexography was also born in England about 1890 in an attempt to keep up with the speed of paper bag making machines. English inks were fast drying "aniline" types and were applied to resilient rubber blocks or plates from which they were printed. By 1905, a bag machine and press had been combined for an in-line operation.

During the 19th and 20th centuries, photographic techniques were invented whereby printing plates could be prepared and chemically etched. Process color separation by photography lowered costs of plate preparation and reduced the number of colored inks required. By the turn of the century (1900) four color lithography was common place and six colors were possible.

Modern high-speed rotary presses with their automatic controls all spring from these earlier beginnings.

PACKAGING AND CONVERTING MACHINERY

It is difficult to establish dates for the various developments in packaging and converting machinery. The packaging industry was so busy growing it forgot to record its accomplishments.

Earliest machines were confined to making hand operations easier. Foot treadle-operated or steam-powered, they each contributed to increases in output and to lower costs. As time went on several separate machines were combined to produce a still more efficient in-line operation.

Laminating equipment was most likely an offshoot of waxpaper making equipment where a second web could be applied to the first web by heat and pressure. However, laminators may also have borrowed features from box gluing machinery and label applicating machinery, all of which were developed during the decades just before and after World War I. One of the first automatic gluers of labels for setup boxes was marketed in the United States just after 1918 by the New Jersey Machine Corporation.

Extruders, which are widely used today, perhaps originated in the manufacture of extruded foods such as spaghetti or macaroni, or building materials such as brick and tile. We know one early extruder was patented before 1840 to extrude lead pipe. Between 1840 and 1860 extruders were used to apply gutta-percha rubber to electrical cables and wires. The first extruders were ram types and therefore discontinuous. This was suitable for molded parts, but not for continuous sheeting or tubing. The screw extruder was invented in 1879. In 1912 a longer-screw, heated-barrel extruder was developed for rubber. It had a 10 to one ratio of length to barrel diameter (l/d) and incorporated air cooling. Little success was achieved in attempts to extrude plastics until the late 1950s when screw design was improved to allow efficient and uniform melting and l/d ratios were increased. This was the double channeled screw with separate mixing and melting section and subsequent metering section. Other screwless extruders have since been developed but some time will be required to gain experience before they will seriously compete with the screw types. The extruder has provided the packaging material converter with a means for making a wide variety of combinations of laminated and coated materials, tailored to meet the demands of nearly any food product. In recent years extruders have been developed that coextrude two different plastic melts through one multiple die to produce a sandwich-structured coextruded film.

By the latter part of the 19th century, packaging materials were being made by machine and all of the ingredients for modern packaging machinery were available. Packages were being fabricated completely by machine as with paper bags, or in several separate operations as with cans. One by one various hand operations were studied and replaced by machinery. These included: package setup, product metering, application of package to the product or vice versa, package closing, collecting of packages, preparation (setup) of shipper container, placement of packages in shipper and closing of shippers.

Weighing, metering and counting equipment date prior to 1900, as well as carton setup and gluing and bag-making equipment. In the next decade machines were developed for wrapping chewing gum and over-wrapping cartons. One machine in 1910 would form, cut and twist-wrap candy kisses. Sometime after 1913 machinery was developed to fold biscuit cartons with paper liners. There was a difficult period between the two World Wars, during which food manufacturers resisted the changeover from bulk packaging to highspeed unit packaging. The machinery manufacturers tried to persuade the food manufacturers to adopt the new technology.

Most machines during this period were designed to fit individual requirements; secrecy was the order of the day. It wasn't until after World War II that serious efforts were made to mechanize fully and integrate entire packaging lines. Pouchmaking equipment came on the scene in the late 1940s and early 1950s, and vacuum-packed pouches in the mid 1950s.

Today, few manufacturers of food products are without automated high speed packaging lines. Without them the cost of packaging in a highly competitive marketplace would be prohibitive.

MODERN PACKAGING MATERIALS AND FORMS

Rigid Packaging Materials and Package Forms

Glass Containers.—Glass containers are one of the stalwarts of food packaging. Over 75 billion containers are used annually by food and beverage packagers. Narrow-necked glass containers are called bottles and wide-necked containers are called jars.

Glass is made from limestone, sand, soda ash and alumina. Colorants may be added to the melt or applied later to the surface. Glass is strong, rigid and very chemically inert. It has complete FDA clearance as do many of the standard coatings. Glass does not appreciably deteriorate with age and is an excellent barrier to solids, liquids and gases. It is, therefore, excellent protection against odor and flavor contaminations. It is also a low cost material. However, strength requirements may demand a heavier weight ratio. Its chief disadvantages as a packaging material are weight and fragility. It is also not easily disposable. Consequently glass is frequently designed for multitrip or multiuse. Significant improvements in strength-weight ratios have recently been achieved and new coatings have reduced damage and enabled higher handling speeds.

The transparency of glass provides excellent product visibility and new methods for surface decoration yield highly attractive finishes. Glass is extremely versatile as to size and shape of the final container. New

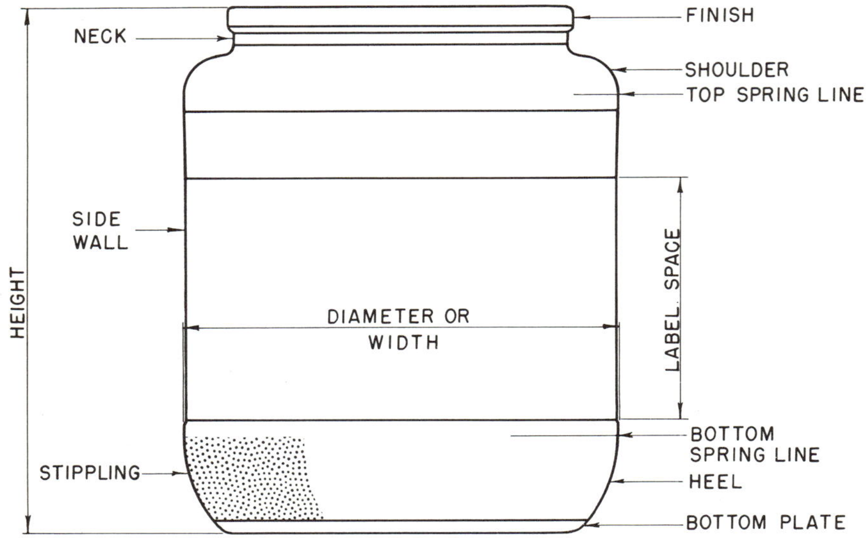

From Joslyn and Heid (1963)

FIG. 1.11. NOMENCLATURE FOR GLASS CONTAINERS FOR FOODS

computerized furnaces have improved production rates and dimensional tolerances.

Glass containers include the following:

Bottles. The bottle is the most extensively used glass container. They may be of many different shapes but the neck is always round and much narrower than the body. The neck facilitates pouring and reduces the size of the closure required. Principal uses are for liquids or small-sized solids.

Jars. The jar is really a very wide-mouthed bottle having no appreciable neck. The opening permits the insertion of fingers or a utensil to remove portions of its contents. They may be used for liquids, solids and nonpourable semiliquids such as thick sauces and pastes.

Tumblers. These are like jars but they are open-ended. They have no neck and no "finish." They are shaped like a drinking glass and are used for products such as jams and jellies.

Jugs. These are large-sized bottles with carrying handles. Necks are usually short and narrow. They are usually used for liquids in ½ gal. (1.89 liters) and larger sizes.

Carboys. These are very heavy shipping containers shaped like a short-necked bottle and having a capacity of 3 gal. (11.36 liters) or more.

Typically they have been used with a wooden crate holder. Other outer protective frames are now finding use.

Vials and Ampoules. These are small glass containers. The latter are principally used for pharmaceuticals. Vials are sometimes used for small quantities of foods such as spices or food colorants.

When selecting a glass container for a food it is important to be careful in choosing dimensions and finish so that the correct volume will be available, the product can be easily filled and dispensed and a proper closure can be selected. "Finish" refers to the type and dimensions of neck and mouth of the container, i.e., thread, lug, friction, snap-cap, roll-on, etc. There are many standard finishes listed by the Glass Container Manufacturers Institute.

Other important factors in selecting a glass container for food are its color which can influence the type of light reaching the food, and its ability to resist thermal shock. Some glasses cannot withstand sudden changes in temperature, i.e., filling a hot product into a container and then plunging it into cold water. Special glasses are available for such purposes.

Metal Cans.—The average American family uses over 350 cans yearly. A large percentage of this amount is for food and beverages. There are over 600 different styles, shapes and sizes of cans to choose from.

Traditionally cans have been made from soldered tinplate steel but more recently aluminum cans have been introduced.[3] Today there are several more choices available: standard tinplate, lightweight tinplate, double reduced tinplate, tin-free steel (coated), vacuum-deposited aluminum on steel and aluminum.

Can bodies can be soldered, welded or cemented. Steel bodies can be combined with aluminum ends. Many new easy-open devices are available for cans ranging from pop-tabs for beverages to complete removal of lids or panels for frozen or meat products.

Can coatings are now regarded as vital components especially for foods and beverages. Coatings must be nontoxic and free from odors or taste. They must not deteriorate or come loose from the can wall during food processing storage. Interior coatings are made from acrylics, alkyds, butadienes, epoxyamines, epoxy-esters, epoxy-phenolics, oleoresins, phenolics and vinyls, depending upon the type of food and process. Outside coatings include acrylics, alkyds, oleoresins, phenolics and vinyls, and are usually pigmented. They are less exposed to food contact but must survive processing and be receptive to further decorative coatings and inks.

[3]It is interesting to note that the taste of tin from the can has been accepted as appropriate in tomato juice. It is so important that aluminum canmakers have thought about adding the tin flavor to juice packed in their containers.

GCMI ITEM NO.	CAP. O'FLOW FL.OZS	WT. MAX. OZS.	A	B MAX.	C	D	E	SPECIMEN FINISH
10-14	4 1/4	3 3/4	3 25/64	2 1/16	35/64	1 49/64	1 3/4	48 – 400
10-20	6 1/4	4 1/4	3 55/64	2 9/32	5/8	2 1/16	1 15/16	53 – 400
10-24	7 3/4	4 1/2	4 1/8	2 25/64	21/32	2 5/16	2 3/32	58 – 400
10-26	8 3/8	4 3/4	4 9/32	2 1/2	11/16	2 9/64	1 15/16	53 – 400
10-27	8 3/8	4 3/4	4 9/32	2 29/64	11/16	2 3/8	2 3/32	58 – 400
10-28	8 3/4	5 1/2	4 11/32	2 1/2	5/8	2 31/64	2 3/32	58 – 400
10-30	9 1/2	5 1/2	4 31/64	2 35/64	41/64	2 9/16	2 3/32	58 – 400
10-36	11 1/2	6	4 25/32	2 3/4	49/64	2 7/16	2 3/32	58 – 400
10-40	12 1/2	6 1/4	4 27/32	2 53/64	49/64	2 1/2	2 3/32	58 – 400
10-48	15 1/2	7	5 1/8	3 1/32	13/16	2 11/16	2 5/16	63 – 400
10-50	16 1/2	7 1/2	5 7/32	3 1/8	27/32	2 47/64	2 3/16	60 – 440
10-51	16 1/2	7 1/2	5 7/32	3 1/8	27/32	2 47/64	2 5/16	63 – 400
10-52	17	7 1/2	5 11/32	3 7/64	27/32	2 55/64	2 5/16	63 – 400
10-53	18 5/16	8 1/4	5 3/8	3 17/64	29/32	2 49/64	2 5/16	63 – 400
10-60	22 3/4	9 1/2	5 7/8	3 29/64	15/16	3 11/64	2 5/16	63 – 400
10-62	24	10 1/8	5 7/8	3 35/64	61/64	3 3/16	2 5/16	63 – 400
10-63	24 1/2	10 1/2	5 61/64	3 9/16	31/32	3 7/32	2 5/16	63 – 400
10-67	27 1/2	11	6 1/4	3 43/64	1	3 25/64	2 5/16	63 – 400
10-71	30 1/4	11 1/2	6 9/16	3 47/64	1 1/32	3 39/64	2 5/16	63 – 400
10-72	31	11 1/2	6 9/16	3 25/32	1 1/32	3 39/64	2 5/16	63 – 400
10-75	32 5/8	11 1/2	6 3/4	3 53/64	1 1/16	3 45/64	2 5/16	63 – 400
10-77	34	12 3/4	6 13/16	3 59/64	1 3/32	3 45/64	2 5/16	63 – 400
10-81	48 3/4	18 1/2	7 11/16	4 13/32	1 1/4	4 17/64	2 5/8	70 – 400

NOTES:

1. WHEN OTHER FINISHES ARE USED, CAPACITY, WEIGHT AND HEIGHT SPECIFICATIONS ARE ADJUSTABLE WITHIN THE REQUIREMENTS OF THE FINISH USED. SEE NOTE 2.
2. HEIGHT (DIMENSION 'A') IS BASED UPON USE OF SPECIMEN FINISH SHOWN.
3. THE SPECIFICATIONS SHOWN MAY VARY MODERATELY ACCORDING TO COMMERCIAL TOLERANCES AND INDIVIDUAL MANUFACTURER'S PRACTICE.
4. 'B' DIMENSION IS VARIED TO MAINTAIN CAPACITY.

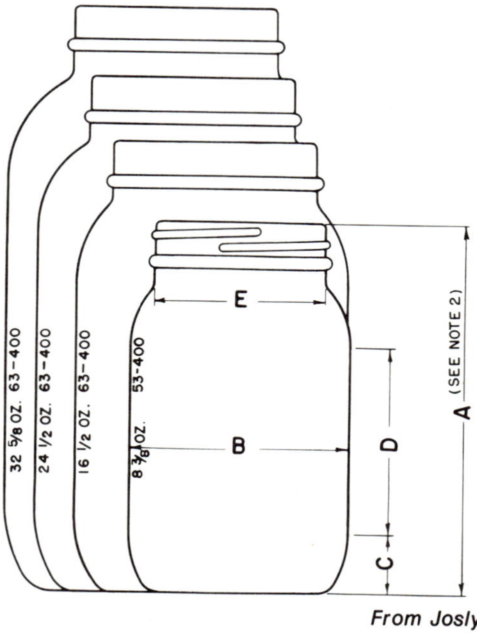

From Joslyn and Heid (1963)

FIG. 1.12. STANDARD DIMENSIONS OF PLAIN ROUND JARS

GCMI ITEM NO.	CAP. O'FLOW FL.OZS.	WT. MAX. OZS.	A	B MAX.	C	D	E	SPECIMEN FINISH
29-16	5	$3\frac{9}{16}$	$3\frac{17}{32}$	$2\frac{7}{64}$	$\frac{11}{16}$	$2\frac{7}{32}$	$1\frac{49}{64}$	48-1740 OR 48-870
29-25	8.2	$5\frac{1}{8}$	$3\frac{7}{8}$	$2\frac{17}{32}$	$\frac{41}{64}$	$2\frac{39}{64}$	$2\frac{5}{32}$	58-1740 OR 58-870

NOTES:

1. WHEN OTHER FINISHES ARE USED, CAPACITY, WEIGHT AND HEIGHT SPECIFICATIONS ARE ADJUSTABLE WITHIN THE REQUIREMENTS OF THE FINISH USED. SEE NOTE 2.
2. HEIGHT (DIMENSION 'A') IS BASED UPON USE OF SPECIMEN FINISH SHOWN.
3. THE SPECIFICATIONS SHOWN MAY VARY MODERATELY ACCORDING TO COMMERCIAL TOLERANCES AND INDIVIDUAL MANUFACTURER'S PRACTICE.
4. 'B' DIMENSION IS VARIED TO MAINTAIN CAPACITY.

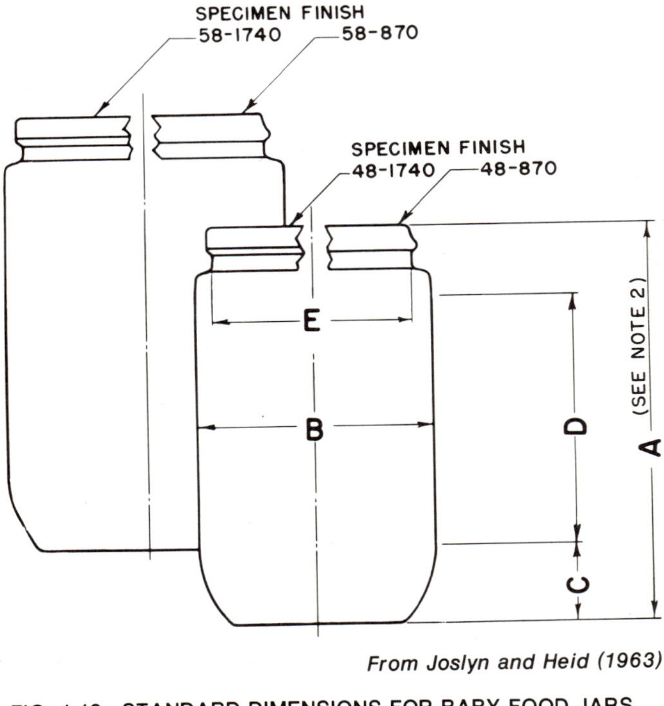

From Joslyn and Heid (1963)

FIG. 1.13. STANDARD DIMENSIONS FOR BABY FOOD JARS

TABLE 1.3. GENERAL CLASSES OF FOOD PRODUCTS AND TYPES OF STEEL BASE REQUIRED[1]

Class of Foods	Characteristics	Typical Examples	Steel Base Required
Most strongly corrosive	highly or moderately acid products, including dark colored fruits and pickles	apple juice berries prunes cherries pickles	type L
Moderately corrosive	acidified vegetables mildly acid fruit products	sauerkraut apricots figs grapefruit peaches	type MS type MR
Mildly corrosive	low acid products	peas corn meats fish	type MR or MC
Noncorrosive	mostly dry and nonprocessed products	dehydrated soups frozen foods shortening nuts	type MR or MC

[1] From Ellis (1963).

In 1959, aluminum was used for the first time commercially for beer in 7 oz (198.5 g) cans. Aluminum cans were supplied as the conventional cemented side seam three-piece can, shallow draw, deep draw and impact extruded varieties. Today most are made by a draw-redraw, or draw and iron process.

Conventional cemented side seam three-piece aluminum cans may be square as well as round; there is no limitation on diameter and height. Side seams are sealed with a special cement. Usage of these cans is limited to nonprocessed products because the cement is not capable of withstanding processing after the can is filled. An interior coating is usually necessary with aluminum cans. However, with some foods, such as orange juice, a coating is not required.

Shallow-drawn aluminum cans are limited in size but can be fabricated in irregular shapes as well as round. They can be used for packing processed foods. Some products successfully packed in this type of can include cod, roe, tuna and sardines.

Deep drawn aluminum cans have been made only in round shapes. They can be used for packing processed foods. Products packed include beer and soft drinks.

Metal cans are advantageous as packages because of superior strength, high speed manufacturing, and easy filling and closing. Disadvantages of metal cans include weight and difficulty in reclosure and disposal.

Composite Containers

A composite container is a container made from two or more constit-

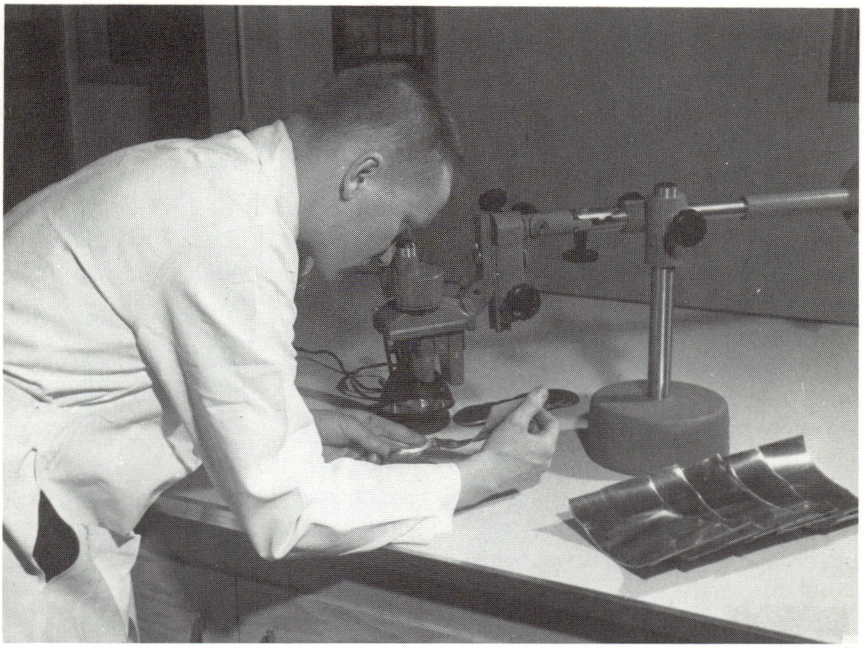

Courtesy of American Can Co.

FIG. 1.14. ANALYSIS OF CAN COATING
Enameled can bodies from packed cans are examined under magnification for coating and tinplate performance.

uent materials. It usually consists of a paperboard body with metal or plastic ends.

Two basic types are available. Spiral-wound containers are made in cylindrical shapes where two or more plies of board are glued together around a mandrel. Convolute-wound composites are produced by straight winding. Body materials used are chipboard and kraft paper. Linings used are vegetable parchment, wax laminates, aluminum foil, glassine and polyethylene-coated paper. Other linings can also be used depending on the product to be packaged.

Composite cans are closed by either a snap-on lid, plug-in lid or a lever lid. In the nondetachable type of closure, perforated tops and string-opening devices are used as well as double-seamed ends.

Aerosol Containers.—Aerosol containers are used to dispense a product by means of a pressurized gas or liquid that is held in the same container. The basic components of the package are the container, the valve and the protective cap.

The container must be gas-tight and may be constructed from steel cans, aluminum cans, glass, plastics or a combination. Choice of material must be related to safety (ability to hold pressure and lack of fragility), size and product compatibility.

There are several types of aerosols which can dispense products as fine mists, sprays, dusts or foams depending upon the type of valve used and arrangement of the product-propellant ratio. A single phase aerosol contains a liquid product layer and compressed gas propellant layer. A two-phase aerosol contains a liquid product with dissolved propellant layer plus a compressed gas propellant layer. A three-phase aerosol contains a layer of liquid propellant, a layer of product and a layer of propellant vapor. Removal of product causes the liquid propellant to boil and replenish the vapor phase.

Piston aerosols contain a flexible plastic barrier between the product and the propellant gas. Co-dispensing aerosols disperse two products through the same nozzle.

Advantages of aerosols are convenience, premixing, lack of evaporation and exclusion of air even when partly used. Disadvantages are high cost, risk of explosion, breakage or leakage. Aerosols have been used for beverage concentrates, cocktail mixes, cake icings, pancake mixes, syrups, salad dressings and seasonings.

In 1974, two scientists proposed that the fluorocarbon propellants used in aerosol cans were drifting up into the stratosphere and causing a reduction of the ozone layer. This caused a widespread public reaction toward the aerosol package and several states passed legislation effectively outlawing the aerosol can. Although evidence that fluorocarbons from aerosol cans deplete the ozone layer is sketchy and inconclusive (loss from refrigeration is more), the aerosol can suffered a substantial market loss.

Aerosols are actually *not* the only threat to the ozone layer. Half the fluorocarbons in this country have gone into air conditioners, solvents and foam products as well as refrigerators. They then eventually escape into the atmosphere (automobile air conditioners are particularly leaky). Sweden has banned fluorocarbon propellants, and Britain, France and Italy are skeptical. Neither Japan nor the U.S.S.R. appear to recognize that a problem exists.

The theory that fluorocarbons destroy the ozone layer is also simplistic. Measurements taken in Britain and Switzerland over the past 50 years have shown an increase in ozone of between 5 and 10%. The argument that a depletion of ozone will aggravate skin cancer may be true, but during the 50 years ozone has increased, the incidence of skin cancer in the United States has increased by 10-fold among the more susceptible members of the community. It has been proposed that no detectable depletion of the ozone layer will occur with the present amounts of fluorocarbon use. Even at their worst, the effects of fluorocarbons on

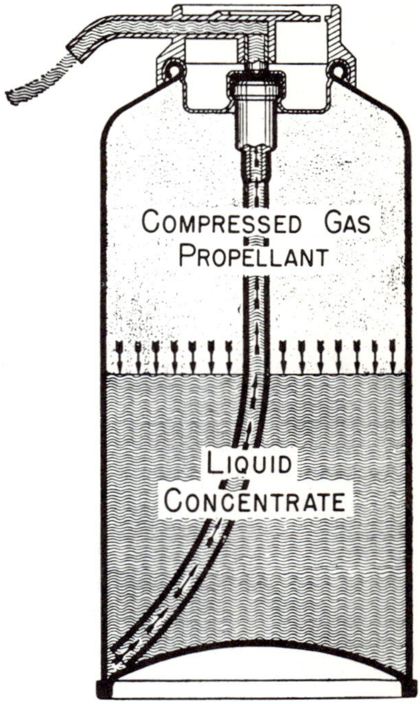

FIG. 1.15. AEROSOL COMPONENTS
One type of aerosol container used for food.

Courtesy of Food Processing and Marketing

ozone are small in comparison to other natural influences already at work. This includes the emission of fluorocarbons during active volcanic eruptions.

On December 15, 1978, the Government implemented a ban on nearly all aerosol products containing fluorocarbon propellants.[4] The effects of this 98% ban are that:

(1) Consumers will use the less effective propellants such as a *pump mechanism.*
(2) There will be a negligible effect on businesses such as DuPont and Allied Chemical (the major suppliers of fluorocarbons).
(3) Other propellants will be used, i.e., propane, butane, nitrous oxide or a high pressure gas such as nitrogen. With the latter, there is a significant reduction in the dispensing force.
(4) Straight fluorocarbons may possibly be used. These do not produce the chlorine atom that starts the chain reaction that does the actual harm to the ozone layer.

[4]Exempted from the December 15 ban were aerosol products for which no acceptable substitutes exist, i.e., contraceptive foams, inhalation therapy drugs, certain electrical cleaning sprays, aircraft maintenance products, some insecticides and coolants used in refrigerants and air conditioners, and those used to produce plastic foam.

Rigid Plastic Packages

A wide variety of rigid plastics can be used in the form of thermoformed, injection-molded and blow-molded containers.

Thermoformed containers are formed by exposing the plastic sheet to heat and forming into a female or around a male mold. Vacuum and pressure forming and several other techniques are used.

Thermoformed trays are used in many food applications ranging from margarine tubs to formed polystyrene foam egg cartons. The plastics used for thermoformed trays are polycarbonate, polyethylene (high density), polyvinyl chloride, polyester, polystyrene, polypropylene, ABS (acrylonitrile butadiene styrene), cellulose acetate and acrylonitrile. The specific plastic selected depends on the food to be packaged and storage requirements. Polystyrene tends to be brittle at low temperatures while linear polyethylene trays are not brittle but tend to appear cloudy in appearance. All thermoformed trays are covered with heat-sealable or snap-on lids made of plastic film or aluminum foil laminates.

Injection-molded containers are used in high volume applications for jars, bottles and tubs. Most widely used plastics are polystyrene, polypropylene and acrylonitrile. A significant advantage in the use of polystyrene is its outstanding clarity. The overall injection-molded container tends to have a thicker wall than one formed by thermoforming techniques.

Blow molding is generally used to produce containers where the neck diameter is small in contrast to the overall container diameter. Plastics used are polyvinyl chloride, polypropylene, polycarbonate, cellulose acetate, polystyrene, polyethylene, polyacetal, acrylonitrile and polyesters. Blow-molded bottles are made of polyethylene, PVC, polystyrene and more recently polyesters (PET). Final selection of the specific plastic is determined by product compatibility, economics and ease in processing. Shallow trays made from polysulfone are being introduced for packaging of frozen food to be reheated in conventional and microwave ovens. These trays appear to be compression-molded.

Advantages of plastic containers are low cost and ease of fabrication. Disadvantages may include lack of product compatibility, low barrier qualities, plastic deterioration, lack of resistance to high heat and fragility at low temperature.

Solid and Corrugated Fiberboard Containers

Solid and corrugated fiberboard materials are used to fabricate shipping cartons and cases used extensively in wholesale and industrial shipping. They are not usually used as direct containers for foods, but are extensively employed as outer shippers for food packages, i.e., cans and bottles.

Both are made from heavy fibrous kraft paperboards either cylinder or Fourdrinier. Solid fiberboard is made by gluing several plies of paperboard together. By using asphalt or a special resin adhesive, such as ureaformaldehyde, enhanced moisture resistance may be built in. Selection of weight, fibrous construction and number of plies is based on the desired strength properties such as burst, tear, puncture and bend resistance, etc. Corrugated fiberboard is made from similar base materials but is generally thinner, as it is then structured by combining facings (flat sheets) and liners (corrugated or fluted sheets) by means of adhesives. There are five major types of corrugated paperboard: unlined, single-faced, double-faced, double-walled and triple-walled. The first two types are used for wrappings of fragile objects or as interior padding of boxes. The latter two types are used where exceptional strength and rigidity are essential. Double-faced corrugated is the most commonly used type for boxes, liners and partitions.

	A Flute	B Flute	C Flute	E Flute
Flutes/foot	35–37	50–52	41–45	90–96
Flute height (in.)	0.185	0.105	0.145	0.085
Thickness (in.)	3/16–7/32	4/32	5/32	5/64

Each flute has particular properties with respect to load support in each of the three possible directions. In double wall construction different flutes can be combined.

Advantages for corrugated board packages are versatility, light weight, strength, disposability and low cost.

Disadvantages include low wet strength; however, newer techniques of manufacture and new coatings are eliminating this disadvantage.

Corrugated containers are now available with easy open-tear strips, self-locking assembly and smooth white liners permitting flexographic printing on the exterior. Special reinforcements can be employed. Where strength is less important, molded pulp has been used for liner material. Other special corrugated boards or fiberboard constructions include foil-laminated facings and plastic foam fillers between paper facings.

Wooden Boxes and Crates.—When lumber was plentiful and inexpensive, many shipping containers were made in the form of nailed or wirebound wooden boxes and crates. With rising costs of lumber, labor and freight rates, and the availability of cheaper substitutes—fiberboards, composition boards and the like—use of wood containers has been on the decline.

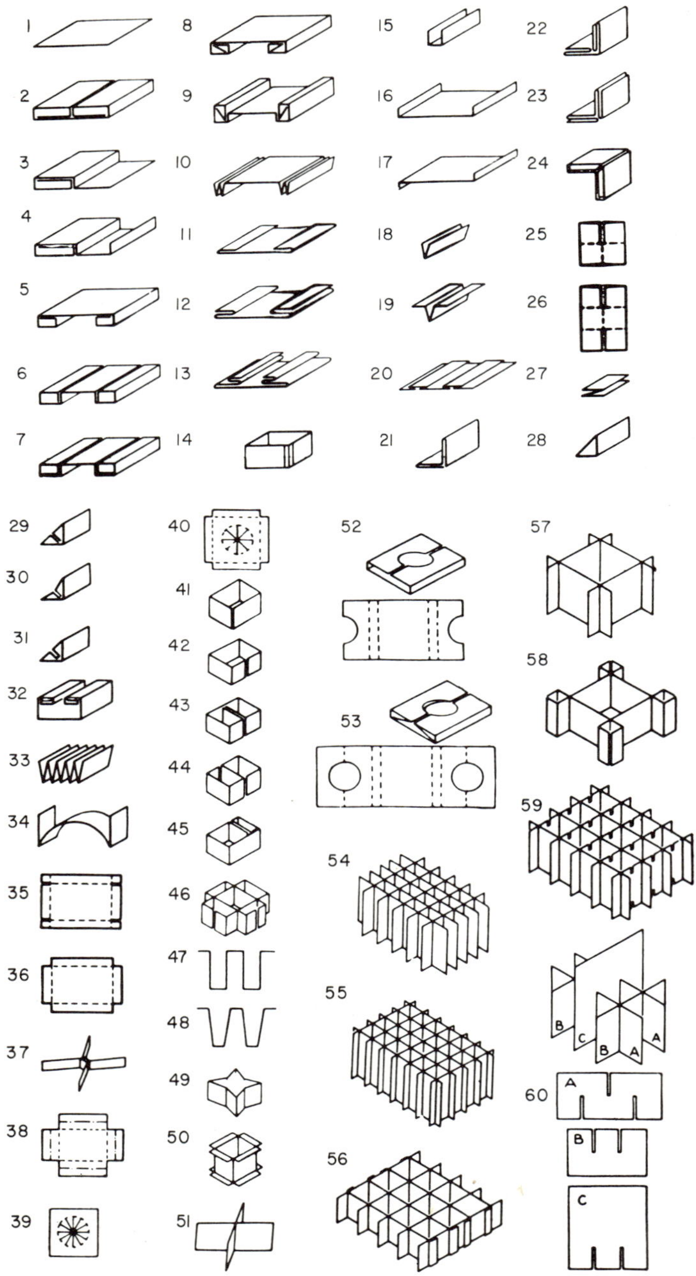

Courtesy of Hinde & Dauche Div., W. Va. Pulp & Paper

FIG. 1.16. STYLES OF INNER PACKING

The type and cure of wood used will affect the weight, strength and ease of fabrication of the container. Soft woods are easier to nail but not as strong as hardwoods. Green lumber is excessively heavy and weaker, and will warp and shrink causing loosening of nails or other fasteners.

Boxes are usually solidly-walled, rectangular-shaped and will vary in construction and in the number of extra cleats and braces used as is required by the load. The top, bottom and sides of a box provide the main structural strength.

Crates are similar to boxes but may be of lighter weight and more open construction—that is, spaces may be left between boards or the crate may be fully enclosed or sheathed. A crate differs from a box in that the frame members carry the load. The sheath merely encloses; sheathing may be corrugated fiberboard, thin plywood or light weight lumber.

Other joining methods may be used for boxes and crates. These include metal fasteners, glues and wires or wire tapes. When using wires, thinner side, top and bottom sheathing can be utilized as the wires add strength. Cleated ends and stiffeners provide the structural strength required.

Advantages of wooden boxes and crates depend on the relative cost, strength and weight ratios involved. Where lumber is cheap or boxes can be reused, they may prove advantageous. However, in most food uses today, wooden containers are being phased out and solid or corrugated fiberboard containers are replacing them. Some wood is still used for reinforcing cleats and bottoms.

Cylindrical Shipping Containers.—Cylindrical containers have high stacking strength and can be rolled in handling. The latter is now less important due to mechanical handling devices. Cylindrical containers may be made from fiberboard, glass, metal, plastic or wood. Glass containers have already been discussed. Plastic containers are principally in the form of liners for other shells made from steel, aluminum, fiberboard or wood.

A barrel is a cylindrical container of greater length than breadth, having two flat ends of equal diameter and bulging at the waist. A drum has straight sides and flat or bumped ends. A pail is a cylindrical or tapered-shaped (truncated cone) container with or without a wire handle or bail. Small pails may be called cans. A keg is a small barrel; a cask is a large type wooden barrel.

Barrels may be made of wooden staves bound together with hoops and may be tight or slack. Metal barrels are made of steel or aluminum. Drums may be made of metal, plywood or fiberboard with wood, metal or fiber ends.

Fiber drums and cans are made from spiral or convolute wound paper or paperboard bodies and may be closed with metal, plastic or fiber ends.

With interior facings of special papers, foils or plastics such as glassine, parchment, aluminum foil, polyethylene or pliofilm they may be used successfully for bulk shipment of foods as the linings prevent contamination of the food or weakening of the container. Other barrier components can be built into the side wall at time of lamination. These may be asphalt, foil or polyethylene to achieve a moisture barrier.

Steel drums are used as single trip or returnable containers. Drum heads may be removable or fixed. Fixed heads may be fastened to the body by brazing, welding or double-seaming. Some drums are formed in two halves and joined circumferentially at the waist.

Drums may be fitted with removable covers which are of the friction lid type or which may be held in place by locking rings or lugs. Tight head drums or pails are furnished with small capped or screw plug openings for ease of pouring. Additional vents and drains may be specified.

Aluminum drums are designed to have similar features to steel drums. Stainless steel drums are available where corrosion resistance or high levels of sanitation are required.

The latest types of drum constructions use blow-molded polyethylene drums as the primary container and steel shell for added protection and rigidity.

Containerization of Freight Shipments.—The basic aim of containerization is to transport goods from the source of manufacture to the sales point safely and economically. The concept uses a freight container which is delivered directly to the factory for loading. In other cases, the goods are sent to a container base to be loaded on the container. At the point of use, the container is directly unloaded. Freight containers are made of various materials including aluminum. Specially designed handling equipment is used to accommodate containers.

Semirigid Packaging Materials and Package Forms

Aluminum Containers.—Containers fabricated from plain aluminum foil have differing degrees of rigidity depending upon thickness, temper, alloy and container design. Some could be called rigid as they are not easily deformed, others are more delicate, hence they are included here as semirigid—that is, a package which is intended to maintain a definite form or shape and is not influenced by the shape or bulk of the contents, but which can rather readily be bent or dented.

Aluminum containers provide extreme convenience in preparation and serving of foods; they withstand extremes of temperatures. Foods may be frozen in the package or cooked in it. Aluminum protects the food against moisture, gases and light. General categories are compartmented, folded-end, ovals, pie plates, rectangulars, rounds, squares and specialty items. Folded-end cartons have wrinkle-free sides and bottoms. Formed containers made with male and female dies have unwrinkled side walls and flanges. Newer techniques of forming are now producing smooth flanges and smooth side walls. With smooth flanges flat-web materials can be hermetically sealed on as closures. Colored containers can be decorated attractively to permit the container to be placed on the table as a server. Soft butter and margarine products are now found in such a package.

Setup Paperboard Boxes.—Four basic components are used to make setup paperboard boxes: paperboard, adhesive, corner stays and covering. Paperboard is selected to give the right weight and smoothness for the size box required. Sheets of the boxboard are cut and scored, the sides are folded up to make a tray and the corner stays are adhered. This is repeated for the cover. Finally the covering material is glued on. Coverings may be colored papers, foil laminates or highly coated, embossed and printed litho papers.

Boxes can be made in almost any shape and with a wide variety of lid arrangements—separate or hinged. Semirigid plastics can be used as lids for better product display or die-cut windows can be similarly employed.

Advantages of setup boxes as packages include convenience, individuality, strength, reusability and excellent product protection and display. Equipment required is minimal and low cost. Boxes are shipped setup, hence no setup time. Small quantities are no more expensive than large quantities, making large inventories unnecessary.

Disadvantages are the generally higher cost in comparison to folding cartons produced in large quantities.

Folding Paperboard Cartons.—A folding carton is a container made from bending type boxboard by die-cutting and scoring it properly to fold into the desired form. It is supplied by the maker as a flat blank, preglued or partially glued, and collapsed. It is erected, filled and closed by the packer. Because they are supplied in knocked-down form (flat), folding cartons are easier to pack than setup boxes and provide economic savings in transportation.

Paperboards used in folding cartons must be capable of being bent and scored. There are many grades and thicknesses ranging from cheap-lined or unlined chipboards to manila, kraft, laminated and clay-coated solid, bleached sulfate boards, the latter being best for high quality printing. Boards may also be coated with plastics such as polyethylene, ethylene vinyl acetate, wax or blends of resins. Foil laminates are used both for aesthetics and for added product protection.

There are two common styles of folding cartons and a large number of special constructions. Tube types are one piece cartons that are bent into a tube (generally square) with a longitudinal glued body seam. End flaps may be glued shut, tucked and with or without selflocking features.

Tray types may be one or two piece with or without a lid. They are shipped flat and are setup and glued to form the tray and/or lid in the packager's plant. Some tray types are glued by the box manufacturer and folded flat along diagonal score lines. They can be snapped open to setup. Carriers for cans or bottles are special types of folding cartons.

Cartons may be printed, embossed or die-cut prior to blanking. Printing may be by letterpress, offset lithography or rotogravure.

Folding cartons are widely used for both solid and liquid foods. Advantages are low cost, ease of automatic high speed setup, filling and closing, good stackability, easy opening and reclosure and excellent graphics.

Molded Pulp Containers.—When a fibrous material is mixed with water and molded, a molded pulp container results. These containers are made of various virgin and chemical wood pulp or wastepaper pulp. The process used closely resembles conventional papermaking techniques.

Molded containers can be formed by pressure, injection or suction-molding methods. They are usually water sensitive but low cost. Uses in the food industry are for egg containers and various produce containers. Water sensitivity can be lowered by application of coatings to interior or exterior surfaces.

Flexible Packaging Materials

Flexible packages are made from combinations of flexible materials. These include the basic substrate, laminating adhesives, protective or decorative coatings, and decorative inks. There are literally millions of possible combinations. The flexible packaging engineer achieves his/her objective by selecting appropriate components and tailor-making a laminate to fit the packaging needs of a particular product.

Paper.—Paper remains an important factor in flexible packaging because it contributes strength, stiffness, smoothness and low cost. In flexible packaging, the basic paper used consists of bond, tissue, litho, kraft, glassine parchment and grease-proof.

Bond papers are uncoated sheets made of bleached chemical pulps in weights ranging from 9.07 to 31.75 kg. They are supplied with a wide variety of finishes and may be purchased with high wet and tear strength and as printing and laminating grades. Machine-finished papers are usually not smooth enough for fine printing. Machine-glazed bond papers are used for gum wraps and printing applications.

Tissues are lightweight papers made of semi- and fully-bleached chemical pulps in basis weights ranging from 3.62 through 9.07 kg with open

TABLE 1.4. SPECIALTY FILMS

Film	Uses	Comments
XT Polymer	Blow-molded tubs	Medium priced, bottle uses possible
Phenoxy	Specialized applications	Expensive
Polyphenylene oxide	Steam sterilizable pharmaceuticals	Expensive
Polyurethane	Motor oil pouches	Expensive
Fluorohalocarbon	Pharmaceutical blisters	Specialized
Acrylonitrile	Bottles, under FDA scrutiny	Future uses widespread
Starch	Food uses, edible	Processing uses in food technology
Cellulosics	Bottle molding	Future bright
Collogens	Sausage casings	Edible
Acetylated monoglycerides	Coatings and dips	Intermediate for food processing
Alginates and pectinates	Thickening agent	Edible
Polyvinyl alcohol	Laundry bags	Various types available

or closed fiber formation. Special treatments may be added such as wet strength and mold resistance. Where a bright, white tissue is required, machine-glazed bleached tissues are used. For applications requiring "bleed," porous tissues are required.

Litho papers are sheets coated on one or both sides in basic weights ranging from 13.15 through 27.22 kg. These papers are especially prepared to produce a smooth printing surface. They are not as strong as bonds, but have a better surface for reproduction of printing and are preferred for label stocks. Two-side coated litho paper is often used in publication-advertising where both sides of the sheet must be smooth for printing. Groundwood-free and groundwood-content paper incorporate a special finish.

Kraft papers are extremely strong and may be used in bleached or unbleached forms. Basic weights range between 11.34 and 36.28 kg. A wide variety of finishes and wet strength levels are available. Krafts are classified as "Southern" or "Northern." Southern grades are made from softwoods grown in the southern part of the United States. They are usually porous, nonuniform in formation and comparatively rough. They also have high tear strength. Northern kraft papers are supplied from the northern part of the United States and are more uniform, less porous and have high tensile and mullen strength values.

Due to the strength of kraft papers (kraft in German means strength), they are most useful in many applications where bonds and lithos are not strong enough. In addition, they are usually much cheaper than other stocks. A high wet-strength natural Northern kraft is used for printed cannister labels where a smooth and strong sheet is required.

Glassine papers are supercalendered chemical pulp sheets in basic weights ranging from 6.8 to 20.4 kg. They have high resistance to air and grease. Genuine glassine pulp is beaten much longer than the imitation grade and therefore the genuine glassines have the highest resistance to grease. Imitation glassines offer greater resistance than bond papers. All glassines are stronger and stiffer than most other papers and have a smooth, glossy surface. Some are fully bleached while others have a greenish, grayish or bluish cast. A pouch paper is made in a similar manner to glassines but plasticizer is added to give the paper softness and machinability. Glassines are used in candy wraps and other food applications.

Parchment papers are made of bleached chemical pulp which is immersed in a bath of sulfuric acid. They have good grease resistance and high wet strength. Greaseproof papers are essentially nonsupercalendered glassines. Both parchment and greaseproof papers have a rougher and more opaque appearance than glassines. They are supplied in grades ranging from 6.8 to 12.3 kg. Applications include butter and oleo wrappers.

Films.—A film is a thin flexible plastic sheeting having a thickness of 0.0254 cm or less. Although solid in its finished form, at some stage of its manufacture it is capable of being formed into shape through the application of heat and pressure, chemical reaction or by casting from a solvent solution.

The first commercial flexible film was cellophane. In 1924, the E.I. DuPont de Nemours Company acquired rights from La Cellophane (France) to produce the film in the United States. Cellophane is manufactured from highly purified cellulose derived from bleached sulfite pulp. The cellulose is treated with sodium hydroxide solution and carbon disulfide to produce "xanthate" which, when dispersed in sodium hydroxide, produces viscose. The viscose is extruded through a slot into acid-salt baths to produce a regenerated cellulose film. It is then washed, desulfured, bleached, softened, dried and wound up as plain, nonmoisture-proof PT film. By incorporating various coatings and modifications, over 100 different grades of cellophane are now available. Nitrocellulose-coated cellophane offers nonblocking, moisture vapor resistance (MVT), slip, flexibility and heat sealability. Saran-coated cellophane offers superior protection from oxygen and better MVT. Most cellophanes are used in packaging baked goods, confectionery, meats and overwraps. Polymer-coated (Saran) varieties are useful for oily and greasy products. Over 90.72 million kg of cellophane are produced annually and its present commercial strength is based on its wide range of coated varieties. Cellophane is fighting for survival in the highly competitive film market. Even though it has been said that modern packaging began with the introduction of cellophane, it has shown steadily declining usage during 1965–1978. Polypropylene is capturing many of cellophane's traditional markets.

Cellulose acetates are derived from cellulose treated with acetic acid anhydride. The cellulose triacetate is partially hydrolyzed. Additives include plasticizer, antiblocking agent and sometimes a uv-absorber. It is used where stiffness, high gloss and dimensional stability are important. Cellulose acetate is also used for window applications in boxes since it has a low static attraction of dusts. It is now seldom used in flexible packaging and laminates.

Polyethylene is the largest volume single film used in the flexible packaging industry. It is a polymer of ethylene and is obtained by two different processes. High pressure polyethylene is obtained by exposing ethylene to temperatures between 150° and 200°C at a pressure of about 1200 atm in the presence of traces of oxygen. This film is known as low density polyethylene. Low pressure or high density polyethylene is produced at temperatures between 60° and 160°C and a pressure of 40 atm with alkylmetal catalysts.

Polyethylene (low density) is a low cost film with moderate tensile strength and clarity. It is a good moisture and poor oxygen barrier and several grades are unaffected by mineral and essential oils. Its greatest benefit is its ability to be fusion-welded to itself to give excellent moisture protection and increased stability to heat.

Polypropylene is produced by the polymerization of propylene. It is more rigid, stronger and lighter than polyethylene. The film has low water vapor permeability, good resistance to greases, high temperature stability and good gloss. The cast-unoriented film has fair low-temperature stability and is a poor gas barrier. Polypropylene is usually used in the oriented form. By stretching the film, monoaxially or biaxially, tensile strength and abrasion resistance are greatly increased.

Polyamides are obtained by polycondensation of ω-amino acids or by polycondensation of diamines with diacids. Various grades of polyamides (Nylon) are available. Nylon-6 offers ease in handling and abrasion resistance. Nylon-11 and Nylon-12 are superior barriers to oxygen and water and have lower heat seal temperatures. Nylon-66 is very high melting and difficult to heat seal.

Polyesters are condensation products of a polyalcohol with a diacid or its anhydride. They offer excellent tensile strength, tear resistance and good aging properties. Formerly more expensive, clear film is now competing with polyolefins. Metallized polyester film is now being used for both coffee and snack food packaging.

Polyvinylchloride films are prepared by polymerizing vinylchloride in the presence of suitable catalysts. By the addition of plasticizers, flexible films are obtained. Vinyl copolymer films are used as shrinkable oriented films for dairy, meat, confectionery and beverage packaging as well as a laminate component.

Polyvinylidene chloride is usually produced as a copolymer with 13–20% vinylchloride. It contains a low percentage of plasticizers, slip agents and stabilizers. "Saran" films are clear, have excellent mechanical resistance and extremely low water vapor and gas transmission rates. The film is used for cheese, meat, sausage and dried fruit wrappers.

Polyvinylacetate is produced by block polymerization of vinylacetate with benzoyl or acetylperoxide, or by emulsion polymerization with potassium persulfate or hydrogen peroxide. Its main use is as a coating on paper; however, a copolymer is used for packaging cheese and meats in direct contact, such as seamless casing.

Polyvinylalcohols are made by hydrolysis of polyvinylacetate and are soluble in water and insoluble in many organic solvents. They are used as a lining on paper for oil and fat packaging.

Amylose Film.—Sold by American Maize Products as "Ediflex" film, it is made from corn starch. Its most unique property is that it is edible.

TABLE 1.5. PERMEABILITY AND CHEMICAL PROPERTIES OF PACKAGING FILM

	Gas Transmission[1]			Water Vapor[2]	Water	Resistance to				
Film Material	Oxygen	Nitrogen	Carbon Dioxide	Transmission	Adsorption[3]	Acids	Alkalies	Greases and Oils	Organic Solvents	Water
Cellophane										
Plain				106–198	44.7–114.8	Poor	Poor	Good	Good	Fair
NC coated	7.8–12.4	7.8–24.8	6.2–93	22–42	45–115	Poor	Poor	Good	Ctg attacked	Fair
PE coated				18.6	—	Good	Good	Fair	—	—
PVDC coated				18.6	—	Good	Good	Good	Good	Good
Cellulose acetate	1813–2325	465–620	13,330–15,500	2480	3–8.5	Poor	Poor	Good	Poor	Good
Polyamide										
Nylon 6	40.3	14	155–186	248–341	9.5	Poor	Good	Good	Good	Poor-Good
Nylon 11	527	53	2370	—	0.27	Poor	Good	Good	Good	Good
Polycarbonate	4650	775	16,660	170	0.35	Good	Poor	Good	Poor-Good	Good
Polyester	46.5–62	11–15.5	232–387	15.5–20.2	<0.8	Good	Poor	Good	Good	Good
Polyethylene										
Low Dens.	7750	2790	41,850	21.7	<0.01	Good	Good	Poor	Good (60°C)	Good
Med. Dens.	3875–8290	1317–4880	15,500–38,750	10.8	<0.01	Good	Good	Good	Good	Good
Hi Dens.	2667	651	8990	4.6	Nil	Good	Good	Good	Good (80°C)	Good
Polyethylene-vinyl-acetate	13,020	6200	93,000	31–46	<0.01	Fair	Fair	Poor	Fair	Good
Polypropylene (cast)	2325–3720	620–744	7750–12,400	10.8	<0.005	Good	Good	Good	Good	Good
(coated-oriented)	8–78	8–31	8–78	<3.8	<0.005	Good	Good	Good	Ctg attacked	Good
Polystyrene (oriented)	3875–5425	—	13,950	108–155	0.04–0.10	Good	Good	Poor-Good	Poor-Good	Good
Polyvinyl chloride-acetate (plasticized)	310–2325	155–930	1085–12,400	77.5–124	Neg	Good	Good	Fair-Good	Poor-Good	Good
Polyvinylidene chloride-vinyl chloride	12.4–107	2–23	59–682	3.1–9.3	—	Good	Good	Good	Fair-Good	Good

Courtesy of Modern Plastics Encyclopedia. 1978–79. Vol. *55* No. 10A. McGraw-Hill, New York

[1] cm^3/m^2/25.4 micron thickness/24 hr/atm at 25°C

[2] g/m^2/25.4 micron thickness/24 hr at 37.8°C and 90% RH

[3] % by weight adsorption of water in 24 hr immersion test (25.4 micron film)

Amylose film is also very soft, slightly hazy, stretchable and water soluble. The film is a poor moisture barrier; however, it is satisfactory as a gas and flavor barrier. It is outstanding in its resistance to oil and grease.

Possible uses for amylose films include an inner wrapper for frozen foods where the film would dissolve during thawing. Other possibilities are its use in portion packs, dehydrated soups and other foods that are dissolved in water.

Ionomers.—The introduction of Surlyn A by E.I. DuPont de Nemours Company marked the development of a new family of thermoplastics called ionomers. Based on low density polyethylene, the ionic bonds in ionomers serve to increase overall bond strengths and yield superior oil, grease and solvent resistance.

At normal temperatures, Surlyn A is tougher than polyethylene and is more sensitive to water absorption. Its high melt strength makes Surlyn A extremely suitable for extrusion coating, vacuum forming and skin packaging. Applications requiring deep "draw" vacuum forming are potential markets. Skin packaging, blister packaging and blow-molded bottles are likely areas of development. Spicy and greasy products can be packaged in Surlyn A laminates since grease hold-out is greater than low density polyethylene. Surlyn A has been used as the inner component of a Tetra-Pack container holding olive oil.

Since Surlyn A has superior hot melt strength as compared to polyethylene, thinner coatings enable a converter to offset part of the higher raw material cost.

Ethylene-vinyl Acetate Copolymers (EVA).—Copolymerization offers polymer chemists a method of obtaining tailor-made polymers with specific properties. Ethylene-vinyl acetate copolymer results from the copolymerization of high pressure (low density) polyethylene and vinyl acetate. EVA is more flexible than polyethylene; however, it is also more permeable to water vapor and gases. Impact strength is excellent and the film is clearer than polyethylene. EVAs are thermally unstable at high heats but very stable and versatile at lower temperatures and have a wide heat-seal range.

A significant disadvantage is EVAs tendency to block due to high surface friction. The polar nature of the side groups present in the copolymer cause blocking characteristics. The addition of slip and anti-block additives partially overcome blocking. In applications requiring high friction, EVA may prove extremely useful. As a coating for bags and industrial sacks demanding stackability, high surface friction is desirable. Future applications include use as a shrink film. Bags can be produced to stretch-wrap products in the form of a contour wrap. The

TABLE 1.6. MECHANICAL PROPERTIES OF PACKAGING FILMS

Film material	Tensile strength[2] (100 p.s.i.)	Elongation[2] (%)	Tearing strength[2] (g mil)	Bursting strength 1 mil. thick (p.s.i.)	Folding endurance/ 1 mil thick (no. of folds × 10^3)	Heat sealing range[2] (°C)
Cellophane-plain	104–186[3]	14–36[3]	275–426[3]	—	—	—
NC coated	70–150	10–50	2–10	55–65[1]	Over 15[5]	93.3–148.8
Saran coated	70–130	25–50	2–15	—	—	107.2–176.6
Polyethylene coated	50 and over[1]	15–25[1]	16–50[1]	40–50[5]	Good	110–148.8[1]
Cellulose acetate	54–139[1]	25–45[1]	2–25[1]	50–85[3]	0.25–0.4[1]	176.6–232.2
Polyethylene						
Low density	13.5–25	200–800[1]	150–350	48[5]	Very high[5]	121.1–176.6
Medium density	20–35	150–650[1]	50–300	—	—	115.5–176.6
High density	24–61	150–650[1]	15–300	—	—	135–176.6
Polyester (Mylar Scotch-pak, Videne)	170–237	70–130	10–27[1]	45–50[3]	20[3]	135–204.4
Polypropylene	45–100	400–600	32–1750[1]	Very high[1]	—	162.7–204.4
Polystyrene (oriented)	80–120	3–20	20–30	23–60[1]	—	121.1–162.7
Rubber hydrochloride (Pliofilm)	35–50	200–800[1]	60–1600	Stretches[2]	10–1000[5]	121.1–176.6
Vinylidene						
Cryovac	60–120[1]	50–100[1]	15–20[1]	—	Over 500[5]	135–148.8
Saran	80–200	40–80	10–100[1]	20–40[3]	—	137.7–148.8
Vinyl chloride	30–110	5–250	30–1400[1]	25–40[5]	250[1]	93.3–176.6
Nylon 6[4]	138–170[4]	200[4]	50[4]	No burst[4]	—	204.4–232.2[4]

[1] Source: Data by permission Modern Plastics Encyclopedia Issue for 1962.
[2] Source: Data by permission Modern Plastics, 1961 all values unless otherwise noted.
[3] Source: Stone and Reinhart (1954).
[4] Source: Miller (1959).
[5] Source: By permission Modern Packaging Encyclopedia Issue 1960. Copyright 1959 Packaging Catalog Corp., New York.

extremely high flexibility of EVA enables it to stretch tightly without the use of a shrink tunnel. In addition, EVAs high impact strength allows the use of thinner gauge films for similar applications than polyethylene.

No pretreatment is necessary to achieve excellent ink adhesion in printing; the resin can be blow-molded, injection-molded and extruded. Stress cracking resistance is good thus allowing EVA to be used for detergent bottle production.

Polypropylene Copolymers.—Polypropylene offers excellent water, gas and strength properties; however, its cold temperature impact strength is poor. In an effort to overcome this defect, a second monomer, usually ethylene, is introduced into the propylene structure. The low temperature impact strength of polypropylene is improved by the addition of ethylene and potential uses include blow-molding applications.

Polypropylene copolymers are particularly useful in shrink wrapping. Due to the reduction in crystallinity, a lower melting point is achieved. Orientation thus takes place at fairly low temperatures. Additional copolymers of polypropylene include the introduction of 1-pentene, 1-butene or 1-hexene. The subsequent films are clear and offer improved impact strength compared with the polypropylene homopolymer.

TABLE 1.7. METALLIZED FILM CHART[1,2]

Film	Volume	Use (present and potential)
Polyester	$12MM; 2.4MM lb	Coffee pouches, snack foods, baked products, vegetables
Nylon	$800,000	Various
Polyethylene	$700,000	Various
Cellophane	$900,000	Various
Polypropylene (oriented)	—	Snack foods, condiments, dehydrated mixes, labels

[1] Potential Market for Metallized Films = 40 MM lb/yr (high barrier uses).
[2] Potential Market for Metallized Films = 5 MM lb/yr (nonbarrier uses).

TPX.—Chemically, TPX is closely related to polyethylene and is based on the polymerization of 4-methylpentene-1. The material is transparent, has the lowest specific density of any plastic material (0.83) and has outstanding electrical properties. Its clarity is excellent and the crystalline melting point of 245.5°C enables it to be used for hot-filling and cooked food applications. The resin can be extruded, blow-molded and injection-molded.

A major disadvantage in TPX is its high permeability to water vapor and gases. The first production facility by Imperial Chemical Industry, Ltd. (United Kingdom) was established in 1967.

Halogenated Polyethylene.—Polyethylene has been produced incorporating chlorine in the monomer. The addition of chlorine enables polyethylene to be printed without any pretreatment and does not effect film flexibility. The film is called halogenated polyethylene. Since the film is nontoxic, it has applications in conventional polyethylene markets. The higher film price must be evaluated relative to potential uses.

Coextruded Structured Films.—Laminations of two films such as cellophane and polypropylene have been available for some time as also have been coated films such as Saran-coated polyesters, nylons and polyolefins. Films called structured films are now being produced by a process of coextrusion (simultaneous extrusion through complex dies). Intermixing of the polymers is avoided through the maintenance of laminar flow throughout the molten phase. Both flat die and blown-structured films are being manufactured. One of the first commercial products was a polyethylene-polypropylene-polyethylene film designed for breadwrap. Another structure available is a copolymer-homopolymer-copolymer polypropylene film. In Denmark and England, a combination of white-black-white pigmented polyethylene is used for packaging milk. It is predicted that about 349.3 million kg of polymer will be used in coextruded film by 1982. Uses include snack foods, trash bags, shipping sacks, pharmaceuticals and meats. Use of coextrusions with metallized films is also considered to be very promising.

Polycarbonate films are heat-sealable and have excellent clarity, high impact strength and good temperature resistance. Although they have been commercially available for several years, their high cost has retarded most flexible packaging applications. The film is resistant to weak acids and alkalies but is attacked by aromatic and chlorinated hydrocarbons. Permeability to gases and water vapor is poor.

Possible markets for polycarbonate films include rigid and semirigid applications such as blister packs and thin-walled bottles for milk. In Japan, a thin-wall polycarbonate bottle is used for packaging salad oil. Another Japanese development consists of a soda siphon which contains plain water and a carbon dioxide charge. Polycarbonates withstand the effect of high internal pressures.

The raw material costs inherent in the production of polycarbonate resin are expensive. Future price reductions will probably not bring polycarbonates into the price range of polyethylene.

Phenoxy Films.—Since phenoxy films are chemically related to epoxy resins, they are extremely hard, scuff-resistant and exhibit superior heat resistance than most other thermoplastics. Bottles, films and extrusion coatings made of phenoxy are very resistant to oils and greases. Permeability to gases, moisture and odors is low. One major problem is

Phenoxy films are subject to stress-cracking with g more than 40% alcohol.

clarity of phenoxy films coupled with good impact gidity made blown bottles a feasible outlet. Several ch as shampoos have already been packaged in blown pheno...

The price of phenoxy resin is fairly high. Most curing type plastics are expensive due to the cost of curing systems. They are unlikely to become significantly cheaper; however, phenoxies are most useful for highly specialized applications.

Acrylic Films.—XT, acrylic copolymer exhibits excellent clarity, oil, grease resistance, gas and odor permeability. Its rigidity is claimed to be four times that of high density polyethylene and the material exhibits good impact strength. It has poor resistance to aromatic and chlorinated hydrocarbons, and alcohol containing products cannot be packed in XT polymer.

Potential markets for this polymer include rigid bottles for foods and cosmetics. The excellent clarity of XT makes direct-printed labeling possible.

Acrylonitriles.—Are copolymers of acrylonitrile, methyl acrylate (or styrene), and butadiene. Examples include Vistron's "Barex" and Monsanto's "Lopac." There is some use of Barex film in the pharmaceutical area.

Polyurethane Films.—The most important properties of polyurethane film are toughness and strength. It is a very soft and abrasion resistant film. Due to polyurethane's excellent oil and grease resistance, it is currently being used as a material for motor oil packaging. Other possible applications include boil-in-bags and flexible meat packages. Solvent resistance is good toward aromatic hydrocarbons and gasoline.

Aluminum Foil.—Aluminum foil is defined as a solid sheet rolled to a thickness less than 0.01524 cm. Plain aluminum foil refers to foil, in either roll or sheet form, that is not combined with another material such as paper, film or cloth. In speaking about plain foil, bare foil, unmounted foil, unsupported foil and free foil are used synonymously. The bulk of converting and packaging applications utilize foil in rolls.

Aluminum foil is available in different tempers and alloys. The high purity grades of alloys (1000 series) are normally used for foil packaging applications. The most widely used alloys for packaging foil are 1235 and 1245. For drawn and/or formed containers, 3003 alloy is used. It has a 1.0–1.5% manganese content.

"O" temper is produced by subjecting the foil to controlled heat followed by controlled cooling. Foil with "O" temper is referred to as

TABLE 1.8. AMOUNTS OF DIFFERENT GASES TRANSFERRED IN 24 HR BY DIFFERENT FILMS AT ROOM TEMPERATURE (cm^3/m^2/25.4 micron thickness/24 hr/atm 25°C

Type of Film	Temp.—R.H. Variations	Nitrogen	Oxygen	Carbon Dioxide	Air	Ethylene	Ethylene Oxide	Sulfur Dioxide
Cellophane (MSAT)[1,2]	—	496	666	1,720	542	—	—	—
Cellophane (MSAD)[4]	2% RH	—	98	574	—	—	—	—
	100% RH	—	3184	5,375	—	—	—	—
Polyvinylidene[1] (Cryovac)	—	48	172	420	79	—	—	—
Cellulose acetate (P903)[3]	25°C	—	5464	29,636	—	8513	302,250	286,900
	0°C	—	2049	12,540	—	—	86,650	121,646
Cellulose acetate (P912)[3]	25°C	—	3875	14,446	—	1747	70,680	84,400
	0°C	—	1747	6,526	—	—	—	—
Polyester[4] 12.7 micron	2% RH	—	289	382	—	—	—	—
	100% RH	—	770	1,341	—	—	—	—
Polyethylene 50.8 micron	2% RH	—	119	—	—	—	—	—
	100% RH	—	241	—	—	—	—	—

[1] Values by Cryovac C., Food Processing *17* (8) (1956).
[2] With moist gas at high relative humidity.
[3] Values by Celanese Corp. on two type of film differing in flexibility and toughness (converted from investigator's values expressed in cc per m^2 per 24 hr per cm mercury pressure).
[4] Values by Nagel and Wilkins (1957) converted from investigator's values expressed in grams per m^2 per hr.

dead-soft and has the lowest physical properties. It has good folding characteristics and is widely used in converting operations.

"H" temper foils are produced by strain hardening the metal during rolling. H-12 or H-14 temper metal are strain hardened by rolling to temper after an intermediate anneal. The strength characteristics of tempered foils are improved as contrasted to soft foil; however, close adherence to certain properties is difficult to control. Full hard foil is H-18 which is fully strain-hardened by rolling. There is no intermediate or final anneal. However, it is halted when specified strength properties are attained. Full hard metal is the direct opposite of dead-soft metal.

Superhard foil is H-19 which is fully strain-hardened by rolling and taken to the maximum physical properties. H-24 metal is an intermediate temper strain hardened by rolling and then partially stress-relieved by annealing. It is used in various container applications. Most foil sold and used by packaging converters is "O" temper.

Flexible Package Forms

Wrappers.—Flexible packages are made from materials whose final shape conforms to and is governed by the product enclosed, as opposed to rigid packages which require the product to conform to the shape of the container. The simplest flexible package is a wrapper which may be loose or tight and may be sealed shut by using heat or various types of adhesives. Some wrappers are left unsealed as is done on stick chewing-gum wrappers, others are partly sealed to hold them in place as is done with carton overwraps. The simplest wrappers are sheets of paper, metal foil or plastic film, with or without decoration. Many films can be heat-sealed, but paper and foil require coatings in order to make them heat-sealable. Early types of paper wrappers were glued shut. The earliest types of heat-seal coating were waxes. Many advances have been made in wax formulations and hot melts so that today some wrappers can be sealed with pressure alone (cold tack adhesives). The advent of plastic resin coatings, such as the vinyls, nitrocelluloses, Sarans and polyolefins, made it possible not only to add heat-seal properties but also to substantially improve strength and barrier properties of laminates of paper and plastic. Metal foil is in itself an excellent barrier. Coatings and other substrates serve only to add strength, decoration and heat sealability.

A wrapper that comes in direct contact with a food such as candy or a loaf of bread is called an "intimate wrap" whereas if it wraps an inner package such as a carton it is called an "overwrap." Wraps may be purchased as roll stock or as precut sheets. The latter are usually used for hand or semiautomatic production.

Other variations of wrappers include labels which wrap only part of a rigid package and serve primarily to identify the contents. Some completely wrapped-around labels on paper containers contribute barrier protection also.

Another variation of a wrapper is a bundling overwrap which combines several smaller packages into one larger unit. Plastic film shrink wraps serve this function. The ultimate is extremely large pallet overwraps where a plastic film stabilizes an entire pallet load of smaller packages.

There are at least 13 distinct types of twists and folds by which wrappers may be closed neatly before sealing. Overwrapping machinery has been developed for each type of wrap and fold and for specific products. Speeds range as high as 1000 units/min for twist wrap candies.

Preformed Bags or Envelopes.—A bag is a tube with one sealed-end and is made from a flexible material. After filling with the product it may be closed by sealing, tying or clipping shut. An envelope is a bag with a flap that can be folded over to close. A paper bag was one of the first converted packages. Paper bags may be made in four standard styles: flat, square, satchel and self-opening. A flat bag has a longitudinal back seam and no gussets. The bottom is folded over and glued or sealed.

A square bag has side gussets which give it a much greater capacity. The satchel bag has no gussets but the bottom is folded into a hexagonal shape and sealed. When filled, the bag will stand by itself. The self-opening bag has side gussets or they can be contour-sealed in order to more closely fit irregular products.

Pouches (Form-Fill-Seal).—A pouch is similar to a bag except it is simpler in construction and is made from roll stock, filled and sealed on automatic high speed machinery. There are many possible variations of pouch sizes, shapes and compositions. A few general types can be described.

Pouches can be defined as vertical or horizontal, depending on how they are formed. Vertical pouches are made by drawing one or two webs downward around a forming mandrel, sealing one or two seals to form a tube and adding a top and bottom seal for closure. When one side seal is used and centered, it is called a pillow type. If the side seal is inserted on one side and a fold on the other, it is called a three-side seal pouch. When two side seals are used, it is called a four-side seal pouch or fin-sealed pouch. The product is added while the pouch is in the vertical position and usually before it is severed from the web. Side seams can be lapped or butted and gussets are optional.

Horizontal pouches are made by drawing one or two webs in a horizontal direction. Usually when two webs are fed they are positioned in the horizontal plane and formed into a four-side sealed pouch. The product is added before the top web is sealed in place. Frequently the bottom web is formed into a pocket. When a single web is fed, it is usually drawn over a plough to fold it into the vertical plane. The bottom may be sealed or left unsealed. Side seals are made and the package is filled through the top, usually after being severed from the web. Again gussets are optional.

When packaging small items, frequently the individual pouches are not separated but left in a continuous strip. Perforations or notches are inserted so that individual packets can be torn off as desired. Odd-shaped pouches can be formed by using contoured heat-seal jaws.

Tetrahedral pouches are made by forming a tube on a vertical form and fill machine and inserting cross seals alternately at 90° angles to the preceding cross seal. The tetrahedron uses the least amount of material for a given pouch volume.

Some pouches or envelopes are sold and displayed as such, however, many are given added protection by inserting them in folding cartons. This is nearly always done for small items or portion packs.

Collapsible Tubes.—The collapsible tube has traditionally been made of soft metal, tin, lead, tin-lead and aluminum. One end of the tube is a threaded dispensing nozzle which can be closed by a screw cap closure. The entire tube is formed by impact extrusion from a slug of metal, then it is trimmed, threaded and annealed. Exteriors are decorated by roller coating and offset printing. Interiors may be lined by dipping, flushing or spraying prior to exterior coatings. Linings may be waxes, vinyl, phenolic or epoxy resins. The product is filled through the open bottom which is then crimped or sealed shut. In use, the closure is removed, the interior seal (if present) is punctured and the product is dispensed through the nozzle by squeezing the tube. Tubes are marketed on cards, in blister packs or in folded cartons.

Blow-molded plastic tubes are being used for some products. Their main disadvantage is lack of dead-fold. They tend to spring back when squeezed making total dispensing of the product difficult. Tubes are best for packaging thick liquids or thin pasty solids.

Also used are tubes made of flexible laminates usually constructed of a complex paper/PE/Foil/PE material in multiple layers. These are widely used for toothpaste. A significant advantage of a laminate tube is that they can be form-fill-sealed on a machine in a users plant. Thus, a user simply purchases rollstock from a converter and fabricates the tubes in-house.

FACTORS INVOLVED IN THE CREATION OF A FOOD PACKAGE

What is a Package?

A food package is a structure designed to contain a food product in order:

(1) to make it easier and safer to transport,
(2) to protect the product against contamination of loss,
(3) to protect the product against damage or degradation, and
(4) to provide a convenient means for dispensing the product.

The addition of printing or other decoration to the exterior of the package serves:

(1) to identify the contents as to type and quantity,
(2) to identify the manufacturer's brand and quality grade,
(3) to attract the buyer's attention,
(4) to persuade the buyer to purchase, and
(5) to instruct the purchaser on how to use the product.

Ease and Safety of Transport.—The first packages were probably invented to make transportation easier. It is impossible to carry a liquid such as water or milk in one's hands and very difficult to carry small solids such as grain, berries or fruits. Open-topped rigid containers such as cups, pails, bowls, urns, baskets and panniers or flexible containers such as sacks and wine-skins make transportation possible. Modern packages are carefully tested to insure they will survive the rigors of transportation.

Product Protection.—The development of wrappers and closed containers was for protecting products against spillage, evaporation or pilferage losses, and against contamination by dirt or vermin. Modern packages also protect against contamination from microorganisms and product degradation from exposure to environmental factors such as heat, light, moisture and oxygen.

Dispensing.—Easy opening and reclosing features make it easier and safer for the consumer to dispense desired quantities of a food from a package.

Selling the Package and Identification of Contents.—Since most bulk shippers and many individual packages are made from opaque materials, some means for identifying the contents is imperative. Over the years this has evolved from simple markings to highly decorative labeling which not only instructs the purchaser as to the quantity, quality and type of contents but also first attracts his/her attention and then persuades him/her to purchase the product.

Birth of a Package

In designing, developing and testing a food package, the manufacturer of the food product today can draw upon a host of new materials and new

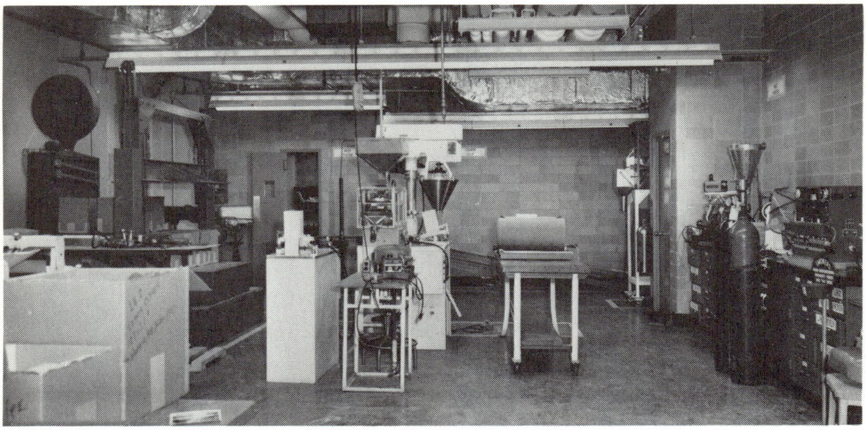

Courtesy of General Foods Technical Center

FIG. 1.17. MODERN PACKAGING LABORATORY
Pilot packaging laboratory with machines for making many types of packages and tests for machine performance.

packaging designs that did not exist years ago. He/she can rely on the talents of his/her own packaging development laboratories or can call upon outside packaging laboratories for assistance. Most of the latter have been established by manufacturers or converters of packaging materials for the purpose of developing new markets for their products. In either case, there are certain steps that must be followed in bringing a food package to the marketplace with reasonable assurance that it will sell.

Idea Initiation

The idea for a new package may be initiated in any one of a number of ways, in the mind of a research scientist, a product or marketing person, a designer, a field salesperson, or a production person on the line. Whatever its source it will be aimed at one of the following:

(1) To modify an existing package for an existing food product in order to improve sales through:
  (a) improved aesthetics of package design,
  (b) improved price due to lower cost materials, to improved manufacturing efficiencies, or to improved handling performance,
  (c) improved product quality due to improved package protection,
  (d) improved product utility due to improved package performance (such as an easier opening device or a better pouring spout) which provides greater convenience.

(2) To put a new food product in a well-tried and proven package in order to improve sales through expansion of the product line.
(3) To develop an entirely new package concept for a proven food product to improve sales through:
   (a) revived customer interest,
   (b) improved product quality,
   (c) improved product utility (convenience).
(4) To develop a new package concept for a new and untried food product.

Typical examples of each of these endeavors would be:

(1) the substitution of a polypropylene film for a cellophane film in order to gain better performance (less breakage) when the product is machined,
(2) the marketing of a new soft drink in a plastic bottle as opposed to a glass bottle,
(3) the marketing of milk in a polycarbonate gallon (liter) container,
(4) the marketing of soft margarines in metal or plastic tub containers.

Regardless of the source of a given idea, the market manager in charge of the food product must decide whether there is a need for such a package and whether it can be sold in sufficient quantity and at a price that will produce the desired return on investment or profitability. The market manager is the one who pushes the "stop" and "go" buttons which get the program started, carry it through to completion and at some point terminate it. In making these decisions there are a number of steps, which when taken more or less in sequence, will develop factual information and help one arrive at a sound judgement.

Initial Market Appraisal

Marketing personnel make an initial appraisal of the market potential for the proposed product and in so doing establish "ball-park" figures on optimum sizes and maximum costs for the product and the package.

Initial Design Concepts

Design concepts are collected from all available sources by the packaging engineer assigned to the project. "Blue sky" thinking and "brain storming" meetings are encouraged at this point. Ideas are then screened to rule out the impractical—that is, those that are drastically out of line costwise or which can not be manufactured by existing technology.

Two lists are prepared. The first list includes all ideas which are feasible with relatively minor changes in materials or manufacturing techniques. The second list includes those which would be excellent packages but which require research to develop a presently not available material or manufacturing technique. In most instances it is decided to go with items from the first list, as these involve less investment in time and money. There may, however, be important reasons for selecting an approach from the second list. A successful long-range development may be much less costly per unit or it may be unique and patentable.

TABLE 1.9. HOW PACKAGE SIZE AFFECTS RATIO OF SURFACE AREA TO CONTENTS

	Package Sizes		
Net wt (g)	453.5	226.8	113.4
Dimensions (cm)	16.8 × 9.2 × 6.4	12.7 × 7.6 × 5.1	10.2 × 6.4 × 3.8
Area (cm^2)	485.2	400.0	254.8
Ratio (cm^2/g)	1.07	1.76	2.25
Ratio (cm^2/cm^3)	0.49	0.81	1.03

From: Brickman (1957B).

Design Concept Testing

Where many ideas abound or where there is some question as to acceptability of a single concept, a consumer panel test can be arranged. Here one or several groups of consumers are asked to look at artists' sketches or at mock-up samples of each idea and to comment upon them. From this interplay of comments helpful conclusions often can be drawn. Care must be taken in intepreting results, however, and skilled professional market research people should be utilized, as it is all too easy to make incorrect decisions based on inadequate data.

Package Design and Performance Testing

The packaging engineer now takes over. This person examines the criteria defined by Markets and makes a selection of packaging forms and raw materials. This selection involves many factors, to name but a few:

(1) package sizes,
(2) nature of product (perishable, fragile, liquid, solid, etc.),
(3) package barrier requirements (air, moisture, grease, etc.),
(4) temperature extremes to be encountered,
(5) does package require easy opening device, reclosure?
(6) will product be packaged before or after processing?
(7) shelf life,

(8) stackability,
(9) in-use conditions,
(10) ease of disposal,
(11) competitive packages,
(12) dealer pack,
(13) shipper,
(14) pallet size or other load unit, and
(15) will package be reused?

Within each package form or category other more specific questions need to be answered; for example, suppose the package is to be a flexible lamination formed into a sealed pouch:

(1) How many plies should the laminate have?
(2) What is the function of each ply?
(3) Which ply should be closest to the product?
(4) What thickness should be used?
(5) How should the plies be joined together?
(6) How should the pouch be formed and sealed?
(7) What outer protection (box, shipper, etc.) will be needed?

After making this selection, the packaging engineer prepares sample packages of each proposed design concept that has survived previous screenings. He/she may introduce several minor variations for each concept and then test the samples to see which ones best meet the performance criteria. These may include tests for strength, product compatibility, heat resistance, permeability, water immersion, resistance to cold and the like. This period of testing will eliminate some design concepts as technically or economically unfeasible and will narrow the choice to a relative few. By this time the selection of the basic concept can be fairly well established and unresolved differences will be relatively minor.

Selection of materials, preparation of samples and performance testing may be a short-term activity or may involve several years of intensive work. This will depend on the type of package development. Minor modifications are short-term problems. In some new package concepts total systems may be required. This may well involve learning new technologies or developing new equipment for manufacture of the package, for filling and closing it or for testing it. For example, in developing a flexible package for heat-sterilized foods, one of the most difficult problems was how to measure the change of temperature of the food in the center of the package, when the package was sealed and was moving in a tortuous path through a pressurized system of steam and air at 121.1°–126.6°C inside a closed heavy steel "cooker." This was solved by placing a radio transmitter in the package which radioed a signal to the antennae mounted on the interior of the retort.

Manufacturing Capability

Now the packaging engineer concentrates his/her attention on establishing manufacturing feasibility (although he/she has not ignored this facet in previous efforts). Trial manufacturing runs are made to get more accurate estimates of cost and to find and "debug" possible problems. Preliminary shipping and handling tests are conducted to help establish design of shipping containers and package performance. Sufficient information is generated at this point for manufacturing to start estimating production costs with some assurance. Plans for consumer placement tests can be begun and test markets studied. Design of the package artwork copy can be brought to a tentative completion.

(Some feeling for the myriad details and decisions required prior to reaching the market place with a new product can be obtained from the following generalized lists:

The Product.—Its specifications; formulation tolerance; analytical techniques for specifications; sampling techniques for specification control; patents on the product itself; toxicity clearance; quality control manual; and uniformity of production from panel evaluations.)

Marketing and Advertising Aspects of Package Design

Not too long ago, the primary aim of a package was to provide protection. Marketing and advertising concepts were generally unsophisticated. Our rising standard of living and the advent of self-service groceries led to a bolder and freer approach to packaging. Packaging now is an integral part of merchandising and every packaging decision has an important effect on sales. The package is now considered to be a top-management marketing tool.

The consumer must choose from a wide range of packages, products and sizes every time he/she shops. In just ⅕ of a second the package must be able to convey the message, "buy me!" No advertisement is ever read as often as a package.[5]

If the package is to sell in this brief period of time, certain factors must be considered:

(1) Does the package have a distinctive shape?
(2) Does it feature a brand name in a standardized location?
(3) Does the manufacturer's trademark or symbol appear plainly? What about the manufacturer's name?

[5]Of the estimated $59 billion spent for advertising in noncommunist countries in 1976, $34 billion or 57% was spent in the United States. Other nations spending at least $1 billion (U.S. dollars) for advertising were Australia, Brazil, Canada, France, Japan, the Netherlands, Spain, Great Britain and West Germany.

Courtesy of General Foods Technical Center

FIG. 1.18. CONTROLLED ATMOSPHERE TESTING ROOM
Interior of controlled atmosphere package testing room with temperature and humidity set to simulate climatic conditions in Jacksonville, Fla.

Courtesy of National Biscuit Company Research Center

FIG. 1.19. CONTROLLED ATMOSPHERE TESTING CABINET
Large controlled atmosphere cabinet with glass doors to permit locating samples before opening to minimize time cabinet is open.

Courtesy of General Foods Technical Center

FIG. 1.20. CONTROLLED ATMOSPHERE SERIES OF TESTING ROOMS
Doors to a series of controlled atmosphere rooms provided with locks to prevent access of unauthorized persons, thus preventing theft and insuring validity of observations.

(4) Is the product name readily identifiable?
(5) What type of decorative technique is to be used?
(6) Are colors appropriate and distinctive?
(7) Is the package attractive and pleasing?
(8) Does the package give adequate and clear information on quality and quantity of contents and how to use them?

Courtesy of General Foods Technical Center

FIG. 1.21. DETERMINATION OF OXYGEN PERMEABILITY
Apparatus used for determining oxygen in head space voids and intercellular spaces in packaged foods utilizing Beckman Oxygen Analyzer.

In addition to these rather obvious questions there is an area which involves the psychology of design. This includes considerations of the implications of color and shape, for example, the connotation of cleanliness and coolness for blue colors, of tranquility for green, of warmth for red and orange, and of strength and cheerfulness for yellow. Strong hues are masculine; pastels are feminine. Ovals and circles are liked by all, but only men like triangles.

Symbolism enters into selection of design motifs, but one symbol may be meaningful to one culture and meaningless to another. Snow flakes have meaning in northern countries but mean little to a resident of

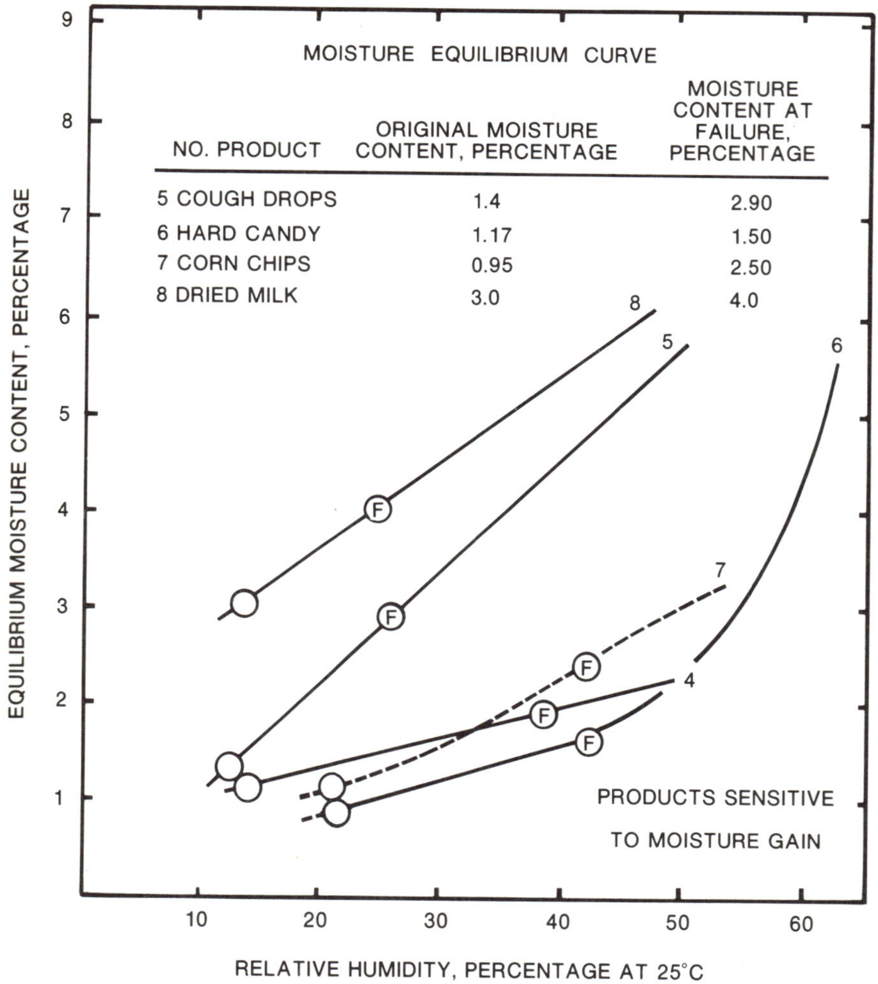

From Brickman (1957A)

FIG. 1.22. MOISTURE EQUILIBRIUM CURVES FOR FOOD PRODUCTS SUSCEPTIBLE TO MOISTURE GAIN
Open circles indicate initial moisture, circles with letter F indicate moisture content at which products become unsalable.

tropical Africa. The toadstool-mushroom is good luck in central Europe, but is meaningless elsewhere. The shape of a design may be detrimental. When a United States designed tunafish can was introduced in Japan, it featured a tuna above the waves with its nose pointing down toward the water. The package sold poorly and questions were raised as to its failure.

TABLE 1.10. COLOR AND SYMBOLISM IN PACKAGING FOR ASIAN MARKETS

Country	Color	Connotation	Symbol	Connotation
China	white	mourning (avoid)	tigers, lions, & dragons	strength (use)
Hong Kong	blue	unpopular (avoid)	tigers, lions, & dragons	strength (use)
India	green & orange	good (use)	cows	sacred to Hindus (avoid)
Japan	gold, silver, white, & purple	luxury & high quality (use)	cherry blossom	beauty (use)
	black	use for print only, prefer gay, bright colors.	chrysanthemum	royalty (avoid)
Malaysia (Population is mixed Malay, Indian, Chinese)	yellow	royalty (avoid)	cows	sacred to Hindus (avoid)
	gold	longevity (use)	pigs	unclean to Moslems (avoid)
	green	Islamic religion (avoid)		
Pakistan	green & orange	good (use)	pigs	unclean to Moslems (avoid)
Singapore	red, red & gold, red & white	prosperity and happiness (use)	tortoises	dirt, evil (avoid)
	red & yellow	Communist	snakes	poison (avoid)
	yellow	avoid	pigs & cows	same as for India & Pakistan (avoid)
Taiwan	black	avoid	elephants	strength (use)
Thailand			elephants	national emblem (avoid)
Tahiti	red, green, gold, silver, and other bright colors.	use		
Arab & Moslem States	white	avoid	animals	avoid
			pigs	religious pollution (avoid)
			Star of David	political (avoid)

Information courtesy of Hygrade Packaging Company (New Zealand).

Market research pointed out that a down-pointing nose on a tuna signified that the fish was dead to the Japanese consumer. As soon as the design was modified, sales increased.

A purple package introduced in Hong Kong was a flop because purple is an unlucky color to the Chinese. On the other hand, red signifies prosperity. When the package color was switched to red, the product was a success. Purple signifies death in some countries, luxury in others. White signifies purity to some and mourning to others. A green-colored wrapper on bread connotes mold, while a brown wrapper connotes "oven-baked freshness." Brown glass containers for milk are popular in northern Europe. Despite the better keeping quality of the milk, United States' consumers rejected the brown glass, perhaps associating it with dirt.

In designing packages for export there are other problems relating to language, for example "gift" in German means "poison."

Consumer Placement Tests

While the initial design concept tests were conducted with panels made-up of consumers, the number of people involved was relatively small. When the product has been brought to the point where technical and economic feasibility appear good and major design features can be fairly well established, it is nearly always wise to prepare a few hundred samples of each minor design variation and to conduct a consumer placement test. In this test, sample packages are coded and given to randomly selected consumers in predetermined communities. Questionnaires are filled out before and after the consumer has tried the product. The consumer comments on the quality of the product, on how it performs, on the convenience and attractiveness of the package, and on what he/she would be willing to pay for the product. These answers serve as guidelines in further narrowing the number of design concepts and in establishing criteria for test markets.

The placement test may also reveal a glaring weakness previously overlooked. This may require further research or further design changes and a second consumer placement test. It may even represent total rejection. It is better to learn what reaction the package will receive at this point than at a later, more costly point in the development program.

Consumer Reactions Are Questionable

It is important to note that even "concept testing" may lead to failure due to the unreliability of results. A new package should not be tested and developed solely via consumer conceptual testing. In our present-day affluent society, people buy psychological pleasures and not biological necessities. It is important that psychological devices be integrated into a concept test. An additional problem inherent in concept testing is that only creative people deal with concepts. The majority of people must relate to discrete "things" and not broadly defined concepts. For these reasons, concept testing can be unreliable. For example, when consumers were asked whether they would want to purchase yellow margarine, over 90% said that yellow margarine was not needed. Yet, today it is still on the shelves and provides a profitable market. Again, when consumers were asked whether they would use instant coffee, they said it was an unnecessary product. Yet, it too is highly successful.

Test Market

When consumer placement test results have been analyzed, selection of the design or designs to be used in the test market is made. Final artwork and copy approval is obtained and manufacturing is asked to produce a sufficient quantity of the product in each package design to fill and sustain the test market areas. Specifications are provided by Packaging. Marketing Research plans the test as to what market areas, what type of stores and distribution, what type and quantity of advertising, etc. Manufacturing establishes the production capability, produces the necessary quantity of packaged product and ships it to the test market areas. Packaging personnel remain in close touch with Manufacturing during the production runs and make quality checks on the product as produced. Packaging engineers participate in the market test by observing package performance in the field.

Marketing Research conducts the test market and reports and analyzes results. If these are favorable, final design, artwork and package specifications are established and the program is turned over to Markets for final exploitation. Manufacturing and Markets work together in establishing full product capability and in developing a national distribution and selling program.

The test market is the final step in the development of a package. It is here that all the predictions and guesswork meet reality in the marketplace. Here consumers must back their words with their money. A test market is expensive and should not be undertaken unless all foreseeable problems have been defined and solved. The package must have a very high level of probable success to justify putting it into the market area. Not the least danger is in allowing competition to become aware of a new development. Competitors have been known to enter the test market area and by heavy advertising and discount prices disrupt the accuracy of the test. It is best to conduct test markets in several areas at the same time to minimize this threat. Package failure during a test market may prove catastrophic. It can be disastrous to a program to have to withdraw a product for performance reasons. If all prior work has been done carefullly and thoroughly, the risk can be minimized.

REFERENCES

ANON. 1969. Packaging and packaging materials with special reference to the packaging of food. Food Ind. Stud. *5*, U.N. Indus. Dev. Organ. New York.

ANON. 1970. Shelf-life data quest quickens, now pays off. Packag. Eng. *15* (10) 60–63.

ANON. 1971. Packaging for food. SPA Mobil Oil Co., Ltd., London.

ANON. 1978. Food packaging: Where it's been, where it's going. Mod. Packag. *51* (7) 17–23, 41.

BINSTED, R., DEVEY, J.D. and DAKIN, J.C. 1962. Pickle and Sauce Making. Food Trade Press, London.

BRICKMAN, C.L. 1957A. Measuring gas transmission of flexible materials in pouch form. Packag. Eng. *2* (1) 21–25, 54.

BRICKMAN, C.L. 1957B. Evaluating the packaging requirements of a product. Packag. Eng. *2* (7) 19–25, 56.

BRODY, A.L. 1970. Flexible packaging of foods. Butterworth & Co., London.

BRODY, A.L. 1971. Food canning in rigid and flexible packages. Ed. Technol. *2* (2) 187–243.

CHESKIN, L. 1966. Selling psychological satisfactions. Comm. Chem. Dev. Assoc. Fall Meet., 80–90, Chicago.

DAVIS, A. 1967. Package and Print. Clarkson N. Potter, Inc., New York.

ELLIS, R.F. 1963. Metal containers for food. *In* Food Processing Operations, Vol. 2. M.A. Joslyn and J.L. Heid (Editors). AVI Publishing Co., Westport, Conn.

FAIN, J.F. and BELGUISE, G. 1970. Odour in foodstuffs (packaged) in cellulose and polyvinyl chloride plastic films. Improm. Nouvelle *187* (October). (French)

FRIEDMAN, W.F. and KIPNES, J.J. 1960. Industrial Packaging. John Wiley & Sons, New York.

GATES, M.Y. and PAINE, F.A. 1969. Food packaging. Rprt. Progr. of Appl. Chem. *54*, 340–356.

GOULD, J. 1969. Packaging an image . . . his? hers? yours? Austr. Packag. *17* (5) 47–49.

GRIFFIN, R.C. 1965. Geometry helps fix size of fin-sealed pouch. Packag. Eng. *10* (2) 92–102.

GRIFFIN, R.C. and MATTHEWS, S.B. 1966. Find material needs for tetrahedral package with geometry. Packag. Eng. *11* (3) 86–94.

HEISS, R. and EICHNER, K. 1971. Moisture content and shelflife Pd. Mf. *46* (5) 53–56, 65.

JOSLYN, M.A. and HEID, J.L. 1963. Food Processing Operations, Vol. 2. AVI Publishing Co., Westport, Conn.

KANE, J.N. 1950. Famous First Facts. H.W. Wilson Co., New York.

KAUFMAN, M. 1963. The first century of plastics. The Plastics Inst., London.

KEFFORD, J.F. 1967. Science in food packaging. Food Technol. Austr. *19* (5) 204–213.

KNIGHT, E.H. 1876. Knight's American Mechanical Dictionary. Houghton, Mifflin and Co., Boston.

LINDOP, R. 1967. Composite containers. *In* Packaging Materials and Containers. F.A. Paine (Editor). Blackie and Son, Ltd., London.

LONG, R.P. 1964. Package Printing. Graphic Magazines, Inc., Garden City, New York.

LOPEZ, A. 1969. A Complete Course in Canning, 9th Edition. Canning Trade, Baltimore.

MAY, E.C. 1938. The Canning Clan. MacMillan Co., New York.

O'BRIEN, R. *et al.* 1965. Machines. *In* Life Science Library. Time, Inc., New York.

PAINE, F.A. 1962. Fundamentals of Packaging. Blackie and Son, Ltd., London.

PAINE, F.A. 1971. The effect of social factors on development in food packaging. Food Trade Rev. *41* (2), 46–49.

RAUSING, R. 1978. How the Tetra Pak idea was born. Paper *190* (8) 474.6 (October).

SACHAROW, S. 1965. The permeability of flexible plastic films. Plastics Design Process *5* (9) 20–23.

SACHAROW, S. 1968. Packaging materials. *In* Encyclopedia of Polymer Science and Technology. H.F. Mark and N.G. Gaylord (Editors), Vol. 9. Interscience Publishers, New York.

SACHAROW, S. 1969. A survey of the new plastics for packaging. Plastics Design Process *9* (1) 14–17.

TUNNICLIFFE, C.M. 1970. A study of the relationship between packaging materials, food, and odour. Food Technol. N.Z. *5* (4) 125–131.

2

Basic Food Processes

> Mankind, says a Chinese manuscript . . . for the first seventy thousand ages ate their meat raw, clawing or biting it from the living animal, just as they do in Abyssinia to this day.
>
> *"Dissertation," Charles Lamb, 19th Century*

EARLY DISCOVERIES

Before the discovery of basic food processes, primitive men had to eat food immediately where they found it. Few foods could be kept for long without spoilage. By trial and error over thousands of years, with no knowledge as to the scientific causes, men perfected methods for processing foods. They separated or extracted the edible portions of foodstuffs from the inedible, or instituted chemical, physical or microbiological changes in the foods to improve their flavor, aroma, tenderness, texture and, above all, their resistance to spoilage.

The earliest discoveries were probably in the areas of selection and separation. It doesn't take much intelligence to learn to peel a banana or to shell a nut to get at the edible portion. Squirrels and monkeys do it. Removal of furry hide or feathers to get at the flesh below is a simple matter of taste.

We have no way of knowing which of the earliest discoveries were accidents and which were derived from intelligent reasoning. Perhaps Charles Lamb wasn't far from wrong in claiming roast pig was discovered as a result of a burned down house in ancient China.

Anthropologists have suggested that the roasting of meats—however discovered—was an epochal event in man's development.

Raw meats are tough and require a great deal of chewing. Cooked meats could be eaten faster leaving more time available for other activities such as the invention of tools.

Many primitive food processes probably could be attributed to intelligent observation of natural or accidental happenings. Fish caught in winter kept longer than fish caught in the summer. Sun dried dates didn't spoil as fast as moist ones. Dampness rotted grain. Smoke from a fire preserved meats.

We can guess that someone made a gruel of pounded grain or allowed some bread to get wet and discovered some time later the pleasantries of beer and perhaps experienced the first hangover! In like manner fermented fruit led to wine, fermented honey to mead and fermented cabbage to sauerkraut.

The accidental churning of milk would have led to the discovery of butter and then later to cheese.

Following animals to salt-licks could have supplied man with this vital necessity. Eating salt in combination with food could have been a matter of taste. Who can say how man discovered salt would preserve meats.

We can only conjecture at the truth and marvel at the ingenuity of these early peoples. Who, for example, was the unsung hero of the South Sea Islands who discovered that heating bitter cassava root destroyed the poison and left edible juices and tapioca starch?

Even in modern times, food processing techniques are of great significance to society. The successful industrialization of China with its 900 million consumers will depend heavily upon a rise in productivity which could be achieved by giving people more time to work. To do this the Chinese want to improve the balance between work outside the home and in the household. Traditional Chinese dining is time-consuming and the time taken for eating and cooking could be curtailed. Ideally, China wants to learn from Japan how to cut up meat and to package it with vegetables so that each consumer can buy prepared food at the stores, bring it home and cook it without spending too much time.

NONLETHAL FOOD "SPOILAGE"

Foods begin to change from the moment of harvest. Some fruits are harvested unripe and continue to ripen during prolonged storage. Others become overripe and start to decay in a very short time. Root crops remain alive during storage for prolonged periods. Dried grains can be stored almost indefinitely. Meats begin to decompose soon after the animal dies.

Foods that are still alive continue to respire and to give off heat. This is typical of fruits and vegetables. Most foods contain enzymes which are responsible for certain types of growth factors. When the food dies or its metabolism slows down, some of these enzymes may cease to function and others may begin to function. For example, in the growth process, enzymes may be converting sugars created by photosynthesis into starches or cellulosic tissues. During the ripening phase other enzymes may take over to convert starches into sugars and breakdown cell structure resulting in a softer, juicier, sweeter fruit.

In addition to the chemical changes induced by respiration, metabolism and enzymes, foods are susceptible to attack by microorganisms. Some times these result in beneficial changes. For example, molds are used to cause desirable changes in meats and cheeses, yeasts are used to make wine and beer, and bacteria are used to make yogurt and vinegar. Much of the time the microorganisms "spoil" the food. A few typical examples of food spoilage and the organisms that cause them are given in Table 2.1.

TABLE 2.1. TYPES OF FOOD SPOILAGE

Food	Type of Spoilage	Causative Microorganism
Fresh meats	Slime	*Pseudomonas* (Bacteria)
	Whiskery mold	*Phizopis* (Mold)
	Spot mold	*Penicillium* (Mold)
	Sour	*Streptococcus* (Bacteria)
Cured	Slime	*Lactobacillus* (Bacteria)
Luncheon meats	Souring	*Straptococcus* (Bacteria)
	Greening	*Leuconostoc* (Bacteria)
Vegetables	Soft rot	*Erwinia cartovona* (Bacteria)
	Gray rot	*Botoytis cinerea* (Mold)
	Sour rot	*Geotrichum candidum* (Mold)
Gravies	Moldiness	*Aspergillus* (Mold)
		Penicillium (Mold)
		Fusarium (Mold)
Fruits	Black rot	*Aspergillus niger* (Mold)
	Soft rot	*Erwinia* (Bacteria)

TYPES OF FOOD HAZARDS

Inherent Poisons

Parts of certain plants and animals are deadly poison to man. Rhubarb leaves contain toxic oxalic acid. Peach pits and the bitter cassava contain hydrocyanic acid. Certain shellfish and certain mushrooms are poisonous. Even the edible potato can be dangerous. Normally grown beneath the earth away from sunlight, it is safe to eat. But when roots are exposed to sunlight the skin turns green and the flesh becomes poisonous.

The principal way to avoid poisoning from poisonous plants or animals is to avoid them. As previously mentioned, sometimes they can be rendered harmless by cooking or boiling in water.

Incidental Poisons

If a plant is sprayed with a poison or is grown in soil containing a poison, such as arsenic or lead, this poison will be ingested by the consumer. Similarly, if an animal ingests such a plant, or an inherently poisonous one, the toxic materials can be passed along. Cows that eat snakeroot plants can pass the poison in their milk. In recent times there has been much controversy over the amount of mercury found in the flesh of swordfish, which has been absorbed from sea water.

Again, the way to avoid poisoning from plants or animals that contain poisons is to avoid them. We rely on careful washing of crops to eliminate insecticidal poisons. We depend upon the farmer or food processor to protect us against crops or meats that contain poisons. Governmental agencies are constantly monitoring foods for hazardous substances. In the past decade two substances were found to be persistent poisons passed on to man via foods—the foods' obtaining them from the soil or water supply. These were polychloro-biphenyls (PCBs) and Kepone, an ingredient of insecticides. PCB was first identified in the flesh of a bird in 1966 by a Swedish investigator. Widely used in industry, the principal chain to foods was probably via printing inks and plasticizers that ultimately became recycled into paperboard and then into waste dumps where it was leached out. Fresh water fish in the Great Lakes were found to have high levels of PCB. Some foods were found to have absorbed it via contact with paperboard packaging. The remedy has been total cooperation of industry to discontinue the use of such compounds. Kepone was first discovered when workers in producing plants complained of illness. Since then it has been traced to the waters and bottom silt of the James River in Virginia and surrounding tributaries. It became necessary for the governor of Virginia to ban all fish and shellfish taken from these polluted waters. Continued monitoring has enabled a partial lifting of the ban, but major concern still exists as to how to decontaminate present contaminated areas and how to prevent expansion of the contamination.

Toxins in Foods

Foods that have been subjected to microbial attack and have nurtured the growth of certain yeasts, molds, fungi or bacteria may become infused with toxic substances produced by these microorganisms. When the food is eaten the toxins cause illness or death. Ergotism is a poisoning caused by eating grain containing toxins produced by a fungus, *Claviceps purpurea*. Poisonous aflatoxin is produced in nuts that are allowed to develop an aspergillus type of mold. The most dangerous of all the toxins is that produced by the bacteria *Clostridium botulinum* which causes the usually fatal botulism.

The remedies against toxin poisoning are:

(1) Identification of contamination and destruction of the contaminated food.
(2) Destruction of the toxin via cooking or heating of the food.
(3) Inhibition or prevention of toxin formation.

Infested Foods

Foods can be infested with certain microorganisms or parasites which do not in themselves render the food poisonous, but which, once consumed, multiply and grow within the human body to produce illness and/or death. A few typical examples of this type of contamination are:

Escherichia coli which results in dysentery or "Montezuma's Revenge."
Staphylococcus aureus which results in nausea, vomiting and diarrhea.
Salmonella (var. sp.) which result in fever, cramps, nausea, diarrhea.
Shigella (var. sp.) which results in headache, fever, abdominal pain, severe diarrhea.
Trichinella spiralis (round worm) which results in trichinosis with several stages of symptoms (muscle pain and fever).[1]
Taenia saginata (beef tape worm) which results in variable symptoms including pain, insomnia, loss of weight, nausea, vomiting, etc.

The remedies against foodborne poisoning from infested sources include:

(1) Identification and destruction of contaminated foods or animals.
(2) Sanitation at all phases of food growth, slaughter, harvest, storage, processing, preparation and serving.
(3) Freezing and/or cooking of foods.
(4) Avoidance of prolonged holding of foods at conditions favorable to growth prior to eating.

TYPES OF MICROORGANISMS AND THEIR BASIC CHARACTERISTICS

In order to understand how to prevent the spoilage of foods by microorganisms it is important to know some of their basic characteristics. The microorganisms associated with foods are members of the plant kingdom falling under the general category of fungi. Within this category can be found mushrooms, molds, yeasts and bacteria. Bacteria and yeasts as a rule are single-celled although these cells may form clusters or chain-like groupings. Molds are multicellular and form rather extensive thread-like or branching mycelia from which may grow "fruiting bodies" that contain spores.

[1]There were 142 reported cases of trichinosis in the United States in 1977. New Jersey led the nation with 50 cases. To prevent trichinosis, a 15.24 cm thick slab of pork should be frozen at −15°C for 20 days or −23.3°C for 10 days, or −28°C for 6 days. Pork should also be cooked to an internal temperature of 76.6°C. Trichinosis, although commonly associated with pork has also been transmitted by the flesh of bears, wild boars, walruses and even horses.

All of these microorganisms require a favorable environment and a specific type of food in order that they may "eat," metabolize, grow and reproduce. In the eating and metabolizing of specific parts of the food several things happen:

(1) Dead microorganisms accumulate (but always a few survive).
(2) Simple or complex metabolic products are manufactured:
  (a) gases such as carbon dioxide or hydrogen sulfide,
  (b) sugars, alcohols, esters, acids,
  (c) pigments, toxins, enzymes.
(3) The physical and chemical composition of the food is changed.

If the accumulation of metabolic products creates conditions unfavorable to growth, the food supply is depleted, or the favorable environment otherwise changes, growth will slow down or cease.

Microorganisms can be destroyed under certain circumstances; however, some are capable of withstanding rather drastic changes in environment by changing from a vegetative cell (capable of growth and reproduction) to a spore. Spores are a dormant form. They don't grow or multiply, and they are much harder to kill. When the environment changes back to favorable conditions the spores then revert back to vegetative cells. Reproduction of vegetative cells of bacteria and yeasts is via binary fission. The parent cell divides into two cells. If this were to occur every 20 min, one bacterium would multiply to more than 68 billion bacteria in less than 12 hr.

FACTORS INFLUENCING GROWTH OF MICROORGANISMS

Assuming the right kinds of foods are present, there are a number of environmental factors that influence the growth of microorganisms. These are:

(1) temperature,
(2) moisture,
(3) oxygen availability,
(4) pH (acidity),
(5) chemical inhibitors, and
(6) biological barriers.

Temperature

For every microorganism there is an optimium range of temperature for maximum growth. At extremes of high temperatures, destruction occurs. At freezing temperatures growth almost stops. One method of classification groups them as to temperature as follows:

	Range	*Optimum Growth*
Thermophilic	>45°C	55°–65°C
Mesophilic	~20°–45°C	30°–40°C
Psychrophilic	<20°–30°C	20°–30°C

Moisture

It is not enough that foods contain moisture. This moisture must be available for use by the microorganisms and not bound to the food or to a constituent of the food. Modern microbiologists refer to this as water activity. In the simplest terms this is the ratio of the vapor pressure of the water in the food to the vapor pressure of pure water. Thus, pure water would have a water activity of 1.00. Most fresh fruits, vegetables and meats have water activities near 0.99 which accounts for their susceptibility to microbial attack. Other typical values are:

Breads	0.95
Ham	0.96
Hard salami type sausage	0.85
Fruit cake	0.80
Salt fish	0.75
Candies and cereals	0.70

Most bacteria require a water activity greater than 0.91. Yeasts will grow in an environment as low as 0.88, while molds require only 0.80 or more.

Oxygen Availability

Microbes can be further classified as to their performance in relationship to the presence or absence of atmospheric oxygen:

(1) Aerobic organisms grow best in the presence of oxygen and will not grow without oxygen.
(2) Micro-aerophilic organisms grow best in a reduced level of oxygen.
(3) Anaerobic organisms don't like oxygen. Strict anaerobes will die. Tolerant anaerobes survive but don't grow.
(4) Facultative anaerobic organisms don't care. They will grow with or without oxygen.

Molds are aerobes. Yeasts are aerobic or facultative anaerobes. Bacteria can be found in every category.

Acidity

Acidity is expressed on a scientific scale called pH which is related to the concentration of hydrogen ions in a solution. On this scale, pure water which is neutral—neither acid nor alkaline— has a pH value of 7. The scale runs from 0 to 14 with values below 7 being increasingly acid and values above 7 being increasingly alkaline. Foods are generally neutral or mildly acidic. The lowest pH value in natural food would be pure lemon juice at 2.2–2.4. On the high end of the pH scale we would find ripe olives 5.9–7.3, clams 5.9–7.1, milk 6.4–6.8 and corn 6.5–6.6.

Optimum growth for almost all microorganisms is at about pH 7.0. Some are more acid-tolerant than others, but in nearly every species growth is slowed as pH drops, and in many instances no growth occurs at all.

Fortunately many of the more dangerous bacteria from the human health standpoint do not grow at pH below 4.3.

	Minimum pH
Escherichia coli	4.4
Streptococcus lactis	4.3
Staphyloccus aureus	4.4
Clostridium botulinum	4.6
Clostridium perfringens	4.5

Since fruits, most soft drinks, vinegar and wine fall below these values, they are relatively free from bacterial spoilage and are sensitive only to molds and yeasts. This is also why a number of historical methods of food processing resulted in safer foods and longer keeping qualities; the processes resulted in acid formation in the food.

Chemical Inhibitors

Some foods contain natural antimicrobial ingredients. These compounds inhibit microbial growth. Examples are:

	Product
Allyl isothio cyanate	Horseradish
Benzoic acid	Cranberries
Cinnimic aldehyde	Cinnamon
Eugenol	Cloves
Lactenin	Fresh milk
Lysozyme	Egg white

For many years chemicals have been added to foods to prevent microbial growth. There is a great amount of controversy over the relative value of these additives *vs* their possible harmful side effects.

A number of consumers will not purchase foods containing such additives. Some even go so far as to insist on foods grown only in soil enriched by "natural" organic fertilizers and without the use of insecticides.

A few examples of such antimicrobial additives are:

	Product
Antibiotics	Poultry
Benzoates	Soft drinks
Nitrates and nitrites	Cured meats
Propionates	Breads
Sorbates	Cheeses, cured meats
Sulfites	Wines, dried fruits

The most controversial at the present time are the nitrates and nitrites. These are under close scrutiny by governmental agencies as possible cancer-causing agents. The cured meat industry has been challenging the allegations and pointing out that natural foods contribute far more nitrates and nitrites to the human diet than cured meats and that the risk of spoilage and food poisoning will be drastically increased if such preservatives are banned.

Biological Barriers

There are a number of natural and artificial biological barriers that exist or that can be created which will prevent or inhibit microbial spoilage of foods.

The skins, husks and shells of foods of animal and vegetable origin are naturally impervious barriers to bacteria. It is only after they are damaged by physical cuts or bruises, or by the destructive action of surface molds or yeasts that bacteria can penetrate.

Bacteria can be physically separated from the food by cleaning and washing processes.

Artificial barriers can be created by coating foods with such materials as waxes or by placing food in impervious wrappers or containers. It is in this final application that food packaging plays a role. What role packaging plays depends upon the type of food and the type of processing it has been subjected to.

BASIC TYPES OF FOOD PROCESSING

In Table 2.2 there are listed a number of food processes and their purpose. In the first category are listed those processes which aim at extracting the edible portion of the food from the inedible portion. In the second category are listed those processes which aim at changing the shape of the food in order to make it easier to handle, eat, cook or process further. The third category lists processes that are performed on the food to institute chemical, physical and microbiological changes which improve its flavor and aroma, its tenderness and texture, or its ability to resist deterioration or spoilage.

TABLE 2.2. TYPES OF FOOD PROCESSING

I.	Separation:
	Picking
	Washing, dry cleaning
	Dehusking, peeling, shelling
	Destemming, deleafing
	Beheading, disemboweling
	Coring, deboning, pitting
	Bleeding, juicing
	Pressing
	Straining, filtering
	Distillation, steeping
II.	Size or Shape Alteration:
	Grading, sorting
	Cutting, slicing, dicing, flaking
	Shredding, chopping, comminuting
	Grinding, milling, powdering
	Pasting, mixing, beating, kneading
	Rolling, extruding, molding
III.	Chemical, Physical or Microbiological Changes:
	Ageing, ripening, enzymatic action
	Smoking, curing, brining, pickling, syruping, spicing, fermentation
	Malting, brewing
	Steeping, extraction, distillation
	Cooking
	Evaporation, concentration, drying
	Condensation, coagulation, flocculation
	Blanching, pasteurization, sterilization
	Refrigeration, freezing
	Acidification, alcoholation, chemical preservation, carbonation

As packaging plays a negligible role in the first two categories we shall not discuss them further. Packaging becomes important as a means for protecting foods that have reached the final stage of their processing during holding, transportation, storage and up to the point of final use. As we discuss these types of food processes we shall note generally what kind of protection is required. More detailed discussions will be found in later chapters.

PRINCIPLES OF FOOD PRESERVATION

Food preservation can be accomplished in several ways: reduction in degree of spoilage by removal or prevention of contamination, delaying spoilage by making conditions unfavorable and elimination of spoilage by destruction of contamination or the factors producing it.

Food Preservation by Removal or Prevention of Contamination

It is possible to minimize the level of contamination in foods, whether in a processing plant, in a food service operation, or in the home by observing fundamental principles of good sanitation and by the removal and isolation of existing contamination.

Removal and isolation of existing contamination involves operations relating to sorting and removal of spoiled, inedible or poisonous components, discarding of offal and garbage, and the washing of the accepted product.

Good sanitation practices apply to personnel, materials, equipment and facilities and are now widely included in industry or government standards for good manufacturing practices.

People working in a food processing plant or food service operation must be clean, healthy, wear sanitary clothing and observe strict personal sanitary practices. All raw materials including dry chemicals as well as food components must be monitored for quality and freedom from contamination. Food handling machinery and equipment must be regularly sanitized and monitored for possible contamination. Buildings and grounds must be so constructed and maintained that they minimize possible contamination from birds, insects, rodents, etc., and do not allow dirt, peeling paint or accumulated product splatter to fall into the food.

Failure to observe these practices may result in outbreaks of food poisoning. Government regulating bodies at local, state and federal levels have the power to close down unsanitary operations or to seize or order recall of contaminated products that might become a danger to public health.

Food Preservation by Delay or Inhibition of Spoilage

Food spoilage can be reduced by inhibiting or preventing the growth of microorganisms or by not giving them time to multiply. The time factor is used throughout food processing. From the moment foods are harvested all processing steps are as rapid as possible.

As mentioned previously, growth of microorganisms can be minimized or halted by altering conditions that favor growth. Processes relating to this principle are:

Temperature control	Refrigeration, freezing, cooking, pasteurization
Moisture control	Drying, freeze drying, salting, evaporation, sugar syruping
Oxygen control	Barrier packaging, vacuum packaging, gas packaging, oxygen scavengers, ethylene oxide storage
pH control	Acidification, pickling, acid fermentation
Chemical inhibitors	Smoking, curing, spicing, salting, sugaring, antibiotics, other preservatives

These processes are frequently used in combinations. For example milk is pasteurized to inhibit spoilage and is then transported and stored under refrigeration. Even so it will not keep indefinitely (see Chapter 5). Luncheon meats may contain smoked meats, chemical inhibitors, spices, sugars and salts. They are packed in barrier films sometimes under vacuum and transported and held under refrigeration (see Chapter 3). Soft semidry pet foods have low moisture content and high sugar content to retard spoilage. Acid fermentation retards spoilage in pickles, relishes, cheeses, yogurt, etc. Vegetables, fruits and meats can be dried to prevent spoilage.

DEHYDRATION

Dehydrated foods are as old as civilization. Vases containing samples of dehydrated foods were found in Egyptian tombs. Marco Polo reported that dried milk was used by the Tartar tribes. In the United States, early settlers found that a staple food of the Indians was "pemmican." It was made of dried meats, such as venison, crushed to a powder and mixed with fat. Colonists adopted various Indian techniques and utilized dried corn ("samp"), dried fish and dried beef ("jerky").

Original drying methods utilized the heat of the sun. In 1795, the first artificial drying process was introduced by Eisen. Vegetables were exposed to hot water before drying. Soon after, two Frenchmen, Masson and Chollet, dehydrated vegetables by slicing and drying them with a controlled warm air flow.

In the Civil War, General Sherman's soldiers used dried foods extensively. During the Boer War, dried compressed vegetables were used by the British Army in South Africa. In 1917, approximately 5.4 million kg of dehydrated vegetables were produced in the United States for war time use.

Between World Wars I and II, the dehydrated food industry was at a low ebb. Although development continued in Europe, most United States firms preferred to import whatever quantities of dehydrated foods they needed. The onset of World War II catalyzed further American development. Soldiers needed light-weight foods and traditional European sources of supply were terminated. After the war, the rise in consumer affluence coupled with the demand for convenience created new interest in dehydrated foods.

What Is Dehydration and Why Is It Needed?

Food dehydration involves the removal of water without destroying the cellular tissues of the food nor altering its caloric values. A variety of processes are used to effect water elimination. The presence of water in a food tends to accelerate: (1) mold and bacteria growth, (2) discoloration and off-flavor, (3) enzyme action, and (4) caking and other physical changes. By reducing the water content of a food below its critical value, using proper packaging and careful storage, increased shelf-life is possible.

The rapid rise in dehydrated food sales has been a direct result of improvements in processing techniques.

Methods of Food Dehydration

Sun Drying.—Sun drying is still used for the dehydration of various fruits. Some varieties of fruits are sulfured to prevent darkening. They are then taken to a drying yard and exposed to the sun for one to 10 days, depending on the climate. Different fruits require different lengths of exposure. Apricots are exposed from one to seven days, bananas one to two days, figs two to four days, etc. In general, the process is rather slow and used mainly for selected fruits.

Spray Drying.—Spray drying is used for some liquids such as milk or dissolved solutions such as brewed coffee. The liquid is finely atomized into a gasous heating medium such as air. As the moisture evaporates, solid particles are formed which float in the air stream. Evaporation occurs so rapidly on the surface of the particles that the temperature of the product never rises high enough to cause thermal damage. A variety of atomizers and driers are available from different manufacturers.

Courtesy of Reynolds Metals Co.

FIG.2.1. DRIED SOUP MIX POUCH
The complete moisture and vapor barrier qualities of aluminum foil enable high quality dehydrated convenience foods to be packed in convenient, inexpensive flexible pouches rather than metal cans. The aluminum foil is combined with plastic film and paper to gain the qualities of each material.

Atmospheric or Hot-air Drying.—This consists of cutting the product into small pieces and spreading them out on trays which have slat or wire bottoms. Hot air is passed over and under the trays in order to effect dehydration. Drying takes place in tunnels about 10.7–12.2 m long and 1.8 m high by 1.8 m wide.

Hot-air drying methods are used principally for vegetables and fruits (which are not sun-dried). Instant potatoes, onion flakes and apple slices are commonly hot-air dried.

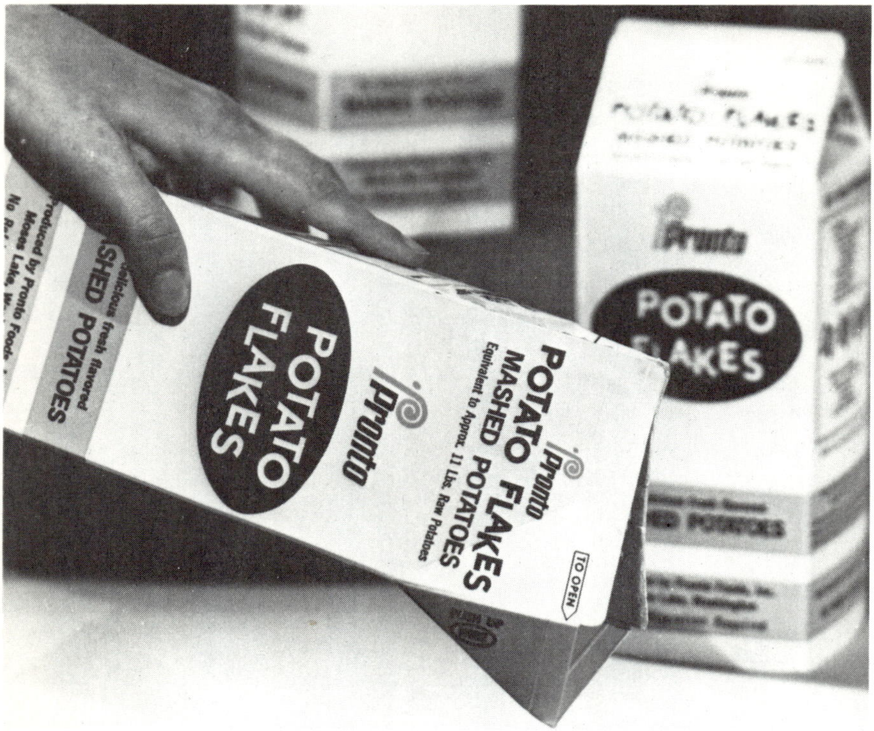

Courtesy of E.I. du Pont deNemours and Co.

FIG. 2.2. ADAPTATION OF MILK CARTONS TO DRIED FOODS
Dried potato flakes packaged in containers like the familiar polyethylene-coated milk carton have been tried in the retail market. Two years of success with this style of carton in institutional sizes made it the choice for the 16 oz (453.6 g) retail package. The carton, coated with Du Pont's "Alathon" polyethylene resin, features easy opening, pouring and reclosing for the consumer, and offers economic and other advantages for the packager.

Foam-mat Drying.—This involves beating a liquid into a stabilized foam, spreading the foam in a thin mat on a perforated belt and passing warm air through the foam mat. Air temperatures are lower than for spray drying and contact times are longer.

Drum-Drying.—Drum drying may be atmospheric or vacuum. In atmospheric drying, the material is applied by feeding devices onto large steam-heated drums 0.6 to 1.8 m in diameter. As the drums revolve, the material dries and is subsequently removed by a scraper.

Vacuum Drying.—Vacuum driers may be of any configuration, that is drum driers, belt driers, kiln driers, tunnel driers, etc. The difference lies in the pressure being reduced below atmospheric pressure. The higher the vacuum, the lower the temperature required to remove moisture and volatiles. This permits drying of heat sensitive foods. Condensation of volatiles may be necessary. Capital investment is higher for vacuum equipment as opposed to atmospheric driers. The absence of oxygen helps avoid food quality deterioration. Some foods, when vacuum dried, are heated so that bubbles form. This makes the product lighter. It may be flaked as is or it may be crushed to reduce bulk. This is somewhat akin to foam drying.

Dehydrofreezing

Dehydrofreezing is a process which combines dehydration and freezing. A certain amount of water is extracted by heat and the product is then frozen. It must be stored under refrigeration and has a few advantages over conventional frozen foods. Since the food shrinks during dehydration, it requires less storage room than conventional frozen foods. Several vegetables, meats and fruits have been dehydrofrozen.

Freeze Drying

The major interest currently in dehydrated foods revolves around a process first used in the early 1920s for medical applications. Blood plasma was preserved by a freeze-drying technique. Although the original process was slow and expensive, it has been developed into a rapid and economical procedure. Freeze drying is an extension of vacuum drying. The product is first frozen and then moisture is removed by sublimation under high vacuum. Just enough heat is applied to accelerate drying without causing the product to decompose. Ice crystals are converted directly into water vapor without passing through a liquid phase. The freeze-dried product retains its original shape and is porous. This porosity allows for more complete and rapid rehydration than is possible with air-dried foods. In addition, the weight is reduced to 10–15% of the original weight. Moisture levels are as low as 2% compared to a minimum 5% with other methods. Large items such as steaks, fish and chicken can be freeze dried without crushing or shredding the product. At present, freeze drying is being applied on an increasing scale to a wide variety of products ranging from soup to vegetables. No refrigeration is required for freeze-dried foods.

Packaging Requirements

All dehydrated foods are subject to two prime causes of spoilage: moisture and oxygen. They are also fragile, sensitive to light, and if fatty, subject to contamination by foreign flavors. Dehydrated fruits and vegetables are especially subject to insect attack.

Moisture Sensitivity.—The problem of absorption of atmospheric moisture is more critical for some products than others, depending on the desired shelf-life. This is particularly so for freeze-dried foods because of their open structure. Even the slightest degree of moisture absorption by the product is harmful and undesirable. Even though the caking of a powdered dehydrated food may not constitute complete spoilage, it is unacceptable to a consumer and must be avoided.

Courtesy of U.S. Army, U.S. Army Natick Lab.

FIG. 2.3. MILITARY PACKAGING SHOWS ADAPTABILITY OF FLEXIBLE PACKAGING
Experimental ration utilizing flexible packaging for both dehydrated foods (potato pattie) and processed foods (chicken loaf).

Oxygen Sensitivity.—Although moisture is the common enemy of all dehydrated foods, many of them are highly susceptible to oxidation. Oxygen adversely affects products containing high levels of fats or oils and those having a high carotene content. Fats and oils turn rancid and carotene develops an odor like newly mown hay.

Freeze drying, while offering a superior product in comparison to other dehydrated foods, yields a product highly susceptible to oxygen. Both vacuum and gas packaging are used to eliminate oxygen from a package.

Light Sensitivity.—Many dehydrated foods are sensitive to light. Bleaching occurs on some products and others darken due to the effect of light rays. Light also accelerates the development of rancidity by means of oxidative spoilage.

Mechanical Abrasion.—Although all dehydrated foods are subject to mechanical damage, this problem has been reduced with compressed foods. Powders, shreds and thin slices of air-dried foods cannot be damaged by mechanical handling to any great degree. In freeze-dried food, the product is extremely friable and brittle. Mechanical abrasion inside a loosely packed can containing freeze-dried foods can crumble and break-up large items, such as shrimp, meat or mushrooms, into unusable granules or powders. Freeze-dried shrimp or steak should be put in a package with a degree of rigidity with some cushioning or some means of immobilizing the product. Materials must be strong and flexible enough to resist being punctured by sharp granular or chunky items.

Flavor Contamination.—If dehydrated foods are fatty, they can be contaminated by foreign flavors extracted from packaging materials or from adjacent external sources. A good grease barrier and a good vapor barrier is needed to prevent this.

Insect Attack.—Insect attack is a problem with dried fruits and vegetables. Fumigation against insects is of no value after the material is in the final container. If insect eggs are present, they are unlikely to hatch in the dried food because of its low moisture content, but they will if the package is broken. Even a minute pinhole in the package will attract insects and encourage entrance.

Packaging Materials and Form

Three different types of packages are used for dehydrated foods: rigid containers, flexible pouches and lined cartons. All fulfill the requirements imposed by the product and an eventual choice is based on economics and sales appeal.

Vacuum and Gas Packaging.—If it is desirable to remove most of the oxygen from a package to achieve prolonged shelf-life, the package can be evacuated and sealed under vacuum or the air can be replaced by an inert gas.

Vacuum packaging presents certain problems with dehydrated foods in flexible pouches. Products with sharp corners tend to rupture the material causing vacuum loss. Many soft products are themselves crushed by the application of high vacuum. The unappealing shape of a vacuum pack makes it difficult to merchandise. In addition, the use of vacuum packaging demands heavy materials in order to resist the strains imposed by the process. This adds cost. These problems are not encountered with rigid containers.

Gas packaging is a more practical process for oxygen removal. The air is replaced by flushing or flooding the package with an inert gas. A vacuum may be pulled and then gas added, or the interior of the package may be flushed prior to sealing, or the package may be filled and sealed within an atmosphere of inert gas. All of these procedures reduce the oxygen level to between 1 and 2%. By back flushing with an inert gas, the product may be under sufficient pressure to prevent movement within the package and yet avoid the damaging stresses of a full vacuum package.

Three gases can be used for the gas packaging of dehydrated foods: nitrogen, carbon dioxide and "Vitagen."

Nitrogen is more commonly used and is truly inert. Carbon dioxide may be absorbed by certain foods and create a partial vacuum after a period of time. Vitagen is a product produced by the burning of natural gas in generators.

FREEZING

Drying, smoking, pickling or canning of foods alter the physical appearance and the flavor and aroma of the product. Eskimo tribes invented a method of food preservation which avoided these changes. They caught fish in subzero weather and the fish froze immediately upon exposure to air. Many months later, upon thawing, the fish were just as edible as when first caught. These procedures were not recognized or discovered by others until relatively recent times. The use of low temperatures for food preservation goes back to the mid 19th century. The first patent issued for a freezing technique was awarded in 1842. In 1876, frozen meat, using ice and salt, was shipped from the United States to England. In 1912, berries were slow-frozen in barrels in Puyallup, Washington. These early products were not quick-frozen and suffered from structural alteration.

On freezing, the water contained in food is converted into ice crystals at a temperature between -5° and 0°C. With slow freezing processes large crystals are formed which may distort and rupture the tiny cells. After thawing, valuable nutrients are lost through the punctured cell walls. In 1912–1915, Clarence Birdseye, while in Labrador, observed what the old Eskimos had long known that animal products which had been killed many months earlier were edible after being exposed to the Arctic weather. He decided that the secret lay in the rate at which the product was frozen. With quick freezing, small ice crystals are formed which have less effect on the structure, flavor and texture of the product.

In 1928, 5000 kg of fruits and vegetables were frozen in the United States. Today, sales of frozen foods account for over $9 billion annually and the growth rate is rapidly increasing. In 1978, the United States per capita consumption of frozen foods amounted to 3.6 kg. Per capita consumption in Europe ranges from 12.56 (Sweden) to 7.26 kg (Italy). Many areas of the world have not yet been exposed to frozen foods. In Portugal and Spain, sales are negligible. The eastern European countries are just emerging as large-scale users of quick-frozen foods.

Courtesy of E.I. du Pont de Nemours and Co.

FIG. 2.4. ANOTHER FROZEN CONVENIENCE FOOD
Here a complete pizza item is prepared for the consumer and frozen ready to bake in the oven. The "Elvax" hot-melt coated carton provides adequate protection against moisture loss.

Method of Quick Freezing.—Four basic methods are used to quick-freeze food products. In all four methods, temperatures below −34.4°C are utilized. The product may be packaged prior to freezing for convenience and protection against dehydration.

Blast freezing employs a current of very cold air directed against the product. An alternate method utilizes a tunnel in which the product is moving in the opposite direction. This process is well-suited for individual fruits and berries and irregular-shaped items.

Plate freezing consists of refrigerated plates on which the product is placed. By compressing the plates together in order to make close contact coupled with the use of an internal circulating refrigerant, at a low temperature, rapid freezing is obtained. Rectangular-shaped items in thin cartons are rapidly frozen.

In immersion freezing, the product is immersed in a liquid refrigerant such as glycerol, alcohol, brine or sugar syrup. It is useful for large products such as turkeys and other types of poultry.

Cryogenic freezing is usually done before packaging using liquid nitrogen or powdered dry ice at very low temperatures (−195°C).

After freezing, foods should be kept at a storage temperature of −17.7°C or lower until ready for distribution to the retailer. By the controlled use of packaging and retail storage temperatures of −17.7°C or lower, frozen foods can be stored for long periods of time. Small size foods such as peas and lima beans may be quick frozen while moving, thus preventing the individual products from freezing to one another. These individually quick frozen (IQF) foods will pour readily from a package if not allowed to thaw and refreeze.

Package Requirements.—Over $800 million in packaging materials are used by the frozen food industry. A wide range of materials account for this volume.

Frozen foods are specifically designed for long-term storage. A suitable packaging material must protect the product from dehydration and oxidation.

Under the dry conditions maintained for freezing and cold storage, foods tend to lose moisture. Quick-frozen meat, poultry and baked goods are particularly sensitive. A material having a low moisture vapor transmission rate is needed.

The development of rancidity in fatty and oily foods is due to oxidative deterioration. Flavor losses occur coupled with the fading and discoloration of various pigments. In order to prevent oxidation, a suitable oxygen barrier is needed. Oxidation occurs at a slower rate at low temperature ranges; however, it is still a factor in the packaging of frozen foods.

If dehydration and oxidation are allowed to occur, "freezer-burn" results. A discoloration of the surface of the food coupled with texture loss results. In addition, a suitable package must not permit the loss of organic volatiles important in the maintenance of flavor. The intermingling of different food odors also needs to be prevented. A gas barrier is necessary to prevent the transfer or pickup of odors.

Courtesy of Chemische Werke Hüls Aktengesellschaft

FIG. 2.5. BOIL-IN-BAG PACKAGE
This frozen Hungarian Goulash meal is quick frozen in a film pouch and can be reconstituted in boiling water.

A final package must be able to withstand the condensation of atmospheric moisture on the exterior. If water vapor from the atmosphere permeates the package, excessive frost formation will occur on the interior. In order to prevent such condensation, air cavities within the package must be minimized.

All materials used in packaging frozen foods should maintain their integrity at temperatures as low as −40°C, or as low as −195.5°C for cryogenic freezing. They must be capable of being properly printed for retail sales and easily stacked in either a vertical or horizontal frozen food cabinet. In addition, a suitable material may have to withstand any abrasion or puncturing caused by the sharp edges or points present in many frozen foods.

COOKING

Cooking is a general term covering a wide variety of heat treatments that change the characteristics of raw foods making them easier to eat and more palatable. A secondary advantage derived from cooking is the inhibition of spoilage.

There are three general categories of cooking: moist, semimoist and dry. Moist cooking involves the immersion of the food into hot or boiling water or into steam. It includes such procedures as: blanching, poaching, simmering, stewing, boiling, steaming and pressure-cooking (steaming).

Moist cooking adds moisture to foods. Because it may also leach out water-soluble nutrients, water volume should be kept as small as possible. Time depends upon the food and the result desired. Temperatures of the heating medium range from about 82.2° to 100°C in uncovered unpressurized vessels up to about 115.5°C in a pressure cooker.

Semimoist cooking involves the heating of the food under such conditions that its natural moisture content is trapped and generally contained. Little or no water is added. Losses due to evaporation are minimized. It includes such procedures as: sauteing, frying, deep fat frying, braising and pressurized deep fat frying.

It frequently involves a sealing off of the exterior of the food to entrap the moisture.

This is accomplished in braising by a preliminary searing or browing of the meat in hot fat. Deep-fat (french) frying accomplishes the same objective by browning or crisping the exterior, as in french-fried potatoes or the batter as in french-fried onion rings, clams, shrimp, chicken, etc. Temperatures in semimoist cooking range from 121.1°–204.4°C depending upon the food and the result desired. More moisture is lost in open-pan frying than in pressurized deep-fat frying. After the initial browning, braising is a long, moist cooking in a lidded container at low oven temperatures.

Dry cooking involves the heating of a food in an environment that drives off moisture as in baking, roasting and broiling (or grilling). Temperatures range from as low as 148.8°C for a pancake on a griddle to as high as 315.5°C or more for a pizza oven or a charcoal broiler for grilling a steak. In baking, the objective is to drive out moisture and lightly brown the exterior. Time and oven temperature are controlled to avoid overbrowning the exterior before bringing the interior to the desired dryness.

In roasting, the objective is to minimize moisture loss and brown (and seal) the exterior while bringing the interior to the desired temperature (degree of doneness). Basting the exterior with fat and or juices helps reduce moisture loss.

In grilling or broiling, higher temperatures are utilized. Doneness is achieved in as sufficiently short a time as needed to minimize moisture loss.

Cooking Equipment

A wide variety of utensils and appliances have been developed to assist man in cooking foods by the methods described. No attempt will be made to list them here. It is sufficient to say that materials of construction have included pottery, leather, copper, wood, iron, steel, aluminum as well as modern ceramics and plastics.

Heat has been derived from wood and charcoal fires, kerosene and naphtha stoves, and gas and electric ovens, the latest development of all being microwave ovens.

Conventional ovens are enclosed heated chambers that transfer heat to the food by radiation (infrared), by conduction from the surrounding hot gases or contact with hot shelves, or by convection. Convection is a more efficient form of conduction whereby the hot medium is moved with respect to the food so as to remove temperature depleted layers and replace them with fresh hotter layers.

The latest example of this is the consumer convection oven where a circulating fan is used to create a forced draft. Conventional ovens by their nature are hot. The heat they generate warms the outer layer of the food and penetrates to the center by conduction. Penetration by the infrared radiant energy is minimal. Most is absorbed on the exterior surface.

Microwave ovens are enclosed chambers that are built to contain and reflect the microwave energy generated by an electronic device called a magnetron. Microwave ovens by their nature are cold. The microwave energy traveling at the speed of light penetrates the food for a depth of several centimeters and is absorbed by causing the polar molecules of the food (such as water) to reverse their polarity about 2½ billion times a second. This creates mechanical friction within the food which generates heat. The advantage of microwave heating is its speed. A 12 oz (340 g) frozen entree that might take 30 to 40 min to heat to serving temperature in a conventional oven could be heated in a microwave in ½ to ⅓ of that time. Also, the microwave oven is ready at the push of a button. It does not need 10 to 15 min of preheating.

The disadvantage of microwave heating is its selectivity. Different foods heat at drastically different rates. Because ice is a poor absorber of microwave energy and thawed food is a good absorber, trying to cook a frozen food too rapidly can result in one component being overcooked while another is still frozen. Another disadvantage is load sensitivity. In a conventional oven the reheating of several 12 oz (340 g) frozen entrees would not take much longer than one—perhaps 10% longer. In a microwave oven, doubling the load would require almost twice the time. With larger and larger loads, microwave ovens are less efficient than a conventional oven. Perhaps this is why microwave oven interiors are made relatively smaller.

Still another disadvantage of the microwave oven is the difficulty in achieving browning of foods. Special infrared devices have to be employed to achieve browning in most consumer ovens.

Nevertheless, despite these disadvantages consumers are buying microwave ovens. It has been suggested that by 1980 one out of four United States households will have a microwave oven. Food processors must bear this in mind in the packaging and marketing of their products.

Frozen entree dinners, single entree or multicomponent meals are now being test-marketed that were designed both in package and in food to be reheatable either in conventional or in microwave ovens.

By taking away preparation time and shortening cooking and serving time the combination of frozen convenience meals and microwave ovens has eased part of the burden of household duties for the dual wage-earning family.

Cooking Films

Another product that has been around for a while is the ovenable cook-in film bag or tube. This product, made from heat resistant polyester or polyamide films, was designed to trap and hold heat and moisture both in conventional and microwave cooking. This not only speeds up cooking and conserves energy, but also traps volatile flavorants and prevents spattering of oven or utensils.

Cook-in Consumer Bag.—For household use the consumer can purchase several sizes of cook-in plastic bags at the supermarket. Food is prepared with seasoning and placed inside the bag which is then tied shut with a heat resistant (and microwave resistant) plastic tie. Several small punctures are made in the top of the bag to prevent pressure buildup. Oven temperatures can be set a little higher than usual because evaporation losses are small. Cooking time will be shortened 25–40% depending upon the food. Roasts will be self-basting and the juices can be collected and retained for subsequent use, not burned onto the pan. Dry foods such as oven heat-and-serve rolls can be cooked without using a pan and will be softer and moister when served. Foods like hashes, meat slices and gravy, or beef stews can be placed in the bag in a skillet and simmered on a range top on low heat settings without spatter leaving no pan to wash.

Food Service Film.—In restaurants and institutional kitchens, chefs have a choice between the consumer type bag and a heavier gauge seamless tube sold on rolls. The latter can be pulled off, cut to length and drawn tightly around the food securing both ends. The use of ovenable film to contain moisture permits the "waterless" reheating of frozen vegetables. It takes longer than in water but the result is better color and flavor and less leaching out of soluble nutrients.

Ovenable Board and Plastic Containers

In recent years containers have entered the market that are made from a polyester-coated paperboard. These are advertised as capable of surviving the reheating of frozen foods in both conventional and microwave ovens. One of the latest food lines being tested in the marketplace has an ovenable film sealed to the rim of the container and is sold in a printed, folded carton. Instructions for cooking entrees with 9–11 oz (255–311.8 g) of meat and gravy are essentially as follows:

Conventional Oven	*Microwave Oven*
Preheat oven to 375°F. Do not remove film cover. Place on cookie sheet on middle shelf. Heat 30–40 min or until hot. Remove film and serve.	Do not remove film cover. Put in oven. Heat 4–6 min on high setting. Raise film, separate meat slices or stir gravy. Replace cover. Rotate dish ¼ turn. Heat 3–4 min. Remove film and serve.

Instructions for cooking entrees with 11–12 oz (311.8–340 g) of a pasta product such as lasagna are essentially as follows:

Conventional Oven	*Microwave Oven*
Preheat oven to 375°F. Do not remove film cover. Place on cookie sheet on middle shelf. Heat 45 min. Remove film and serve.	Do not remove film cover. Put in oven. Heat 5 min on high setting. Let stand 5 min. Rotate dish ¼ turn. Heat 5 min. Let stand 5 min. Remove film and serve.

Another dual package being test-marketed is a line of frozen entrees containing a portion of meat, a starch, a vegetable and appropriate sauces. The entrees are in a shallow rigid heat-resistant plastic dish and are covered with a gold-colored foil tuck-wrap. A semirigid high density polyethylene snap-on "domed" lid has two features—a flange that serves to prevent shifting within the carton and two scored vents. Instructions for cooking are essentially as follows:

Conventional Oven	*Microwave Oven*
Preheat oven to 325°F. Remove plastic lid. Bake at 325°F 40–45 min. Remove foil and serve.	Remove plastic lid. Remove foil tuck wrap. Open vents in lid. Replace lid. Microwave 3½–4 min high setting. Rotate dish. Microwave 3½–4 min high setting. Let stand 5 min. Remove lid and serve. (When cooking two dinners add 6–7 min microwave time at high setting.)

With these meals retailing in the price range of from \$3 to \$5 per lb (0.45 kg) of food, it can be seen that they are being aimed at a different market segment than the familiar inexpensive TV dinner.

Food Preservation by Destruction of Spoilage Organisms

If microorganisms in the food are destroyed and the food is protected from recontamination, it will remain sterile indefinitely. Even the most heat-resistant bacterial spores can be destroyed by the application of sufficient heat or by ionizing radiation. More sensitive microorganisms can be destroyed by less heat or by ultraviolet radiation.

The use of moist heat to destroy microorganisms in food was discovered by Nicolas Appert who began experimentation in 1795. Neither he nor leading scientists of that day understood why foods spoiled nor why his process worked.

Appert's first experiments and first commercial processes were far from efficient. There were unpredictable and occasionally resulted in high rates of spoilage. Imperfect closures leaked. Sometimes his processes were inadequate. But it was a beginning. Peter Durand's tin can was another step forward. All through the 19th century men struggled to contribute to this growing industry. There were many tries, many successes and many failures. As the development of the tinplate can reached closer and closer to an acceptable level of performance, the development of pressure cookers and of satisfactory processes became a commercial reality. One can find success stories in the middle and late 1800s, but the food canning industry really didn't become strong until about the beginning of the 20th century. By then the work of Pasteur in France and Koch in

Courtesy of the Senior Author's Collection of Napoleonica

FIG. 2.6. NAPOLEON BONAPARTE
The prize given by Napoleon for the discovery of a method to better preserve food was won by Nicolas Appert.

Germany in relating bacteria with food spoilage had left the laboratory bench and had become a technology which could be used by industry. Canning became less and less an art and more and more a science. Today, the principles laid down by Appert and refined by those who followed him are used to process hundreds of different foods and total production is measured in terms of billions of kilograms. Pasteur found that microorganisms can be destroyed or weakened by moist heat. Later workers proved that different organisms in different foods required different amounts of heat to destroy them. They also found that many desirable properties of food were destroyed by the long cooking time necessary to destroy the microorganisms. The search for better methods led to higher temperatures and shorter cooking times. This search continues.

Courtesy of Jewel Food Stores

FIG. 2.7. GENERIC PRODUCTS
Generic type packages have captured the imagination of the budget-minded consumer in 1980.

Where Appert processed bottles of food for many hours, later workers found that at higher temperatures shorter times could be used to achieve sterility with less lowering of food quality. Modern high temperature short time (HTST) sterilization processes of liquids like milk use times as short as a few seconds.

Aseptic processing also uses heat to destroy microorganisms. It differs from canning in that the food and the package are sterilized separately (and by different methods) and then the package is filled and sealed under sterile (aseptic) conditions.

Sterilization by ionizing radiation has not met with much commercial success because of undesirable alterations in many foods by the heavy doses of radiation required to achieve complete sterilization.

FOOD CANNING

Canning is defined as the processing of food in a hermetically sealed container with heat in order to reduce the population of bacterial contamination to a commercially safe level.

Preparation for Canning

After foods have been graded, inspected and separated from their inedible or defective portions, they are put into the desired shape for canning. Vegetables and fruits may be canned whole, halved, quartered, sliced, diced, in chunks or in pieces, or they may be precooked and pureed. Blanching with steam or hot water is done to soften tissues, eliminate air and retard or destroy enzymes. Meats may be in chunks, diced or slices. Some small items such as small birds or sardines are canned whole. Other products such as hams are deboned and pressed into a shaped can. Still others are chopped or ground and formed into shapes such as patties, sticks or balls.

Containers are cleaned prior to filling. As they are shipped from the manufacturer in an essentially clean state, new containers usually require only an inversion and a blast of clean air to blow out solid particulates. Recycled beverage bottles on the other hand have to be thoroughly washed and rinsed prior to refilling. Filling may be done by hand, automatically or by some combination. This depends upon the product and economics. Hand filling may be best in a low labor-cost society. For some products there is no automatic machinery available. The amount of solid and liquid products added must comply with food laws governing the fill of containers.

The final step before closure is exhausting. It is done by introducing steam to drive out air and preheat the contents. Proper exhausting creates a partial vacuum in the final closed container.

Container closure involves putting the lid in place and seaming a can or application of a screw-on or crimpson lid to a glass jar or glass bottle.

TABLE 2.3. GENERAL CLASSES OF FOOD PRODUCTS AND TYPES OF STEEL BASE REQUIRED

Class of foods	Characteristics	Typical examples	Steel base required
Most strongly corrosive	Highly or moderately acid products, including dark colored fruits and pickles	Apple juice Berries Cherries Prunes Pickles	Type L
	Acidified vegetables	Sauerkraut	Type MS
Moderately corrosive	Mildly acid fruit products	Apricots Figs Grapefruit Peaches	Type MR
Mildly corrosive	Low acid products	Peas Corn Meats Fish	Type MR or MC
Noncorrosive	Mostly dry and nonprocessed products	Dehydrated soups Frozen foods Shortening Nuts	Type MR or MC

Heat Processing

For acid foods with pH of 4.0 or lower heating briefly to 100°C and filling hot is sufficient. Low acid foods (those with pH greater than 4.5) must be processed at temperatures substantially above 100°C to render the food "commercially sterile." New regulations have been published on foods of intermediate acidity, pH 4.1 to 4.5. Those that combine an acid food, such as tomato sauce, with a low acid food, such as meat, in a combined entree type recipe require processes to be calculated on the assumption that the food is low acid.

Other intermediate acidity foods require careful evaluation of process, as they can be processed for shorter times or at lower temperatures. The amount of processing will vary with the type of food and degree of acidity.

In the early days of canning the temperature of boiling water in unpressurised tanks was raised by adding salts to the water. The pressure retort makes it possible to heat water above 100°C without addition of salts. Sealed containers are placed in the retort, the retort is closed, air is displaced by water and/or steam, and "cooking" begins when the retort reaches the desired "cook temperature."

When the food has been given the necessary heat treatment, the retort is opened and the containers are removed to a cooling canal where temperature is reduced. With tinplate cans temperature is brought to about 37.7°–48.8°C average internal temperature in order that the remaining moisture on the outside will evaporate without causing rust. Additional air cooling may be used. (With aluminum cans rust is no problem.) If

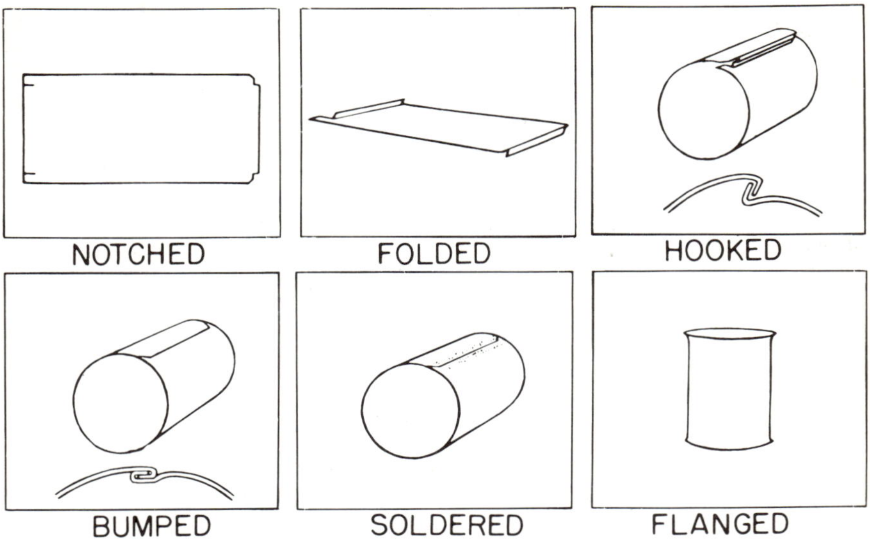

Courtesy of American Can Co.

FIG. 2.8. SEQUENCE OF CAN MAKING OPERATIONS

cooling is inadequate food continues to cook and quality suffers. This is known as "stack burn." Some cans, most glass containers and flexible packages must be cooled in the retort under pressure to avoid buckling, breaking or leakage. Jar lids are not built to withstand internal pressures. Pressure cooling is done by maintaining pressure with air while cooling water is introduced.

Heating of food products in containers is, in general, a slow and inefficient process compared with heating in a heat exchanger. However, the problems encountered in manipulating and filling some types of foods, such as asparagus or pickles, and the inability to pump other types of food products through heat exchangers due to their physical consistency plus history and tradition are reasons why most foods are sterilized in the container after packaging.

The rate of heating of food products in containers is a function of the geometry of the container, physical properties of the food product, heat transfer characteristics of the heating medium and the heat transfer characteristics of the container.

The nature or consistency of a food product, the presence of particles of food and the use of starch and sugar in the covering liquids are some of the factors that determine whether the product heats by conduction or convection.

Courtesy of National Canners Assoc.

FIG. 2.9. PREPROCESSING OF FRUIT
Peach halves are rinsed with water prior to canning operation.

Determining Process Times

In order to determine the process time required (at a nominal retort temperature) to render a food commercially sterile it is necessary to select a suitable test organism and to determine how many minutes are required at the specified temperature to reduce a known population of the test organism in the food to a safe level.

It has been proven through extensive experimentation that the destruction of microorganisms by moist heat is logarithmic in nature. In simple terms a given unit of time at a specific temperature will reduce a population of microorganisms by 90%, leaving 10% survivors. A second equal unit of time at the same temperature will give a second 90%

Courtesy of American Can Co.

FIG. 2.10. AUTOMATIC BODY FORMING MACHINE WHICH FABRICATES CAN CYLINDERS FROM FLAT METAL SHEETS

reduction or a surviving population of 1% of the original population. The unit of time we have just described is called a D value (decimal reduction). Hypothetically, total destruction is impossible. Practically, it has been decided that a process which achieves 12 decimal reductions (12 D) will render the food commercially safe. The number of minutes at a given temperature to achieve a decimal reduction varies with each microorganism, with each food and with temperature. By determining several D values at different temperatures it is possible to make a plot of temperature *vs* the logarithm of D. Here again we find the relationship is a straight line so that we can determine another value called Z. The Z value for our selected organism in our selected food is that number of degrees increase in temperature required to reduce the D value by a factor of 10. For example, if it takes 10 min at 110°C to achieve one decimal reduction (1 D value) and it takes 1 min at 120°C to achieve the same amount of "kill," then the Z value is 10°C (120°−110°C).

It has been established that the D value at 121.1°C for *Clostridium botulinum* is about 0.21 min. Thus a 12D process at 121.1°C would be 2.45 min. Because *Cl. botulinum* is so hazardous to work with and because further safety factors are desirable, a spore forming "cousin" called *Cl. sporogenes* has been widely adopted as a test organism. This

Courtesy of National Canners Assoc.

FIG. 2.11. CAN UNSCRAMBLER
Retorted cans of food are removed from jumble-packed retort baskets and unscrambled (foreground) and sent on to labelers and carton packers. Each can is coded for identification.

organism is much harder to kill. It has a D value at 121.1°C of about 1.0 to 1.5 min depending upon the environment (food). After much testing it has been accepted generally that a process that will achieve five decimal reductions (5 D) on *Cl. sporogenes* will render the food commercially

Courtesy of National Canners Assoc.

FIG. 2.12. FILLING OF PEAS INTO CANS
Peas are filled into open-ended prewashed cans and then sent on to next operation, all automatic and high speed.

sterile. This would be a process equivalent to 5 to 7.5 min at 121.1°C.

But in a pressure retort the food does not instantly jump from its initial temperature up to the sterilizing temperature. The heat of the retort atmosphere is transferred through the package to the outer layers of the food and from there it penetrates by conduction or convection to the coldest point near the center of the food. As the food warms up it will pass through all of the intermediate temperatures. As the food reaches a temperature sufficiently high to begin killing bacteria the process of achieving sterilization is begun. As we have noted, at each temperature the sensitivity of the bacteria is different and the time to achieve a decimal reduction is different. The rate of heat penetration is also different for every food and must be determined. There are mathematical equations available which enable the food processor to determine how much of the total requirement of lethality was achieved during the period the food is "coming up" to the process temperature and how much during the initial part of the "cool down." If necessary the food processor

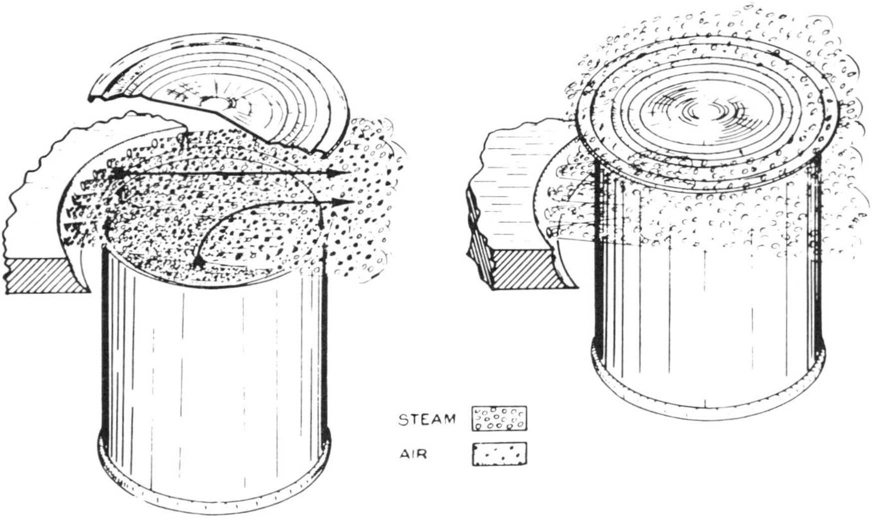

Courtesy of American Can Co.

FIG. 2.13. OPERATING PRINCIPLES OF STEAM FLOW CLOSURE

"holds" the food at process temperature the added time needed to reach the targeted lethality.

There are microbiological techniques that assist in the confirmation of a thermal process's effectiveness. Inoculated test packs are used to help select the required time of retorting. Individual sampling of production batches is used to confirm process adequacy by microbiological examination after incubation at optimum growth temperatures.

There are many variables that affect the operation of a retort and can have a critical influence on the thermal processing of a food. Because of the potential hazard to human life, government regulations have been published as guidelines to "good manufacturing practices." These can be found in the Code of Federal Regulations published in the *Federal Register* and in separate pamphlets.

Title 21, Chapter 1	contains regulations promulgated by the Food and Drug Administration of the Department of Health, Education and Welfare
Subchapter B	relates to Foods for Human Consumption
Part 113	relates to Thermally Processed Law Acid Foods Packaged in Hermetically Sealed Containers—Good Manufacturing Practice
Subpart C	relates to Equipment

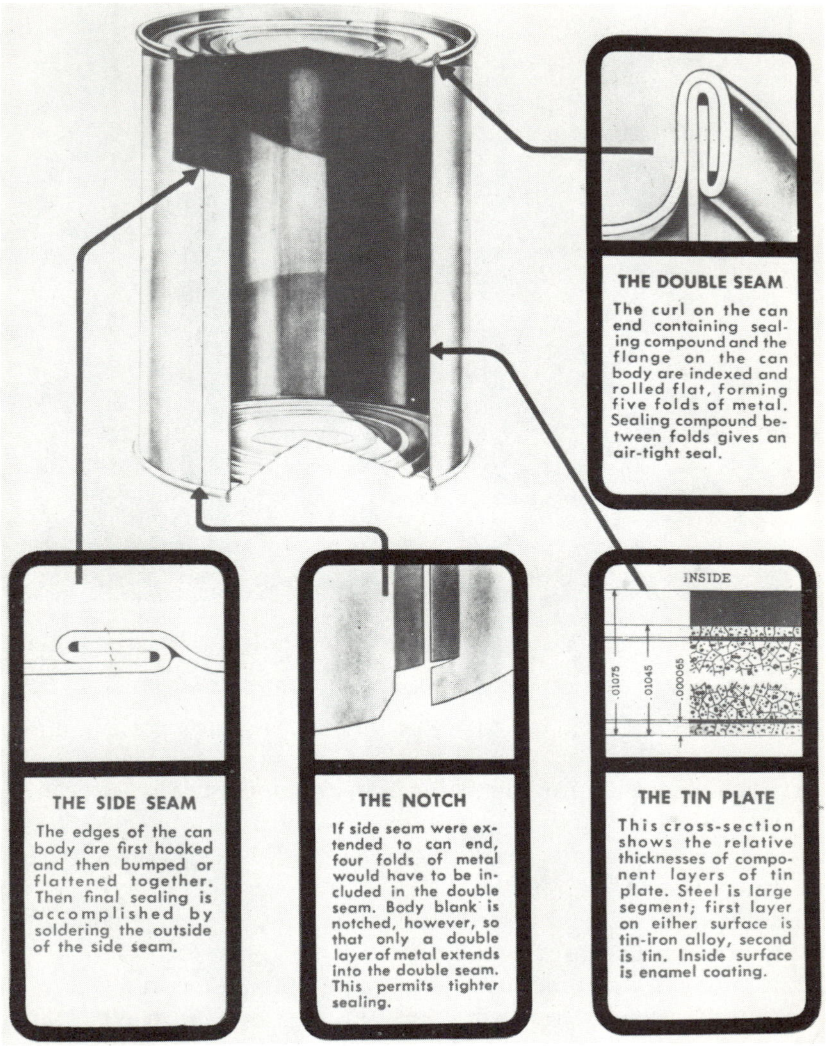

Courtesy of American Can Co.

FIG. 2.14. CONSTRUCTION FEATURES OF THE SANITARY TIN CAN

Within Subpart C there are the following headings relating to retort operations:

(a) Equipment and procedures for pressure processing in steam in still retorts.

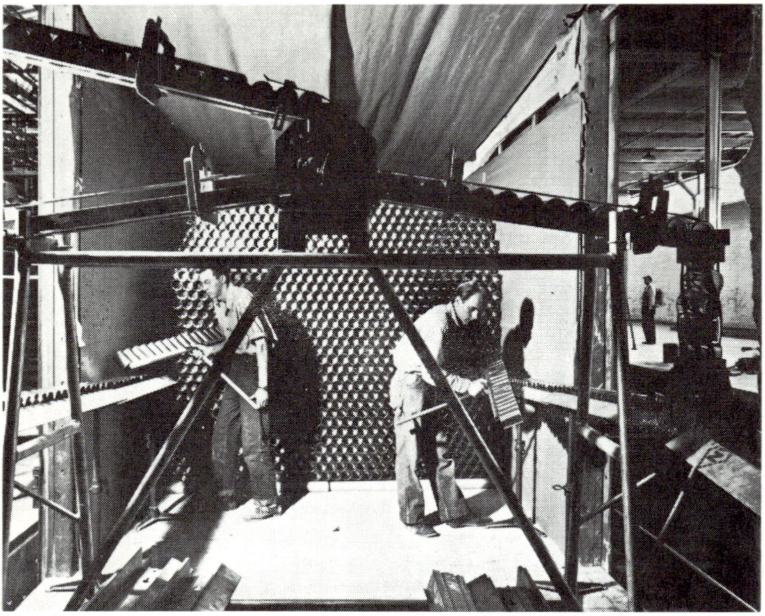

Courtesy of American Can Co.

FIG. 2.15. BOXCAR LOADING OF SANITARY CANS FOR SHIPMENT TO FOOD PROCESSOR

(b) Equipment and procedures for pressure processing in water in still retorts.
(c) Equipment and procedures for pressure processing in steam in continuous agitating retorts.
(d) Equipment and procedures for pressure processing in steam in discontinuous agitating retorts.
(e) Equipment and procedures for pressure processing in water in discontinuous agitating retorts.
(f) Equipment and procedures for pressure processing in steam.
(g) Aseptic processing and packaging systems.
(h) Equipment and procedures for flame sterilizers.
(i) Equipment and procedures for thermal processing of foods wherein critical factors such as water activity are used in conjunction with thermal processing.
(j) Other systems.

Types of Processing Equipment.—*Still Retorts.*—A still retort is nothing but a pressure tank that can be filled with containers, closed, heated,

TABLE 2.4. FOODS OF THE FUTURE[1]

1—	Decrease in frozen foods due to high energy costs.
2—	More dried products (martinis, champagne, soy sauce, butter, vinegar) on market.
3—	More "nutrified" snack foods (high protein cola drinks).
4—	Increased growth of vegetarian foods.
5—	Decreased volume of animal protein consumed.
6—	More fabricated foods "identical" to natural foods developed.
7—	Meat lower in saturated animal fat; development of a "polyunsaturated cow."
8—	Leaner chickens and larger turkeys.
9—	More fish and seafood because of commercial development of aquaculture.
10—	More and more use of fabricated foods.
11—	Growth of produce with electrical energy by establishment of regional greenhouses.

[1]More and more of the future food supply will be controlled by the processors. The majority of people will not think about what they are eating nor care about the food they eat. People will probably think of "food as fodder."

cooled and opened. Both horizontal and vertical types are available. Heating can be done in the presence of steam or water. Care must be taken to purge air by means of bleeders and to distribute steam uniformly (via headers). Packages must be so loaded as not to prevent free circulation of the heating medium. Traditionally, cans are jumble-packed into perforated crates or baskets.

The "Malo" system eliminates the basket and automatically loads cans through the top of a vertical retort and discharges them through the bottom. Steam enters through the top and forces out the hot water introduced at the start to cushion the impact of loading. This is still a batch system, but it is automated and drastically reduces labor required.

Agitating Retorts.—Some batch retorts have been designed wherein the atmosphere is circulated or the packages are moved or both. Such motion improves heat transfer, particularly when the heating medium is not pure steam.

Other agitating retorts have been designed as continuous processors. The FMC Sterilmatic conveys cans through locks and a spiralled conveyor system which rotates the cans. This rotating action improves heat transfer and substantially reduces the processing time needed. Latest models are capable of temperatures as high as 143.3°C and speeds of 550 #303 cans per min. The same type of equipment can be operated at lower temperatures for canned beverages. Glass containers or more delicate metal containers, can be loaded into carrier shells and then conveyed through the retort.

The Rexham Hydrolock conveys containers through a warm water immersed lock into a pressurized chamber. The packages continue through the warm water which is now under pressure and then enter a circulating atmosphere hot chamber. Here steam, or steam air mixtures, or heated water sprays may be used to sterilize the food. On leaving the chamber the packages reenter the warm water which cools them under pressure

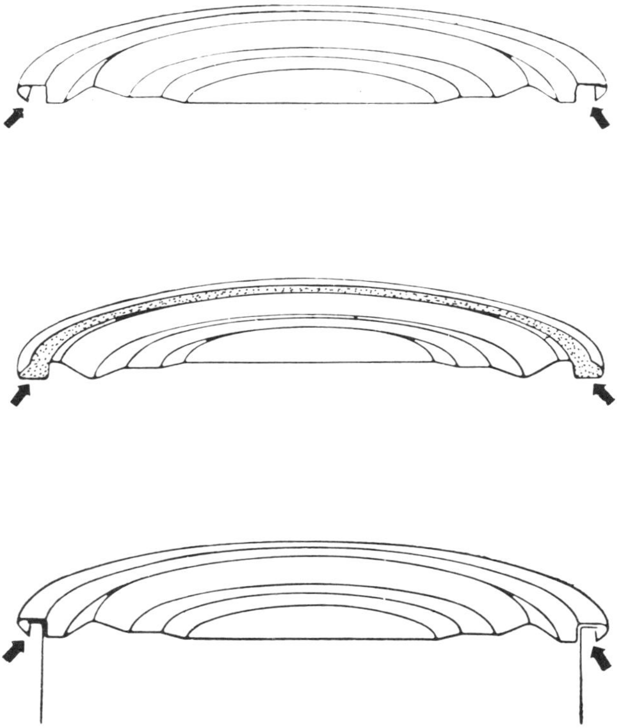

Courtesy of American Can Co.

FIG. 2.16. INSIDE CURL OF THE CAN END IS LINED WITH SEALING COMPOUND, A RUBBER GASKET MATERIAL

sufficiently to permit their safe exit through the lock. Final cooling is done at atmospheric pressure. Carriers can be mounted on the conveyor chains which permit rotation of the packages.

Hydrostatic cookers are open-ended permitting continuous conveyance of containers in and out. The pressure in the sterilizing section is counterbalanced by the head of water in the vertical legs leading in and out. Containers may be rotated as with the Hydrolock.

Aseptic processes differ from conventional processes in that the heat of sterilization in the latter case is applied after the product is sealed in the package, whereas in the aseptic system the product and container are sterilized separately, then combined and the container closed in such a way as to prevent recontamination. Hard to heat conductive foods can be heated much faster in such systems through the use of heat exchangers.

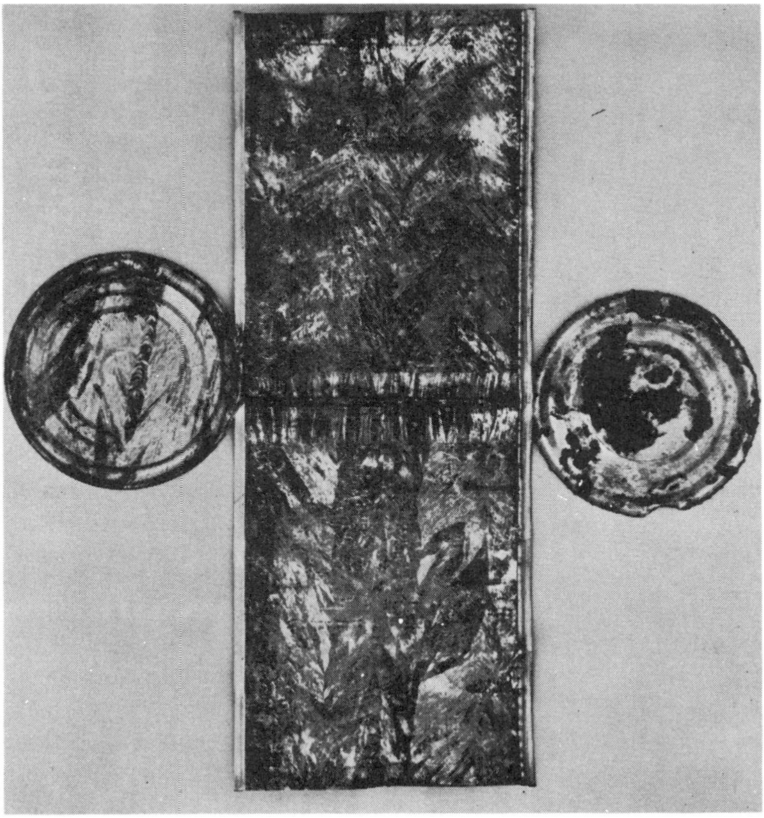

From Desrosier (1963)

FIG. 2.17. DETINNING OF INNER SURFACE OF SANITARY CAN AS MAY OCCUR WITH ACID FOODS

These may be plate type, tube type, steam injection type, or moving film or scraped surface type.

The "Steriflamme" process was developed in France by M. Beauvais, H. Cheftel and G. Thomas. This uses gas flames to bring the preheated, rotating container to sterilization temperature. Heat penetration rates are extremely high and processing is very fast for convection heating foods.

RETORTABLE FLEXIBLE CONTAINERS

For over 20 years a great deal of effort has been expended in the development of the retortable flexible pouch. Extensive studies have shown that aluminum foil is an essential ingredient. Rigid or flexible plastic containers are available that will survive retorting but these have not provided adequate barrier properties for long-term storage of foods at room temperature. Properties of food that deteriorate on exposure to light and oxygen will do so much more rapidly at room temperatures than at refrigerated or frozen conditions. The barrier properties of aluminum foil prevent the entry of light and oxygen. Methods of filling and closing are available which will insure that initial air content is at a sufficiently low level.

Flexible laminations for retortable packages are comprised of an outer plastic film which contributes strength and can carry the decoration or labeling information, aluminum foil and an inner plastic film which contributes some strength, inertness to product constituents and heat sealability. Lamination of these components is accomplished by adhesives which must develop a high level of adhesion and maintain it even under retort conditions. Adhesives are usually curing-type adhesives that develop maximum bond strength after a period of ageing. Primers may be required to achieve desired bond strengths. A typical laminate might be polyester film-primer-inks-adhesive-primer-aluminum foil-primer-adhesive-polyolefin film.

Most retortable pouches are fabricated as fin-sealed pouches with filling being done through the top prior to making the final seal.

In making packages, it is essential that care be taken to avoid contaminating final seals with large particles of food. Minor contamination can be tolerated provided the type and thickness of interior film are carefully selected.

Since flexible pouches will expand if internal pressures are allowed to exceed external environment pressure, processing must be done with overriding pressure from air or steam during heating and with air during cooling.

While this package has been marketed in Europe and Japan for over 10 years, in the United States the Food and Drug Administration and Department of Agriculture have been keeping a close watch on its development. Strict regulations have beeen promulgated as to the possible migration of film or adhesive components into the food at the temperatures of retorting. These regulations caused a delay in commercialization of the pouch as from mid 1976 manufacturers collected data to prove their products were in compliance.

The government has exhibited concern over the integrity of the pouch seals, particularly the final seal, because if improperly closed these would represent sites for entry of contaminants. The development of machinery for the filling and closing of pouches with perfectly made, uncontaminated seals and with means for excluding residual air has been concurrent with the development of the pouch material. There are two

Courtesy of U.S. Army, U.S. Army Natick Lab.

FIG. 2.18. AUTOCLAVABLE POUCH
U.S. Army sponsored research has been underway for a number of years to develop a flexible package which can survive retorting. Illustration shows an experimental package on the left and the comparable can on the right.

basic approaches—preformed pouches and the in-line pouch form-fill-seal machine.

The latter present, some potential economics, but in practice the preformed pouch approach permits an inspection step that culls out a high percentage of defectives. Experience in the Japanese market has been less than 1 leaker in 2000 pouches when using the preformed pouch. It has been suggested rollstock would result in at least four times more defects.

The most popular method of air elimination is by steam injection. Just prior to making the final closure steam is flushed into the top of the pouch sweeping out most of the headspace air. After the pouch is closed the steam condenses and a partial vacuum is created. Some machines use a mechanical compression to force the food inward closer to the seal. This eliminates air to a large extent but increases the chance of contaminating the seal.

In Japan, Spain and England some companies have chosen to thermally process in mixtures of steam and air (to achieve the overriding pressure).

In the United States most prospective processors have opted for processing in water-steam-air retorts.

The equipment development program is still very much in a hectic state and in the proverbial "chicken and egg" dilemma. Food companies looking to economics are seeking efficient lines at minimum capital investment. Machine manufacturer's are caught between the cost of development of high-speed machinery and the lack of an existing market.

It seems likely that first installations will be with multiple units of slower machines using preformed pouches. Full automation will be deferred until markets have been firmly established.

The potential United States market for retort pouched foods is variously estimated from 70 to 221 million pouches per year by 1982. The U.S. Army anticipated delivery of 40 million pouches of rations during 1980.

The chief selling points for the retortable pouched foods are convenience, quality and energy savings. Significantly less energy is consumed both in the delivery of the pouch to the consumer and in the evaporation of the food in the home as compared to frozen food yet quality levels are competitive.

Many major food companies in the United States are studying the retort pouch's possibilities; some have already test-marketed products.

REFERENCES

ANON. 1961. Aseptic bottling by four methods. Dairy Ind. *31*, 997.

ANON. 1969. Regulations and petitions on radiation preservation of foods and packaging materials. Food Irrad. *9* (3). (insert pages)

ANON. 1970A. Package forms for frozen foods. Food Mr. *45* (9) 47–50.

ANON. 1970B. Pack for dehydrated potato powder. Packag. Technol. *16* (110) 15 (January).

ANON. 1976. Energy consumption in the freezing process. Quick Froz. Foods Int. *17* (3) 90, 190.

ANON. 1977. Retort tray opens new vistas for institutional food packer. Canner/Packer *146* (8) 40–43.

BRENNAN, J.G. 1978. Wrap it up the role of packaging in maintaining good food quality. Nutr. Food Sci. *55*, 10–12.

COOK, G.D. and MARTON, S. 1968. Thaw pack bids to eliminate paper can for frozen fruits. Quick Frozen Foods *30* (11) 35–40.

DAVIS, R.B. and MAUNDER, D.T. 1967. New system for aseptic pouch packing. Mod. Packag. *40* (14) 157–163.

DESROSIER, N.W. 1963. The Technology of Food Preservation, 2nd Edition. AVI Publishing Co., Westport, Conn.

DOMICE, K. 1970. Automatic packing and preserving of freeze-dried products. Verpack-Rdsch. *21* (4) 514–526.

FLEISCHMAN, F.R. 1970. Aluminum packaging contribution to development of frozen foods. Quick Frozen Foods *33* (5) 103–109.

FODA, Y. *et al.* 1967. Effect of dehydration, freeze drying, and packaging on the quality of green beans. Food Technol. *21* (7) 83–86.

GEDDES, J.P. 1969. New package keeps raisins fresher by far. Food Eng. *41* (9) 54–55.

GHOSH, K.G. *et al.* 1968. Lightweight flexible packaging for accelerated freeze dried (AFD) meat. J. Food Sci. Technol. *5* (1) 12–16.

GRIFFIN, R.C. *et al.* 1968. Use of computer derived tables to calculate sterilizing processing for packaged foods—Part I. Application to straight line heating curves. Food Technol. *22* (4) 129–140.

GRIFFIN, R.C. *et al.* 1969. *Ibid.*—Part II. Application to broken-line heating curves. Food Technol. *23* (4) 121–126.

GRIFFIN, R.C. *et al.* 1971. *Ibid.*—Part III. Application to cooling curves. Food Technol. *25* (2) 36–45.

GRIFFIN, R.C. and SACHAROW, S. 1972. Principles of Package Development. AVI Publishing Co., Westport, Conn.

HEIDELBAUGH, N.D. and KAREL, M. 1970. Changes in pouched heat-processed foods. Mod. Packag. *43* (11) 80–90.

JOSLYN, M.A. and HEID, J.L. 1963. Food Processing Operations, Vol. 2. AVI Publishing Co., Westport, Conn.

LEONARD, S. *et al.* 1976. Determination of a flame sterilization process for apricot halves in 303 × 406 cans. J. Food Sci. *41* (5) 1222–1224.

LONDAHL, G. 1977. Yes, frozen foods do have a future. Food Mfg. *52* (1), 19, 57.

LOPEZ, A. 1969. A Complete Course in Canning, 9th Edition. Canning Trade, Baltimore.

MEADOWS, V.S. 1969. The frozen foods explosion. Food Mfg. *44* (8) 33–35, 56.

MERMELSTEIN, N.H. 1976. An overview of the retort pouch in the U.S. Food Technol. *30* (2) 28–37.

MITTEN, H. 1976. Advances in frozen dessert processing. Amer. Dairy Rev. *38* (11) 18B–18F.

PRESTON, L.N. 1967A. Quick frozen foods. Food Packag. Design Technol. *June*, 33–37.

PRESTON, L.N. 1976B. AFD foods. Food Packag. Design Technol. *July*, 21–28.

ROGERS, J.L. and BINSTEAD, R. 1972. Quick Frozen Foods—The Commerce and Technology of Processing, Packaging and Distribution, 2nd Edition. Food Trade Press, London.

ROSS, V. 1976. Recent advances in packaging systems reflect changing needs of FF industry. Quick Froz. Foods *38* (12) 21–24.

SACHAROW, S. 1970. Converting for the dehydrated food industry. Flexography *15* (8) 14–16.

SACHAROW, S. and GRIFFIN, R.C. 1970. How to package frozen foods. Quick Froz. Foods *32* (8) 56, 180 (March).

SACHAROW, S. and GRIFFIN, R.C. 1973. Basic Guide to Plastics in Packaging. Cahners Books (Now C.B.J. Publishing Co.), Boston.

TRESSLER, D.K., VAN ARSDEL, W.B. and COPLEY, M.J. 1968. The Freezing Preservation of Foods, 4 Vols. AVI Publishing Co., Westport, Conn.

WILES, R. 1977. Aseptic packaging and processing of products. J. Soc. Dairy Technol. *30* (3) 151–156 (July).

3

Red Meats

> The primary functions of a package for meat and meat products are to protect the product against physical damage, chemical changes, and microbial contamination, and to present the product to the consumer in the most attractive manner.
>
> *John M. Ramsbottom*

Meat was one of the earliest of man's foods. When he learned to make fire (about 750,000 years ago), he learned to roast meat.

Up until comparatively modern times meat was eaten promptly after it was killed. The seller of meat purchased live animals and kept them near his shop to be killed when a buyer was at hand. There were problems in keeping live animals through the winter as grain supplies might be needed for human life. In cold climates animals were slaughtered at the approach of winter and great feasts were held. Much experimentation was done to find ways of keeping meat longer. Even in Biblical days spices and salt were used to pickle or brine the surplus. Other techniques included drying, smoking and dry-salting. The resulting tough, salty, stringy meat encouraged the development of stews, meat pies and sauces or gravies.

In the 17th century, pirates in the West Indies were called buccaneers since they sold and used dried or "boucaned"beef, making them one of the earliest meat processing industries in the New World. Other early meat packers in the United States were New England farmers who preserved meat with salt.

Commercial meat packing developed with the concurrent development of direct-expansion ammonia refrigeration in the late 1800s. This replaced natural ice refrigeration methods and enabled meat packing to become a year-round operation. Scientific research into meat science began in the 1890s when many large meat packers organized research programs to study processing and packaging.

In the days of immediate consumption or alternative preservation of meat, packaging was limited to that necessary for protection against contamination and for ease of transportation and storage. The development of the canning process in the early 1800s was quickly adapted to meat and meat products. Wm. Underwood Co. of Boston was one of the first to can meats in 1845. Tinned canned beef was introduced in 1868 by the Libbey brothers, and the canning of meat soups and stews by Campbells came in the 1870s. Despite the popularity of these canned products, fresh meat and some cured meat products were still purchased from the local butcher and carried home in any suitable wrapper. Later, special juice-resistant butcher's wrapping papers were developed. In Victorian and Edwardian England, toy butcher shops were a common child's possession.

Not until the development of supermarkets in the 1930s and 1940s did any significant change occur. They introduced the prepackaged retail cut of meat, already weighed and priced, in order to reduce labor costs. Without cellophane, which enabled the buyer to see the meat, this would not have been possible. In more recent years, the supermarkets have demanded packages which would provide increased shelf-life and reduce their losses due to spoilage. This has required increased attention on the development of packages and packaging materials.

Different packaging principles, dependent on inherent biological requirements of the product, are involved in packaging fresh, cured and processed meat. In the latter category the various methods of processing each have their own peculiar packaging requirements. In recent years, frozen red meats have steadily been increasing in sales volume.

FRESH MEATS

In the United States, over 3.5 million kg of fresh meat are packaged annually. In 1977 this amounted to a per capita consumption of 87.5 kg. In most supermarkets fresh meat is packaged manually as an in-store operation with automatic labeling equipment, cut-off devices and film dispensers serving to increase speed. The most commonly used package is a film-tray combination.

Characteristics

Fresh meat is highly perishable and biologically active. Even with packaging to reduce moisture loss and refrigeration to reduce the attack of microorganisms, a shelf-life of only 2 to 3 days can be expected. Change of color is the first cause of reduced sales appeal.

Color.—The pigments present in fresh meat are the conjugated proteins, hemoglobin and myoglobin, which form complexes with the oxygen inspired by the animal. Hemoglobin transports oxygen in the blood; myoglobin is a storage mechanism for oxygen in the cells. The purple-red color of a freshly cut surface is due to the concentration of myoglobin. Upon combination with oxygen, oxymyoglobin is formed and a bright red color results. The conversion of myoglobin to oxymyoglobin is completely reversible and involves an oxygenation process. When the bright red surface is exposed to air for several days, an additional reaction forms the brown pigment, metmyoglobin. Except for the presence of reducing coenzymes which convert it to myoglobin, metmyoglobin is highly stable and its formation is essentially nonreversible.

Both oxygen reactions occur at a rate proportional to the partial pressure of oxygen; however, the rate of oxidation of myoglobin to metmyoglobin reaches a maximum at a partial pressure of oxygen of only 0.0013−0.0018 atm. The rate falls as the oxygen pressure is raised and reaches a constant value over the range of 0.04 to 1.0 atm. If a freshly cut piece of meat is wrapped in a highly oxygen-permeable film, a sufficient amount of metmyoglobin is formed to cause the color to deteriorate in 2 to 4 days.

The formation of metmyoglobin is also dependent on other factors. Oxygen utilization by meat occurs first through a solubilization of oxygen in the surface fluid and subsequently by a diffusion process into the interior. Maintaining low temperature storage conditions enhances oxygen solubility. Under poor sanitation conditions, microorganisms utilize the available oxygen, causing the conversion of oxymyoglobin to metmyoglobin. Bacterial spoilage is also accelerated in the higher pH range, which is generally between 5.3 and 6.0 depending on the feeding and handling procedure prior to slaughter.

Another type of surface discoloration, a dark reddish-brown color, occurs with loss of moisture. As dehydration takes place, the concentration of pigment increases on the surface of the meat. Interior moisture, containing dissolved pigments, then migrates to the surface and evaporates causing pigment to concentrate further.

Literature contains conflicting evidence about the effect of light on fresh meat color. Some say ultraviolet light accelerates desiccation and oxidizing of myoglobin, others that light is almost without effect. Tests run under controlled conditions on beefsteaks prepackaged in films and stored for various lengths of time at −1.1°C in a dark environment and then in a lighted display cabinet support the former contention. The color score of the steaks remained stable during storage in the dark but display under light caused color degradation.

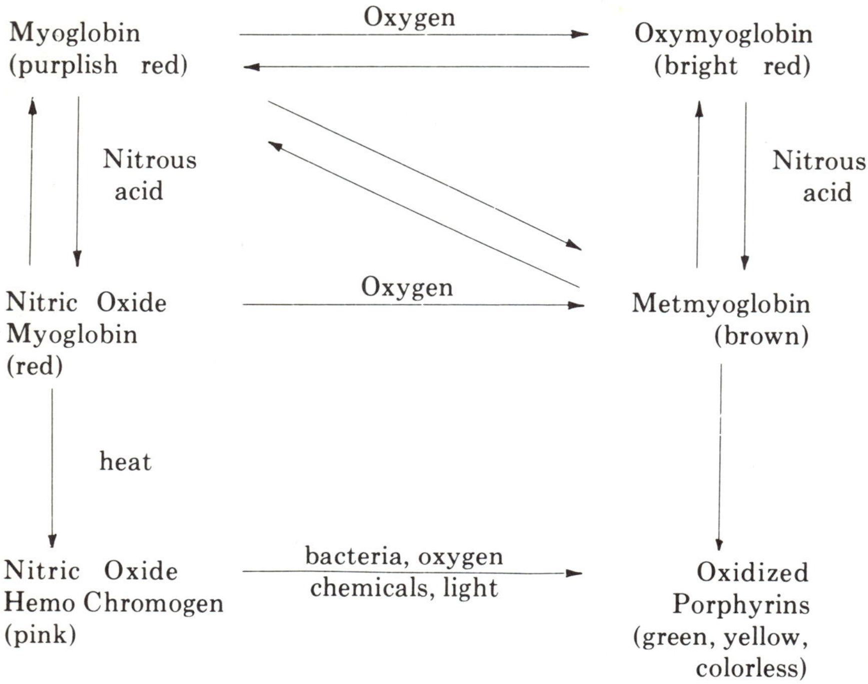

Courtesy of the American Meat Institute Foundation

FIG. 3.1. COLOR CHANGES IN FRESH AND CURED RED MEATS

Flavor, Odor and Texture.—Undesirable flavors, odors and textures can occur in fresh meat if the actions of enzymes, molds, bacteria and oxygen are not properly controlled. Tenderness in texture is achieved by ageing freshly slaughtered carcasses of beef and lamb for up to 2 weeks at 1.6°C, giving them an "aged flavor." (Pork is not aged because the meat is already tender and because of the rancidity problem.) The better grades of carcasses do not lose much weight during ageing because the good fat covering minimizes moisture loss and surface contamination. Lower grades not do have this protection and therefore are seldom aged. Ageing at higher temperatures can be accomplished if special methods such as ultraviolet radition or bacteriostatic agents are used to prevent surface spoilage from molds and bacteria.

Packaging Principles

The package is not intended to prevent bacterial contamination of fresh meats. They are handled so frequently prior to retail packaging that contamination is inevitable. The development of flavors and odors due to spoilage from bacteria or molds is prevented by control of temperature. Except during ageing, fresh meats should be stored as close to 0°C as possible and never above 10°C. The principal role of the package is prevention of moisture loss, the exclusion of foreign odors and flavors, and the moderation of oxygen transfer. A relative humidity of 85–90% is needed to prevent desiccation. Unless the package material is a perfect barrier, the atmosphere outside the package should be kept at this level. A good barrier material will prevent absorption of odors and flavors from external sources. Control of oxygen permeation requires a compromise between development of ideal color and prevention of oxidative degradation reactions. Oxygen is needed for "bloom" in red meats, but it will promote rancidity in fats, particularly in pork fat. Usually refrigeration is used to control rancidity as much as possible and a moderate oxygen permeability is sought. The package also should be capable of resisting tearing and puncturing from normal handling in a retail outlet, and it should offer a pleasing appearance to the purchaser.

Package Types and Materials Used

In supermarket packaging, most fresh meat products are placed in rigid trays and overwrapped with a transparent film. The nonshrink films are now being challenged by heat-shrinkable films which produce a tighter package.

Trays.—Until the late 1960s, most fresh meat trays were made from molded pulp and paperboard. These trays were absorbent, economical and fairly rigid. Disadvantages were loss of strength due to decomposition by excessive moisture absorption, adherence to the meat when frozen and poor visibility. Polystyrene foam trays offer an aesthetically appealing white background for red meat and have a "quality" feel. The use of blotters underneath the meat has served to absorb meat juices not absorbed by the tray.

Recent years have witnessed a growing concern by the consumer for full visibility in fresh meat packages. Molded pulp, paperboard and foam trays are still used but clear-formed polystyrene trays are common. The "Peek-A-Boo" bill introduced in New York City provided legislation for clear, fully transparent meat packages in supermarket retailing. Although the law requiring these packages is operative only in New York City, many other localities have their own regulations. Clear plastic trays offer improved visibility but are fairly expensive; the sealing of the film overwrap on the back of the tray reduces visibility. Meat juices may also create a poor sales effect. The use of blotters eliminates excessive juice accumulation.

Courtesy of Borden Chemical Company

FIG. 3.2. FILM OVERWRAPPING
Worker is overwrapping wholesale cut of fresh red meat for distribution to supermarket and eventual retail sale.

Transparent Films.—The earliest transparent film used for wrapping fresh meat was a special grade of cellophane. In order to retain the red "bloom" of a freshly cut piece of meat, a film with an oxygen permeability equal to 5000 ml O_2/sq m/24 hr/atmosphere at 23.8°C with 100% rh inside and 52% rh outside the package is required. The cellophane used was one-side nitrocellulose coated in order to be permeable to oxygen but impermeable to water vapor. The uncoated side was placed next to the surface of the meat. The moisture in the meat caused the one-side coated film to become saturated, and since wet cellophane is more permeable than dry cellophane, its oxygen permeability increased while the nitrocellulose coating on the outside prevented excessive moisture loss to the atmosphere. One-side coated cellophane is still widely used for fresh meat packaging. Two-side coated film cannot be used because of its low oxygen permeability which accelerates the formation of metmyoglobin. An additional grade of cellophane used for packaging heavy and irregular-shaped cuts of meat is one-side polyethylene coated film, which is extremely durable, punctureproof and permits tight contour wrapping.

The introduction of synthetic plastic films was widely hailed by the fresh meat industry. Softer films, capable of packaging meat, have captured a large segment of cellophane's market. The first noncellulosic plastic film used by meat packagers was rubber hydrochloride. Originally introduced in 1934, the self-service meat packaging industry became one of its major markets. While cellophane wrapping generally is done with sheets, most plastic film systems utilize rolls and sealing wires. Rubber hydrochloride is more durable than cellophane and may be pulled tightly around the package. It was withdrawn from the market by Goodyear in 1976.

Low density polyethylene film may also be used for fresh meat packaging. At gauges lower than 0.00254 cm, it is sufficiently permeable to oxygen and provides a suitable moisture vapor barrier. One problem, however, has been the condensation of moisture on the inner surface of the polyethylene. In order to avoid this, one variety has been produced with minute holes. Other types are available whose surfaces have been treated to prevent condensation. Polyethylene, however, is not widely used because of its extensibility, low strength in thin gauges, marginal clarity and gloss. By modifying polyethylene with vinyl acetate, a good clear film can be made which has suitable oxygen permeability.

The most widely used transparent plastic film for fresh meat is polyvinyl chloride (PVC). It accounts for an estimated 95% of all applications. Most observers believe it will control virtually all of the red meat market by 1981. It has cut the instance of rewraps in half and holds a seal excellently. Costing about 10¢ per package, its cost-benefit ratio still makes it a useful film. Gauges as low as 0.00778 cm may be used and a "soft" package results. The film is also highly plasticized in order to provide oxygen permeation and contour wrapping. Advantages include excellent gloss, clarity and package "cling."

Making a strong bid to replace PVC film is an Ethylene Vinyl Acetate (EVA) film. Problems still remain in sealing and visibility, but should PVC loose its share of the market, EVA stands to be the winner in the future. Most observers feel that PVC will lose out to another film by 1985.

Shrink Films.—The larger and more irregular cuts of fresh meat are usually wrapped in shrink films. The advantages of shrink film include neat appearance, ease in handling and a contour fit which requires less film. At present, shrink wrapping accounts for about 10% of the fresh meat market.

The trend toward the use of heat shrinkable polyvinyl chloride has been increasing. Cellophane-wrapped packages are about 25% more expensive. Other shrink films used include polypropylene and "Cryovac" film.

Package Forming and Filling.—Most supermarket packages are wrapped manually in a film-tray combination. The overturned tray is positioned centrally on the film with its edges diagonal to those of the film. The two pairs of opposite corners of the film are folded over the bottom of the tray and sealed. The entire wrapping operation is fairly simple and 10 to 15 packs may be produced per minute. The film is generally unprinted because of the many different sizes of cuts. Sealing is accomplished by heat or by means of pressure-sensitive labels. In a large volume operation, semiautomatic and automatic machinery become a realistic necessity.

Several systems are based on semiautomatic machinery involving sleeving the film around the tray manually. The package is then fed into the machine for weighing, labeling and pricing with speeds up to 20 packs per minute possible. Fully automatic machinery involves a roll-fed mechanism. The filled tray is inserted on an in-feed conveyor system. Cut-off length, weighing and subsequent operations are fully automatic with speeds up to 35 packs per minute possible.

If the cut of meat is large, irregularly shaped and requires a shrink film wrap, the package may not contain a tray. Hot air tunnels or hot water dips are used to effect a tight wrap.

Centralized Prepackaging

Most supermarkets have many problems in packaging fresh meat. Standardization is difficult and product turnover serves to complicate

the overall operation. Rewraps for torn packages,
products account for about 10% of all packages prepa
store recovers only about 50% of the retail price on rew
These problems plus the shortage of skilled butchers, over
special service orders have focused considerable attention on the
aging of fresh meat in central locations. But in spite of moves towa
boxed beef programs, most retail packaging is still done at the store level by store butchers. Perhaps this situation will change in the 1980s.

The principal factor involved in a centralized prepackaging operation is proper shelf-life. If the conventional 2 to 4 day shelf-life is retained, the area over which the product can be shipped is severely limited. By increasing the final shelf-life to 7 days or more, centralized meat cutting and packaging becomes practical. Close temperature control, excellent sanitation and color maintenance through oxygen control increase the shelf-life of fresh red meat to 7 or more days with advantages for the supermarket in better distribution, more uniform cutting methods, quantity discounts and reduced overhead. Pork has been distributed in wholesale cuts for several years and many other meats are now centrally packaged. Primal cuts of fresh meat are supplied to institutions and retailers. They are then cut into consumer portions as an in-store operation.

A successful centralized operation should be situated fairly close to the retail store so that meats can be cut specifically for certain geographical areas and shipping costs reduced. After hanging the carcass at a temperature of −1.1° to 1.1°C for several days, the meat is prepared. The cutting room is maintained at 15.5°C and "bloom" occurs within 15 to 20 min. Control of shipping temperatures is necessary for the operation to be successful.

Many United States firms have examined central consumer packaging systems but few have adopted them; most tried have not been successful because of poor shelf-life. In European supermarkets, however, centralized packaging is a distinct reality with nearly 100 firms preparing meat in central locations.

Vacuum Packaging of Fresh Meat

An alternate method of packaging fresh red meat is vacuum packaging, but it is difficult to introduce on the retail level because the purple-red color is not aesthetically appealing. Since vacuum packaging provides a barrier to the product from oxygen and moisture, it is suitable for periods up to 3 weeks. During this holding period, the enzymes in the meat continue the total tenderizing process called ageing. Meat "blooms" satisfactorily at the end of the storage period. Major retail cuts of beef now vacuum-packed by the packer are boxed for ageing, storage and handling on pallets and are used by hotels and restaurants. They are sold as "family packs" to the retail consumer. Wholesale cuts vacuum-packed can be opened in the supermarket, recut and repackaged. Large size briskets, roasts and corned beefs are vacuum-packed on automatic machinery using deep-draw films.

Vacuum packaging might well be integrated into a centralized packaging operation. Economic advantages include more efficient use of labor, reduced freight and storage, and holding of the trimmings at a centralized location. Quality advantages include more uniformity, better ageing and reduced bacterial contamination.

The most commonly used film for fresh meat vacuum packages is polyvinylidene chloride. It offers low oxygen permeability and shrink characteristics, so that large cuts may be held for up to 21 days with minimum loss of moisture. Other materials useful include polyesters, polyamides, and cellophane-polyethylene, polyester-polyethylene laminations or nylon laminates.

FRESH SAUSAGES

Fresh meat sausages are made from comminuted and seasoned meat blended with various cereal binders, spices, preservatives and even wine. Their packaging requirements are similar to fresh meat, however, they contain a comparatively heavy bacterial population. They are highly perishable and the use of a preservative does not appreciably reduce perishability. There are probably more varieties of sausage known to man than any other food, i.e. cervelat, salami, etc.

Fresh sausages are subject to oxidative color changes, bacterial spoilage, dehydration and on occasion sensitivity to high intensity illumination.

Prior to the use of packaging materials, fresh sausages were sold unwrapped or bound together in multiples.[1] The first transparent film used to wrap fresh sausages was coated, heat-sealable cellophane. The film was designed to maintain a proper balance of moisture. Protection from moisture loss is necessary, however, a truly moistureproof wrap would maintain too high a relative humidity within the package. This would encourage mold growth and slime development.

Other transparent films are used for prepackaging fresh sausages including polystyrene, polyethylene and cellulose acetate. Polystyrene does not heat-seal readily and does not have a high enough permeability to

---

[1]One of the early makers of sausage machines was Mye in England. His model of 1859 could be recognized by the long nozzle which fitted over the normal mincer outlet. Lengths of sausage skin were pulled onto this, gathered together like a stocking about to be pulled up. As the tube of minced meat came out, suitable lengths of skin were twisted over it to make the sausages. Another model by Lyons could cut up 57 lb of meat in 25 min.

water vapor to offer maximum shelf-life. Cellulose acetate is expensive and also is not easily heat-sealed. Polyethylene in the perforated form has been used to some extent. Since it is a limp film, it does not machine well on high speed packaging machines.

Courtesy of Cryovac Division, W.R. Grace, Co.

FIG. 3.3. A CHUB PACK USED FOR A GROUND BEEF PRODUCT

A small proportion of fresh sausages are wrapped by hand in precut sheets of printed film. For automatic packaging, two types of equipment are used. A direct wrap may be applied to the collated sausages and the ends folded and heat-sealed around the pack, or a film may be overwrapped on a paperboard or rigid polystyrene or polyethylene tray. Speeds of up to 75 per minute are possible on this equipment.

Fresh sausages are mostly sold in natural casings. Cellulose-based casings and alginate formulations are used to a lesser extent. In the skinless variety, the sausages are linked in a natural casing and the meat is coagulated by a heat treatment. The casing is removed prior to packaging. Sausage casings are normally preserved in salt.

PRESERVATION OF RED MEATS

Frozen Meats

Although frozen cuts of red meat have until recently found little consumer acceptance, much meat is frozen in the home. The advantage of the frozen product is that each item can be prepared and frozen under the ideal conditions for preserving fresh texture and flavor. Because frozen meat is prone to dehydration and loss of surface texture, it requires high protection against moisture loss and temperature fluctuation. A barrier against oxygen and light is also desirable on fatty products that are subject to rancidification. Freezing with liquid nitrogen is the preferrred method.

A frozen meat film must accommodate expansions and contractions that may occur during freezing and thawing.

During the late 1970s, frozen red meat finally started to appear in the frozen food cabinet. Frozen hamburger patties wrapped in polypropylene film are now being sold by six different meat packers. The barrier properties of the film slow-up oxygen interchange between the packed interior and the outside, making freezer burn less likely. Over $250 million of unprepared frozen red meat was sold in supermarkets in 1977.

Irradiated Meats

The irradiation of fresh meat is still highly experimental. The U.S. Food and Drug Administration withheld approval for irradiated hams and withdrew prior approval of irradiated bacon. The problem of off-odors and off-flavors that occur in radiation sterilization of fresh meats can be largely overcome by irradiating the product at −80°C because such low temperature inhibits the associated chemical reactions. It is best to use metal cans or an impervious flexible laminate to protect the product from recontamination.

Dehydrated Meats

Many meats are successfully preserved by dehydration. They are very well suited to military rations because of their lighter weight and because

TABLE 3.1. RANCIDITY DEVELOPMENT IN PACKAGED FROZEN PORK CHOPS DURING STORAGE AT −15°C[1]

Packaging Material	Storage Time to Beginning of Rancidity,[2] months
Vinylidene copolymer (Cryovac)	Over 14[3]
Cellophane (moistureproof)	3–4
Vinyl chloride-acetate copolymer and nitrile rubber base film	3–4
Rubber hydrochloride base film	2–3
Locker paper (wax controlled)	2–3

[1] Lowry and Nebesky (1953).
[2] Criterion for rancidity not given.
[3] End of test.

Courtesy of Reynolds Metals Co.

FIG. 3.4. ALUMINUM FOIL CARTON OVERWRAPS
Frozen chopped beef steaks are packaged in a paperboard carton overwrapped with a printed aluminum foil-paper lamination.

they do not require refrigerated storage. All dehydrated meats require protection from moisture, oxygen and mechanical damage.

Hot-air drying has been attempted on cooked ground meat and pork fat. It is not suitable for uncooked meat or cooked chops and steak. The process is too slow and causes surface hardening.

Freeze drying is a more acceptable method. All types of meats can be freeze dried, however, their stability heavily depends on the method of packaging. Many commercial firms are using freeze-dried meats in the production of soup mixes and as intermediates for other food formulations. Freeze-dried meats and products containing freeze-dried meats can be packaged in tinplate cans or heat-sealable laminates containing one or more layers of aluminum foil. Representative laminates are polyester-polyethylene-foil-polyethylene and cellophane-polyethylene-foil-polyethylene. Since freeze-dried products are friable, the use of an intermediate layer of polyethylene serves to give the pouch better resistance to crushing. Also, many freeze-dried products are gas packaged rather than vacuum packaged, as vacuum packaging tends to crush the fragile product. Freeze-dried beef, gas packed in carbon dioxide appears to give a more stable red color than when nitrogen is used.

Cured Meat

At present, virtually all of the cured meat market uses vacuum packed packages. Over $12.5 billion of cured meat was sold in 1978. United States consumers spent $2.7 billion on bacon alone, and ⅔ of all pork is eaten in the form of cured meat.

Meat Curing

Meat curing was developed to prevent spoilage and reduce the need for refrigeration. The original curing process consisted of treatment with salt and saltpeter. The carcass was immersed in brine and the resultant product yielded an attractive pink color. Cured meat may be ground, comminuted or consist of specific cuts. Most meat curing is confined to beef and pork.

Present day curing practices utilize salt, sugar, sodium nitrate and sodium nitrite. Salt is generally used as a flavor enhancer and secondly as a preservative. "Mild cures" involve a reduced concentration of salt. Sugar is added for flavor and both sodium nitrate and sodium nitrite are color-fixation agents.[2] The product may be immersed in the curing solution or the curing agents may be added in the dry state. For sausage, dry curing agents are added in the dry mix. Hams and bacon may be immersed in the solution or the solution may be pumped through the vascular system of the meat.

Corned Beef and Pastrami.—The curing of corned beef is a simpler process than that used for pastrami. The classic cut is beef brisket and the premium portion is the wide, meaty first cut. As a brisket tapers

[2]If nitrite were eliminated from cured meats, bacon would be salt pork, frankfurters would be bratwurst and ham a salty, warmed-over roast pork. There would be no kosher meat products, corned beef, deviled ham, pastrami, canned ham, chopped luncheon meat, cold cuts, Vienna sausage or salami.

TABLE 3.2. UNITED STATES PER CAPITA CONSUMPTION OF BEEF AND PORK (IN KG[1])

Year	Beef	Pork
1977	57.1	27.9
1976	58.7	26.4
1975	54.5	24.7
1974	53.1	30.2
1973	49.7	27.9
1972	52.7	30.6
1971	51.3	73.1
1970	51.6	30.1
1965	45.1	26.6
1960	38.6	29.4
1950	28.8	31.4
1920	26.8	28.8

[1] Figures, from the U.S. Department of Agriculture, represent carcass weight—how much the meat weighed at the packing-house. The National Livestock and Meat Board estimates that, allowing for bone, fat and cooking losses, we actually ate about 20 kg of beef and 10 kg of pork per person last year.

toward the back, slices become narrower and more interlaced with fat and connective tissue.

The meat may be cured (or corned) by two different methods. The dry cure is done with salt, sugar, sometimes spices and saltpeter, and always preservatives such as sodium nitrite and sodium nitrate saltpeter which are rubbed onto both sides of the meat and then layered in barrels. For the wet cure, the same curing mixture is dissolved in water to create a brine in which the meat is soaked. The dry-curing method takes longer and is preferred because the meat cures only in its own juices, absorbing no water and resulting in a firmer, more flavorful product. Time for curing varies from 1 to 3 weeks depending upon the thickness of the meat and the degree of cure desired.

Once cured, corned beef is washed or soaked briefly to remove excess salt and is then boiled before being served. One desirable ingredient in the brine is pickling spices, which give corned beef a particularly subtle and piquant flavor.

There is more variety and complexity in the making of pastrami, generally considered a Rumanian invention. The first variable is the cut of meat used. Since both corned beef and pastrami are the traditional products of kosher butchers, all cuts used come from the forequarters of steers. In keeping with kosher law, no hindquarter cuts may be used; what began as a religious requirement has become a custom. The most frequently used cuts of pastrami are the lean, wide firm shoulder cut known as deckle, and the smaller, juicier, fatter cut from the chest known as plate or navel. Other cuts used are small, very lean sections of shoulder or chuck, especially for true Rumanian-style pastrami.

Pastrami is cured exactly like corned beef, but it is then dried and rubbed with a heavy mixture of coarse black butcher's pepper, other spices and often with crushed fresh garlic. It may be left to absorb those flavors for a few hours a day, and is then smoked over hardwood sawdust and possibly over a pan of water to create steam. Smoking time varies from 2 to 12 hours depending upon the heat, the size of the pieces of meat and the effect the producer is trying to create. The longer, slower method results in a richer, more subtle flavor, but it also causes greater shrinkage, a more important consideration to wholesale producers than to delicatessen owners who sell pastrami directly to the public.

After it has smoked and cooled, pastrami is cooked by steaming, which also keeps the meats hot behind the counter until they are sliced. To be fully appreciated, both corned beef and pastrami should be eaten hot immediately after being sliced because flavor and texture deteriorate rapidly as the meats cool.

The techniques described are the most old-fashioned; mechanized shortcuts have been devised that alter the final results to varying degrees. The first shortcut involves the curing process. Instead of either the time-consuming wet or dry methods, a brine solution is injected into the meat. Injection curing may take from a few hours to 2 days, but the meat it produces is tougher and not as finely permeated or ripened as that cured by the older method.

Faster smoking of pastrami is another carefully calibrated shortcut that results in firmer, less ripened meat. If this process is accomplished by brushing the meat with liquid smoke flavor instead of using sawdust, the pastrami takes on a medicinal or acrid aftertaste.

If the pastrami and corned beef are to be strictly kosher, the raw, uncured cuts are first cooked and salted. The entire process must be done under rabbinical supervision. Otherwise the meat is kosher style.

Nitrite–Nitrate Controversy

"Meat products cured with nitrite and/or nitrate and processed or cooked for consumption should not contain confirmable concentrations of nitrosamines." This is the policy of the U.S. Department of Agriculture (USDA) as indicated in the October 18, 1977, issue of the *Federal Register*. Producers of cured meats have been asked to demonstrate that their products do not contain nitrosamines and to find alternatives to using nitrate and nitrite in the curing of meats.

The apprehension surrounding the use of nitrates and nitrites has been that they may combine with certain amines present in the meats, fish and poultry being processed to form nitrosamines, which have been shown to cause cancer in laboratory animals. The results of the recent study conducted for the U.S. Food and Drug Administration (FDA) by Dr. Newberne of the Massachusetts Institute of Technology (MIT) indicate that nitrite itself is carcinogenic, without reference to the formation of carcinogenic nitrosamines. The FDA is now faced with the difficult choice between a Delaney Clause ban and the danger of botulism.

Prior to June 1978, a typical curing mixture for hams contained 24% sodium chloride, 2.5% sugar, 0.1% (1000 ppm) nitrate and 0.1% (1000 ppm) nitrite. The function of each of these ingredients is summarized in Table 3.1. After curing, the product should contain 2–3% salt, 0.75–1.25% sugar, 10–40 ppm nitrite, 0.3% phosphate and 200–300 ppm ascorbate.

As of June 15, 1978, the regulation issued by the USDA required bacon processors to use only 120 ppm nitrite with 550 ppm ascorbate or erythrobate. The ascorbate enhances the color reaction and acts to block nitrosamine formation. Furthermore, the USDA issued a proposal, to take effect in 1 year, to require the use of 40 ppm nitrite with 0.25% potassium sorbate and 550 ppm ascorbate in bacon, providing the combination is shown not to be a health hazard.

If nitrate and nitrite were to be banned immediately, several problems would develop. The cured flavor, as presently known, would be lost. The color of sausage would be greenish instead of gray. The rate of fat oxidation would increase. The shelf-life of some sausage products would be diminished by as much as 50% with increased mold growth. According to the National Food Processors Association, the decision to remove nitrite from cured meat products "would result in a dramatic increase in the incidences of botulism in the United States and a dramatic change in the type and character of canned cured meat products available to the American public."

Of primary concern is the lack of effective alternatives for nitrite. The FDA and the USDA have begun reviewing alternative systems to use during the transition period during which nitrite and nitrate use is to be phased out. These alternatives include refrigeration and freezing, irradiation, dehydration, chemical nitrite substitutes and thermal processing.

To insure against the outgrowth of *Clostridium botulinum* and toxin formation, cured products must be maintained below 4.4°C. As nitrite provided a margin of safety in refrigerated cured meats, it would be necessary to freeze meat products cured without nitrite. The capital investment required to increase refrigerator and freezer capacity to handle this extra load would be large with concommitant increases in energy costs. Consumers will have to be educated as to the proper handling procedures of foods formerly treated with nitrite. Even though refrigeration and freezing may substitute for the use of nitrite as a microbiological preservative, the flavor and color characteristics of the foods will be different.

Another method to prevent growth of *Cl. botulinum* is to destroy the spores by the use of ionizing radiation; however, the capacity for sterilizing foods by irradiation is now very low. Furthermore, it would be almost impossible to demonstrate the safety of irradiated products to obtain food additive clearance. At the present time, irradiation is an impractical alternative to nitrites.

Spore outgrowth can also be controlled by lowering the water activity of the food product by such methods as freeze drying. The dried products are easily degraded by moisture and oxygen and would require costly hermetically sealed containers. The dried products are not sterile and abuse after rehydration is a principal source of danger. Upon rehydration the quality of the products may be greatly impaired.

Over the years, more than 700 chemical components have been evaluated as substitutes for nitrite. Salts, at a concentration of 10% has been demonstrated to prevent toxin formation by *Cl. botulinum* in cured meats in the absence of nitrite. However, such salt levels render the meat unpalatable. Monsanto Company has demonstrated that 0.26% potassium sorbate, with or without 40 ppm nitrite, has an inhibitory effect on *Cl. botulinum* growth and toxin formation in vacuum-packed bacon. Further work is needed to evaluate the effectiveness of sorbate alone in bacon and other products. National Food Processors Association Research Services have indicated that the use of 500 ppm sulphur dioxide without nitrite will extend the shelf-life of pork or poultry up to 3 months.

Another alternative is thermal processing. Presently canned cured meats include pasteurized-refrigerated meats such as canned hams, shelf stable products such as luncheon meats, and commercially sterile products such as meat spreads, potted meat, vienna sausage, and canned beef hash. The stability of canned cured meats is dependent upon the synergistic inhibitory interaction of nitrite, pH, water activity and sodium chloride in combination with mild heating. Due to the antimicrobial properties of nitrites, canned cured meats require a significantly shorter process time and temperature than noncured low-acid meats. Stable canned cured meats can be produced with a heat process that has a sterilization value (F_o) of 0.05−0.4 min, a salt concentration of 3.5−4.0% in the water phase, initial nitrite levels of 75−150 ppm, yielding a residual level of 20 ppm nitrite and a low incidence of *Cl. botulinum* spores. Therefore, if the present shelf stable process (F_o=1.28 min) used for the 0.68 kg ham were to be increased to the commercially sterile process (F_o=6 min), the process time would have to be increased from 130 min at 110°C to 205 min at 110°C, an increase of about 58% in process time. However, such an increase in process time results in an organoleptically unsuitable product. In the case of emulsion meat products such as frankfurters the heat required to kill residual spores would render the fat and destroy the taste and odor qualities of the meat. Furthermore, to convert all canned cured meat products to commercially sterile products without nitrite, 2.5 to 3 times the number of retorts presently in use would be required.

Cured meat is not the only nor the major source of nitrite in human diet. The FDA and USDA estimated that only about 20% of the dietary nitrite comes from cured meat with most of the remaining 80% coming from conversion in the saliva of nitrates occurring naturally in vegetables, such as beets, lettuce and spinach (1000 ppm nitrate), water (up to 10 ppm nitrate legally) and other foods. According to Dr. S.R. Tannenbaum of MIT, the adult daily exposure to nitrite as a food additive is only 3% as compared to 15% from salivary nitrite and 82% from intestinal nitrite. He concluded that "the risks arising from the use of nitrite as proposed by the USDA would be miniscule compared to those resulting from the body's natural processes."

FDA Commissioner, Donald Kennedy, had been quoted as saying that final action of *Federal Register* proposals could not come before September 1979. Prior to banning nitrite the USDA and FDA must draw up proposals for various classes of products based upon *Federal Register* notices. Once this was completed (now completed) a 120-day comment period followed. The agencies next examined the comments and wrote a final order. Furthermore, under the law, no action could be taken until the 96th Congress convened. At the present time the push is for a gradual phase-out of nitrite allowing for time to find alternatives that safeguard the public from botulism and to minimize disruption of food production and distribution systems while industry and the public adjusts to the nitrite ban.

Frankfurters.—Frankfurters[3] are made from a wide variety of ingredients. The meat formulation is usually 40–60% beef and pork. Filler meats such as tongues and snouts may also be used (up to 20%). Nonmeat ingredients such as cereals, milk solids and corn syrup solids are also commonly used. Flavor is obtained by various spices and seasonings, i.e., pepper, nutmeg and mustard. Curing is obtained by salt, sugar and curing salts.

The meat cuts are reduced in size by grinding. Comminution of the mixture is then necessary. In this process, the beef components are

[3]Frankfurters or hot dogs are said to have been named by the sports cartoonist, T.A. Dorgan, who once drew a German sausage as an elongated dachshund in a bun, the latter being a facetious symbol for things German in the early 1900s. Other stories trace the name to that used by Harry M. Stevens, caterer to the New York polo grounds.

chopped first with the salt, spices, sugar, cure and seasoning plus water and ice. Pork is then added and the entire mixture is chopped to final texture. A colloid mill is usually used to emulsify the mixture. One passage is normally enough; however, in order to prepare a fine emulsion, two machines may be used in-line.

A vacuum is applied to remove large bubbles of trapped air. This is done either by placing the truck containing the emulsion in a vacuum chamber or using a vacuum mixer with an agitator.

The sausage is then stuffed into casings by a hydraulic or air driven piston machine. This machine extrudes the emulsion from a stuffing horn at a specified pressure.

Linking is performed by hand or machine. If ascorbic acid is added, processing follows immediately after stuffing. In the absence of ascorbic acid, the sausage is held for a period of time in order to cause cure to proceed.

Three phases of the heating and smoking operations are usually employed. First, the sausage is tempered by mild heating (32.2°–103°C). Then a smoke is applied and the temperature is raised (73.8°C). This allows the cure to develop and the protein to denature which yields a firm product. Finally the product is cooked with a hot water spray (76.6°–82.2°C).

The product is then cooled by cold water showering. The entire processing time may take between 65 min and 2½ hr, depending on the curing reaction and overall formulation.

Coloring may be added by colored cellulosic casings or by a dye added to the hot water shower.

For frankfurters processed in cellulosic casings, peeling is done by hand or machine, the latter being usually employed.

The final product is then commonly vacuum-packaged in plastic films in two layers on automatic form-fill machinery. Standard Packaging's 6-14 machine employs a thermoformed bottom web of nylon-polyethylene or nylon and ionomer with either a polyester-polyethylene or a nylon-polyethylene top web. Vacuum is drawn up to 58.4 cm prior to sealing the top web in place. Institutional units consist of paperboard boxes with the layers separated by wax paper or parchment inserts.

Characteristics.—Unlike fresh meat, which has a shelf-life of only 2 or 3 days, cured meats will survive for several weeks depending upon the level of cure. Bacon, frankfurters and smoked links will last about 2 weeks; luncheon meats and Canadian bacon may last up to 4 weeks. The general trend is a demand for longer shelf-life; consequently more attention is being paid to the use of vacuum, gas and the more expensive barrier type packaging.

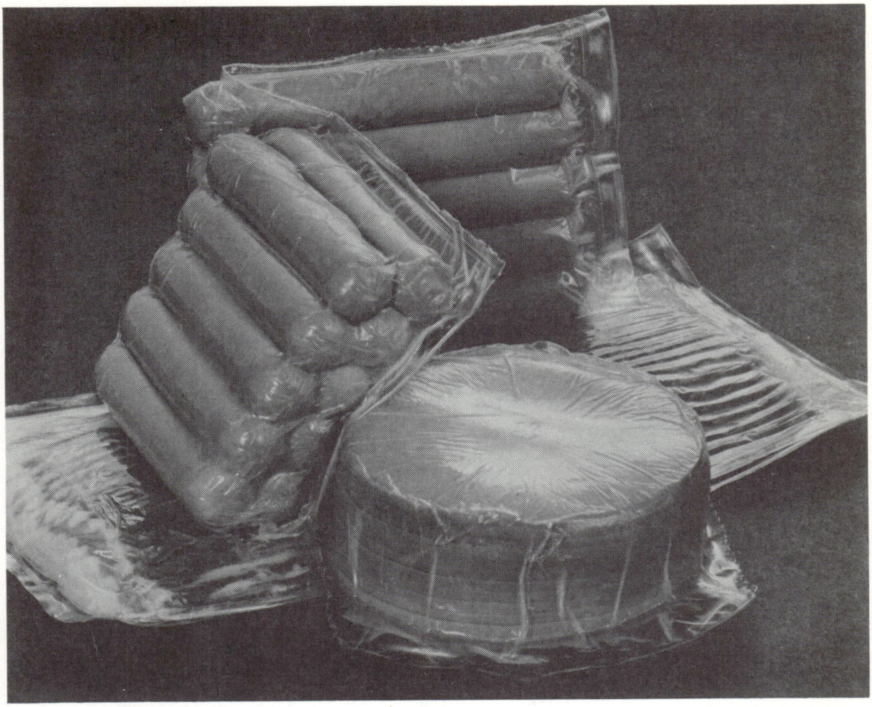

Courtesy of Nixon-Baldwin Div., Tenneco Chemicals, Inc.

FIG. 3.5. STRUCTURED FILM LUNCHEON MEAT PACKAGE
Luncheon meats are packaged in structured films. The formed web is a combination of ionomer and nylon; the flat web a combination of ionomer and nylon; the flat web a combination of ionomer and polyester.

Color.—The attractive pink color present in cured meats is due to a modified pigment called nitroso-myoglobin. This pigment is formed by the reaction of myoglobin with the nitrites present in the curing solution. Although nitroso-myoglobin is more stable than oxymyoglobin, it is readily oxidized to metmyoglobin. The stability of the nitroso pigment is dependent on the partial pressure of oxygen in the atmosphere to which the pigment is exposed. The reaction is also subject to an increase in rate with a rise in temperature. Raising the pH inhibits the tendency of nitroso-myoglobin to undergo oxidation.

All of the green and blue visible spectrum as well as ultraviolet light catalyzes the breakdown of nitroso-myoglobin by oxygen into metmyoglobin and nitrates. The use of colored filters in supermarket display cabinets is unattractive and not completely effective. It has been reported that fading was reduced in products containing free nitrite and sulfhydryl groups, but if the product lacked one of these components, reduction of fade was not noticeable. Most meats contain free sulfhydryl groups; 2% nitrite dip is an effective way of retarding or restoring light fade. Even if faded, cured meats recover some of their original color upon exposure to darkness. It is good practice for supermarkets to turn-off their display lights when not in use to permit color restoration.

The intensity of light in display cabinets for cured meats should not exceed 7.6 m candles; for ham the intensity should not exceed 4.6 m candles. Since the color fade reaction is related to the presence of oxygen, prepared or cured meat products can be vacuum-packaged to prolong color life. The presence of oxidizing chemicals will of course also tend to fade color.

The development of a brownish color on the outside surfaces of a cured meat is usually due to dehydration; however, excess nitrite can cause the same off-color. If insufficient is present, meats will have a weak color that readily fades to gray. Green streaks may be due to excess nitrite (nitrite burn) or to the action of bacteria such as *Lactobacillus viridescens*. Certain fats have high concentrations of organic peroxides; pork fat is particularly high. These fats can become rancid even under frozen storage; the result is not only a bad flavor and odor, but also a color fading on the surface.

Bacterial Spoilage.—The development of a slimy surface is related directly to heavy growth of bacteria or of yeasts. This can be prevented by lower temperature storage, but once encountered the meat is beyond recovery. Some bacteria produce carbon dioxide, others produce sour flavors or odors. Since molds are strictly aerobic, vacuum packaging is an effective means of control.

The bacteria responsible for spoilage in cured meats can largely be determined by the curing method. Excessive cures or reuse of cover pickle cause the growth of halophilic flora. Bacterial counts are sharply reduced by curing and smoking. Lactic acid bacteria may be present in commercial curing brines and may survive inadequate heat processing. These tend to grow in the interior of the product during refrigeration. Green cores may be caused by *Lactobacillus viridescens*. Souring is usually caused by *Streptococcus faecium*. Gas development is caused by heterofermentative strains. If the meat has been properly cured, recontamination is still possible during handling and packaging. Since the surface contains a high salt concentration, a barrier exists to bacterial penetration even though large numbers of lactobacilli and microbacteria may have been deposited on it.

There is conflicting evidence in literature on the effect of vacuum packaging on shelf-life extension. Most researchers feel that vacuum

TABLE 3.3. UNITED STATES MEAT CONSUMPTION PER CAPITA (1977)

Meat	Consumption (kg)	%
Beef	57.1	62.7
Pork	27.9	30.6
Veal	5.3	5.8
Lamb	0.8	0.9

packaging does prolong the shelf-life of cured meats. However, a study of sliced cured meat on the Swedish market discovered no difference in total bacterial count between vacuum-packed and nonvacuum-packed products after storage.

Vacuum packaging helps to establish anaerobic conditions, thus encouraging the growth of anaerobes and facultative spores. The possible existence of *Cl. botulinum* has been the source of much research, but this is unlikely to grow in cured meats at normal refrigeration temperatures. Also, the concentration of salts in the meat tends to inhibit its growth. Cured meat products contain few putrefactive anaerobic spore-forming bacteria. Even if they are added during slicing and packaging, they would not grow in competition with the lactic acid bacteria that are developed during normal storage.

Packaging Principles.—As with fresh meat, the package for cured meats is not expected to perform as an absolute barrier to bacterial contamination, however, some protection is inherent in any barrier. The principal role should be to minimize light fading by preventing the entry of oxygen and the loss of moisture. Spoilage is expected to be reduced by refrigeration or freezing, therefore, the package must be able to survive these conditions as well as present an attractive appearance.

Package Types

Some special cuts of cured meats are sold in the same case as fresh meat cuts and are packaged accordingly. For example, smoked pork chops are packaged in a film overwrapped tray in the same display as fresh pork chops. Chub packs, which are tubes stuffed with a soft product, twisted, tied or clipped at each end, are used for some ground meats and sausage preparations. Other smoked sausages may be vacuum packaged in film envelopes with or without trays or cardboard inserts. Sliced meats are sometimes loosely packed in a film pouch, but more frequently they are shingled or stacked and vacuum-packed. Some plastic films can be drawn into cavities which make a neater package for stacked slices or other shapes. Most hams are overwrapped and shrink-wrapped. Some are boned and sold in cans (see section on Cooked Meats).

Package Materials

Overwraps may be plain film or foil laminations. Film overwraps can be selected from a variety of materials. Plain or coated cellophane, polyethylene, polypropylene, polyvinyl chloride, polyvinylidene chloride (PVDC) or various combinations. Foil overwraps are usually combinations of aluminum foil and paper. The selection of a suitable overwrap is based on economics, shelf-life and machine availability.

Overwraps do not offord long term shelf-life with cured meats. Tightly fitting overwraps are obtained with PVC and polyvinylidene chloride. Also, these films are available in reel form for ease in in-store handling. Cellophanes are used in sheets and may be more difficult to apply. Foil laminates are useful for light protection where the supermarket lights are not turned off at night. They also offer consumer aesthetics. Chub packs can be made from polyethylene alone. Sometimes they are a combination of two films where one is a cellophane and the other a polyethylene. Shrink films are used for packaging trayed items and large cuts of cured meats. Those meats that are packaged in bags or pouches may be nitrogen gas-flushed or vacuum-packed. The use of vacuum or gas packaging is necessary for longer shelf-life. Under these conditions the film is likely to be a high barrier combination such as cellophane-PVDC-polyethylene, or polyester-PVDC-polyethylene. Drawn cavity packages are usually made from combinations of polyethylene, PVDC and either polyester or polyamide films. The polyamides are softer and will give better draw, thus are capable of packaging large size units such as 1 lb (0.45 kg) bologna. Most luncheon meats and bacon are now packaged under vacuum. The standard bacon package is a nylon-PE laminate, vacuum-sealed with a coated overprinted paperboard overwrap.

Package Making, Filling and Closing

When automatic pouch-making machinery was initially developed, most packers preferred to use preformed pouches. Cellophane-PVDC-polyethylene and cellophane-foil-polyethylene were the most frequently chosen materials. Handfed, manually operated filling equipment was generally used. At present most large packers still use preformed pouch type units for small volume products and many smaller packers rely on preformed pouch machines for almost all their production, but the trend is toward automatic, vacuum- or gas-packaging machinery which form, fill and close a drawn cavity package from roll-fed stock at higher speeds and with less material consumption.

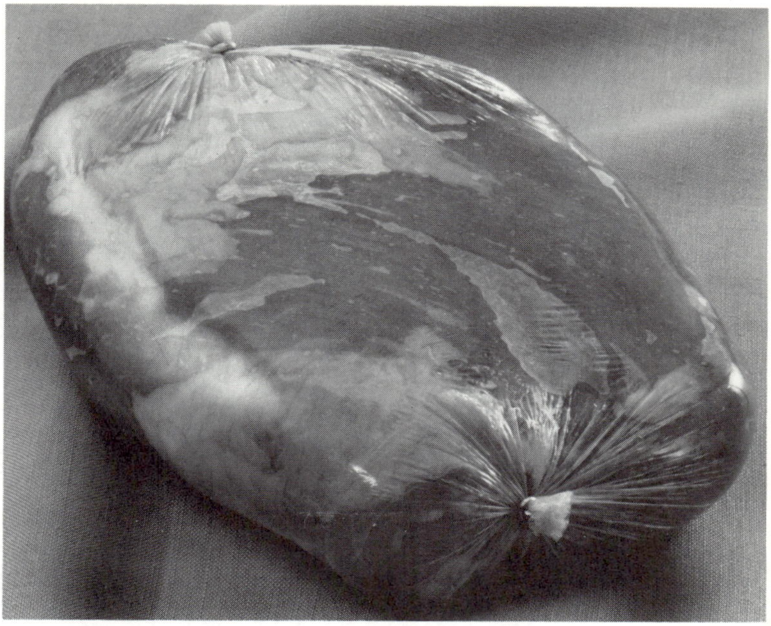

Courtesy of Cryovac Div., W.R. Grace Co.

FIG. 3.6. SHRINK-FILM TIGHT WRAP
A cut of corned beef is wrapped in a PVDC-PVC copolymer film. After closure the package is immersed in hot water to shrink the film tight.

The basic concept for pouch units incorporates an air pressure equilization principle. By means of a chamber, air pressure is equalized so that no collapsing or distortion occurs when a product is vacuum-packed and a stretching mechanism prevents top seal wrinkles during evacuation. Hand-loaded models are available with single or multiple chambers, and pouch machines can vacuum and/or gas flush packages. For some sliced meats, such as chipped beef and liverwurst, only gas flushing is feasible because vacuum packaging causes adhesion of the slices to one another. Nitrogen or carbon dioxide can be used to create an oxygen-free environment. The packaging materials are the same as for vacuum packaging. The meat absorbs some of the gas and creates a partial vacuum in the package. Gas packaging is not widely used by the meat industry due to poor efficiency and lack of specifically designed meat packaging equipment.

Most vacuum machines evacuate a package to an absolute pressure of 0.03 atm Hg using a rotary piston pump. The pump should be closely situated to the machine and additional outlets eliminated. It is very important to obtain the highest degree of evacuation possible. If gas-flushing is desired to render the package oxygen-free, it can be accom-

Courtesy of E.M. Todd Co., Inc.

FIG. 3.7. VACUUM-PACKAGED BACON IN A PRINTED CARTON

plished in three ways. One concept vacuumizes the package and then gas-flushing is conducted. Another flows nitrogen through a tube to displace air from the package. Still another conducts the total operation of filling and sealing in a nitrogen atmosphere.

The basic concept for drawn cavity packages is the use of two webs of material. The "formed web" is thermoformed into varying size cavities while the second "flat web" is used for sealing. The package is evacuated or flushed in a chamber or through a small opening in the final seal, which is then sealed shut. Automatic integrated lines operate at about 40 units per minute per cavity. Typical package sizes are 12 (340) and 16 oz (453.6 g) for frankfurters, 8 (226.8) and 16 oz (453.6 g) for bacon, 6 (170), 8 (226.8) and 12 oz (346 g) for sliced luncheon meats. The trend is toward larger packages and deeper draw materials.

Shrinkable vinylidene chloride copolymer films are used for packaging cooked hams. The package is evacuated prior to shrinking. A hot water immersion or hot air tunnel is used to effect shrink. In addition, the product is pasteurized by exposure to heat. This process is used for other large cuts of cured meats such as bacon, corned beef and various sausage mixtures such as the Polish (Kielbasa) sausage. It is also possible to stretch-wrap large cuts of cured meats with tight contour films.

Cured sausage meats are usually marketed in natural or synthetic casings. A newer trend is the use of a vinylidene chloride copolymer as the casing. The meat is stuffed into the casing automatically and sold as an overwrapped film unit.

TABLE 3.4. CURING INGREDIENTS AND THEIR FUNCTION

Salt—Decrease water activity (a_w), reduce microbial growth, increase shelf-life, improve flavor.
Nitrate—Serve as a source of nitrite for the curing process (in longer cures, bacteria reduce the nitrate to nitrite).
Nitrite—Provide antimicrobial properties, flavor, color, antioxidant qualities.
Sugar—Reduce the harshness of salt, improve flavor and surface color.
Phosphates—Reduce moisture loss during processing, improve firmness.
Ascorbates—Speed color reaction, increase stability.

Recent Developments

The latest trend in cured meat packaging is towards resealable units. Consumers had long been concerned about the leftovers in an open bologna or salami package. In the early 1970s, acrylonitrile (AN) film started to penetrate the cured meat market. Presently, about 50% of the cured meats sold in semi-rigid packages from supermarket cases are in AN (PVC is the other widely used semirigid material).

There are two currently available resealable packages for luncheon meats. Oscar Mayer is marketing a resealable AN package in certain areas. Made from "Barex 210," an acrylonitrile resin manufactured by Vistron, a division of Sohio, the unit consists of an AN cap that is heat-sealed to a flat base AN sheet. After the cap is peeled away, a circle of adhesive on the base provides a seal when the cap is pushed down on it. Also available is Eckrich Foods unit which is made-up of an AN web which provides a package seal during the products' trip from the packaging plant to the consumer's kitchen. When the consumer opens the package, he/she throws away the overwrap. Resealing is accomplished by pushing the clear AN cap into a notch in the base. Reasons given for the popularity of AN include good barrier, competitive price and also that it remains stiff and holds its shape.

Also on the horizon are resealable packages for bacon. AN film is in the forefront of the race towards the development of resealable packages.

The only factor possibly clouding the total success of AN in the marketplace might be a total FDA ban. Ever since the FDA banned the use of AN for soft drink bottles, most meat packers have been leery of using AN; its future remains undecided.

Another interesting bacon concept is "Saran-Pac," in which plastic film is extruded in line at the packing plant. The final package is composed of PVC-PVDC-meatboard-PVDC-PVC. Factors limiting the acceptance of the "Saran-Pac" system are its high cost compared to a conventional roll-stock machine and that only one size package can be used, which requires a large volume operation.

COOKED MEAT

Cooked meats are intended for consumption without any prior heating, although they can be heated if desired. Many of these products have also been cured to provide the desired flavor. Tinplate or aluminum cans frequently are used as the package, and unless the product is sterilized, the package must be handled and stored under refrigeration. At present tinplate is the standard material; however, aluminum is making inroads. Aluminum cans offer light weight, corrosion resistance, variety of shapes and easy-opening. Both types of cans require coatings to prevent the action of salts on the metal as well as other effects. The coatings used are called meat enamels and consist of modified epons with aluminum pigment. If unprocessed cooked meat is thinly sliced, short term shelf-life is possible with a cellophane or PVDC overwrap, or pouch under refrigerated conditions. Meat pies require a breathing film to allow for moisture escape because the ingredients present tend to deteriorate in a damp atmosphere. An equilibrium humidity of less than 75% insures adequate ventilation. Cellophane and ventilated polyethylene may be used as overwraps. In the United Kingdom, meat pies are readily available most everywhere. The food halls at Harrod's Department Store in London have these pies as prominently displayed as the sturgeon and cavier. Butchers and pork stores make their own pies. It used to be the custom in England to have them delivered right after the milkman left milk products.

Retortable Meat Package

A noteworthy trend in cooked foods has been the development of flexible packages for retortable meat products. If a food product has a pH below 4.5 the very dangerous spoilage bacteria, such as *Cl. botulinum*, will not multiply. Food poisoning from eating improperly processed acid foods is, therefore, extremely unlikely. The growth of harmless yeasts and molds can be prevented by pasteurizing the product at temperatures up to 100°C. Foods with pH above 4.5, on the other hand, must be processed at high temperatures (115.5° to 132.2°C) to insure destruction of the dangerous bacteria. A satisfactory packaging material must be capable of withstanding the processing conditions for the food it is to contain and must also provide both an oxygen barrier and protection against subsequent recontamination by microorganisms.

Most canned meats have unlimited shelf-life; however, 2 years is considered to be the norm. A flexible package should provide a minimum of 1 year storage. To obtain this or a longer shelf-life, a composite lamination must be used. The most functional material consists of a film-foil-film laminate. The outer layer of plastic film provides support and physical strength to the composite. Aluminum foil contributes excellent barrier properties and a heat-sealable innerply completes the structure. Polyester and/or oriented polypropylene are possible outerply candidates. The sealing media may be HDPE, polypropylene or some of the newer copolymer films. An all-plastic laminate may also be used for heat-sterilized meats, but the life of such packages is limited by the permeability of the packaging material. Cooked hams have been marketed in nylon-11 (Rilsan) film after storage for periods of up to 3 weeks.

Courtesy of the Metal Box Co., Ltd.

FIG. 3.8. CURED MEAT SPREADS IN ALUMINUM CANS
All-aluminum cans with easy open tear-out lids are used for cured meat spreads in Great Britain.

Pork luncheon meat, ham fingers and chopped pork are being marketed in Europe in a retort pouch. The meats are vacuum-packed in a polyester film-foil-polyolefin film laminate pouch, retorted and then placed in a printed board carton. The packaged meat products are processed in batch retorts using a pressurized water cook. No refrigeration is needed and ample shelf-life is reported. A Norwegian company marketed a full range of heat-sterilized meat products in polyester-polyethylene laminated pouches. The products were intended to have a maximum shelf-life of 3 months. Since oxidative deterioration is rapid at room temperature without a foilply, these products were sold from chilled display cabinets at 1.6° to 4.4°C.

In the United States, ITT Continental Baking Company is in the throes of reintroducing a line of meat entrees in retort packages. Over 80 food companies have also expressed interest in supplying to the military's Meal-Ready-to-Eat (MRE) program which will support canned rations with retortable, pouched foods. In the camping market, Hormel is now supplying the Boy Scouts with pouched meat products.

In Canada, Christy Brown is marketing four sauce products in a retortable pouch. Also, Swan Valley Foods is preparing a reentry into the market for meat entrees.

Most observers in North America are looking toward ITT Continental's reintroduction during 1979 as a concrete test of the market.

SOLID FATS

Lard

Over 2 million kg of lard are produced annually in the United States. The lard is rendered by heating the fat tissue of swine to melt and separate the pure fat from the tissue. The best lard comes from caul and kidney fat. Fuller's earth, activated carbon, or other suitable clarifying media are used to bleach the product, after which it is filtered. Lard may be sold plasticized (smooth) by precooling to 43.3°C, then rapidly chilling, or it may be sold "grainy" by allowing it to cool slowly. Lard flakes from 3 to 10% may be added to raise the melting point. (Lard flakes are hydrogenated lard.) With the addition of an antioxidant higher melting lard can be stored without refrigeration in moderate climates. There are a number of antioxidants approved by the Technical Services Division of the Consumer and Marketing Service of the USDA. The most effective, however, are BHA (butylated hydroxyanisole) and BHT (butylated hydroxytoluene). BHA together with propyl gallate and citric acid is the most commonly used stabilizer for lard.

When a more bland flavor and odor are required, additional bleaching, deodorizing and refining are necessary.

When properly refined and stabilized, lard needs only protection against soilage; consequently packaging has been principally in paper wrappings or waxed cartons. Nitrogen purging of lard helps further to improve shelf-life. Newer hot-melt coated boards may prove superior to wax-coated boards for this end-use.

Other Solid Fats

Various tallows are obtained by rendering fatty tissue of beef cattle or sheep. Some of these tallows are useful in food products. A great quantity is consumed in the manufacture of oleomargarine. Little if any tallow is sold at retail. Bulk packaging may be in large tins, lined paperboard drums or other convenient package. The main property requirement is grease resistance. In the United Kingdom refined tallow is sold in printed paperboard tubs as "Beef Drippings."

REFERENCES

ALIKONIS, J.J. and ZIEMBA, J.V. 1967. Edible coatings. Food Eng. *39* (12) 78–80.

ANON. 1969. Food keeps indefinitely in boilable pouch. Food Mfg. *44* (10) 5.

ANON. 1970A. Shrinking the meat-distribution gap. Mod. Packag. *43* (1) 60–63.

ANON. 1970B. "Roastnet" stretch-netting for meat packaging. Packaging *41* (481) 48 (April).

ANON. 1970C. The commercialisation of fresh meat. Emballages *40* (273) 244–255 (May).

ANON. 1970D. Swedish co-operative turns to chub packs. Food Mfg. *45* (3) 17–19.

ANON. 1971A. Aseptics take off. Mod. Packag. *44* (1) 46–50.

ANON. 1971B. Cans prefried bacon. Food Eng. *43* (5) 103–104.

ANON. 1971C. Meat packers discover advantages of packaging. Mod. Packag. *44* (11) 56–58.

ANON. 1979A. New packaging ups shelf-life for pork cuts. Food Drug Packag. *40* (2) 1, 10 (January).

ANON. 1979B. Skin packs: A fresh idea in meat packing. Packag. News *March* 7 (supplement).

BARAN, W.L. *et al.* 1970. Effects of carbon dioxide and vacuum packaging on colour and bacterial count in meat. J. Milk Food Technol. *33* (3) 77–82.

BRODY, A.L. 1970. Shelf-life of fresh meat. Mod. Packag. *43* (1) 81–86.

CIANI, L.A. JR. 1972. Bacon buyers will see a slice. Food Prod. Dev. *6* (1) 34–36.

HILL, W.M. *et al.* 1976. Total plate count and sensory evaluation as measures of luncheon meat shelf-life. J. Milk Food Technol. *39* (11) 759–762.

MARRIOTT, N.G. *et al.* 1967. Color stability of prepackaged fresh beef as influenced by pre-display environments. Food Technol. *21* (11) 104–106.

McDOUGALL, J. 1971. The inter-relationship between the micro-climate within the package, shelf-life, and quality. Food Technol. N.Z. *6* (4) 20–24.

PAPADOPOULOS, E.P. 1977. Beef handling in North America. Meat *50* (2) 25, 29.

PERMO, D.A. 1970. Fresh meat packs—challenge and opportunity. Paper Film Foil Convert. *44* (12) 66–68.

PETERS, J.W. 1975. Fabricated steaks: Process, package innovations open markets for fabricated steaks in retail, food service outlets. Food Prod. Dev. *9* (8) 78, 82.

PRESTON, L.N. 1966A. Flexible packaging. I. Fresh meat. Food Pack. Design Technol. *Nov.*, 19–23.

PRESTON, L.N. 1966B. Flexible packaging. II. Fresh sausages and meat pies. Food Packag. Design Technol. *Dec.*, 25–27.

PRESTON, L.N. 1967. Flexible packaging. III. Cured meats. Food Packag. Design Technol. *Jan.*, 14–17.

SACHAROW, S. 1970A. Cured meat packaging. Food Prod. Dev. *3* (8) 60–64 (Dec.–Jan.).

SACHAROW, S. 1970B. The cured meats industry is a big and growing market for flexible packaging converters. Gravure *16* (4) 24–44.

SCHUTZ, F. C. 1969. Plastic meat trays. Spe. J. *25* (8) 79–81.

SIGNOCINI, P.P. 1969. Armour scores a first in the big food packaging market. Plast. World. *27* (3) 24–28, 31.

TAYLOR, A.A. 1978. Getting it right. Meat *51* (4) 55–60.

4

Poultry and Eggs

It is surely one of the prettiest presents the Old World has received from the New.

Brillat-Savarin (about the turkey)

POULTRY

The raising of domesticated flocks of geese, chickens and other fowl was known to many early civilizations including the Chinese, Egyptians, Greeks and Romans. Seventeenth century English feasts loaded tables down with roasted meats and fowl. "Four and twenty blackbirds baked into a pie" was not just an imaginative nursery rhyme. As late as the 1830s lark pie was an English favorite.

Few birds have created a sensation to match that of the turkey, which was known to the Indians of the Americas, first discovered for the Europeans by the Spanish explorers. In 1511, the Bishop of Valencia ordered Miguel de Passamonte to fetch 10 birds for Spain on his next voyage. Within a century the people of France, Italy and Spain were driving flocks of turkeys to market. The Portuguese brought them to Goa and to India. From there they spread throughout the Middle East.

The market for poultry meat is growing rapidly all over the world. Consumption is increasing both over all and per capita, particularly in the United States and Europe. In the United States alone poultry flocks continue to expand and are now at a record 3.5 billion birds with the number expected to show a sizeable increase in 1980. Most of this growth has been in broiler chickens and in turkeys. The per capita consumption of turkey in the United States has doubled since 1948. World production is expanding steadily with the United States the largest producer followed by the USSR, Canada and Great Britain.

Poultry raising has become a large industry and its very largeness has been the main reason why prices have been held relatively stable. Special breeds have been developed for special purposes. The Rock Cornish game hen, for example, is a small roasting chicken produced by cross-breeding the Cornish with the White Plymouth Rock. Birds that are best suited for producing eggs are not necessarily good meat producers and vice versa. The Broiler industry in particular has concentrated on development of fast-growing birds who attain market weight in the least possible time with the most efficient usage of feed. The remarkable success of this industry is evidenced in the decrease in time needed to bring a broiler to 1.6 kg in weight; from 78 days in 1950 to about 37 days in 1962; and by the decrease in kilograms of feed required; from 1.5 to 0.9. Cockerels are no longer permitted to survive as their meat cannot be sold profitably. Likewise, laying hens used to produce hatching stock are no longer a primary meat source but rather a byproduct. Research on the role of vitamins and minerals in the diet of poultry conducted during the 1920s and 1930s, made it possible to formulate feed scientifically and to raise broilers the year-round in confined hatcheries and brooders. Successful close herding was only possible with the introduction of disease resistant stock and new medicines. Automation of equipment brought labor savings, but ultimate cost reduction was made possible only by vertical integration of all operations. In 1934, the U.S. Department of Agriculture (USDA) reported 11,405 chicken hatcheries. By 1978 there were only 2000. Modern chicken farms are capable of producing as many as 1 to 1.5 million broilers per day.

Turkey Production

Twenty major companies account for 83% of turkey production. Compared with the broiler-chicken industry, where only 1% is produced by nonintegrated independents, the turkey industry still has a number of independent producers. But large turkey farms, the ones that produce at least 100,000 birds a year, have doubled in number in 15 years, and the five largest turkey factories now account for 30% of production.

According to the USDA, the biggest producer is Swift & Company. In 1976, Swift produced in its 12 plants about 16 million kg live weight turkeys, 20% of the national turkey output of about 900 million kg. Swift's closest competitors, Armour & Company and Norbest, produced roughly half that amount, followed closely by Cargill Inc. and Lewis Rich.

The big companies are centered in 15 states which produce more than 90% of the nation's turkeys. California and Minnesota lead, followed by North Carolina, Arkansas, Missouri, Texas, Virginia, Iowa and Wisconsin.

"Integration" is the core of the modern turkey factory system, and most typical is the integration of a feed mill with a processing plant. Cargill of Minneapolis is a large feed concern that took over the processing facilities of Ralston Purina in 1975 to produce turkeys. In effect, the feed company subsidizes turkey production by contracting for the finished product at an advance price based on feed costs.

According to the USDA, roughly half of commercial turkey growers today are "contract growers" who raise birds for an integrator. The growers produce from 30,000 to 500,000 birds and are usually based within 50 miles of the feed mill or processing plant.

A lesser percentage, slightly more than one in four, are "owner-integrated" in which the growers themselves own the hatcheries, feed mills and processing plants. Norbest, which is the marketing name for a federation of about eight large turkey-grower cooperatives, is owner-integrated in this way.

To produce a high meat-yielding bird of uniform appearance, commercial growers have gradually eliminated all turkey breeds but one. Of the six turkey breeds native to America—the Bronze, Bourbon Red, Narragansett, Black, Slate and White Holland—only the White Holland survives commercially.

The trend toward a single breed has rendered the USDA labeling of classes and grades for turkey irrelevant to the ordinary consumer. Today, only one class of turkey is generally available to the retail consumer—a young hen, between 12 (5.4) and 20 lb (9 kg), which fattens within 20 weeks.

Distinctions between the three USDA labels for quality (A, B and C) are similarly meaningless to the ordinary consumer, since only Grade A birds are marketed whole.

Consumer Reports has complained that the USDA grading system of turkeys is inadequate to the modern factory bird because "quality" refers exclusively to appearance and health of the carcass in the assembly line; "good conformation" and "wholesomeness" evade questions of flavor and texture and ignore problems of damage and contamination in freezing, storing and hauling the finished product, it adds.

Crucial to flavor and texture is a well-developed layer of fat beneath the skin of the bird's breast, once a USDA requirement which is now ignored. The prototypical factory bird has dispensed with natural fat in favor of artificial vegetable oils and water that are injected into the bird to produce the basted or self-basting turkey, as in Swift's "Butterball."

For the producer, feed remains the most expensive item, costing 26.2¢ to produce 1 lb (0.45 kg) of dressed weight. Therefore, a 5¢ injection of water and oil is a far cheaper method of "finishing" a bird than feeding the bird a mixture high in corn, which would develop a natural layer of fat in the bird's final feeding cycle. Many consumer-minded critics have asserted that the modern basted turkey is, in fact, a deception since the consumer is paying for water at turkey prices.

Characteristics of Poultry Meat

Poultry meat is high in protein, low in calories, and easy to chew and digest. It is a particularly good food for the young, the old, convalescents and weight-watchers. The quality of poultry meat is primarily a function of the heredity of the flock, however, freedom from disease and proper nutrition are critically important. From the moment the bird is killed the meat is subject to bacterial growth, off-odors and discoloration. Since most of the fat is located under the skin, very little runs throughout the muscle tissue. Cooked chicken breast contains only 1.3% fat while a cut of beef contains 13−30%. Poultry fat is higher in unsaturated fatty acids, although not so high as some vegetable fats. Fresh poultry fat is almost tasteless and odor-free. Rancidity occurs more readily because of the unsaturation and is accelerated by oxygen, heat and ultraviolet light. Turkey fat is less stable in this respect than chicken fat. Even refrigerated poultry carcasses are susceptible to absorption of odors and can develop off-odor from either bacterial decomposition or fat rancidification. Shelf-life varies from as little as a few days to as much as 7 days for top grade depending upon type of processing. Frozen poultry is less susceptible to the foregoing but more vulnerable to desiccation or freezer burn. Surface desiccation of skin tissue or underlying meat causes discoloration, toughness and loss of flavor.

PREPACKAGE PROCESSING OF WHOLE POULTRY

Birds are weighed live, shackled, stunned, killed, bled, scalded, defeathered, singed and washed. After removal of shanks and oil glands, they are eviscerated, inspected and decapitated. Giblets are removed, cleaned, wrapped in parchment paper or polyethylene film, and replaced in the body cavity. Carcasses are sorted according to weight and chilled, usually in a mixture of ice and water. USDA regulations specify carcasses must be chilled to at least 4.4°C internal temperature or lower. Chillers may or may not be agitated. Agitated chillers cause more damage and more moisture pickup (up to 12%). Also, they require giblets to be withheld and added later. After chilling, birds are diverted to be cut up, left whole, refrigerated or frozen. Each type of end-product requires a different handling procedure and a different packaging system.

REFRIGERATED POULTRY

Shipment to Retail Market

Wet Shipment.—Birds are packed in containers, covered with about 13.6 to 27.2 kg of crushed ice and shipped. One-trip containers may be wirebound wooden crates, wax-coated corrugated boxes or polyethylene-coated paperboard containers. Metal containers or wire baskets can be reused if decontaminated. Wire-bound crates are popular, easy to handle and stackable, but they are not easy to open, unload or dispose of. Paperboard containers eliminate drip and are easier to unload. They are smaller but not so easy to handle and not as stackable. They do, however, reduce evaporation losses. Metal containers are less costly because of the multi-trip feature, but they require an initial investment and sometimes are lost or stolen. Converters of corrugated boxboard containers supply about 20% of the poultry shipper market. While there has been a general decrease in the use of wirebound containers, these are still widely used for most shipments of whole ice-packed poultry. Corrugated containers are used for 30% of the dry shipments of raw, frozen or chill-packed birds. New trends are toward greater use of these boxes for wet-iced shipments due to improved coatings and improvements in insulative properties.

Dry Shipment.—If chickens are well-chilled and contain a good level of moisture, they can be shipped "dry" in refrigerated trucks. No wet ice is added to the container. For long shipments, sometimes dry ice snow is added. There are differences of opinion on the virtues of wet and dry shipment. Some feel dry shipment causes excessive losses of moisture and is not properly controlled for temperature. The Forest Products Div. of Owens-Illinois Corp. claim otherwise. Their shipper includes a blanket of porous cellulosic material with a polyethylene backing. The moisture draining from the birds is absorbed by the cellulosic material and wicked upward. Since the polyethylene prevents its escape, it recirculates and helps maintain moisture equilibrium. The saving in weight through elimination of up to 27.2 kg of ice offsets added costs and permits 38% more chickens to be shipped per container.

In-store Handling and Packaging

Whole Birds.—Small whole fryers, broilers or roasting chickens may be trayed and overwrapped as with cut-up parts, or they may be bagged. Bags are made from low-density polyethylene and twist-tied or clipped-shut. Their main disadvantage is the collection of bloody fluid which is unsightly and may leak out. With trayed birds, blotters can be used to soak up the liquid.

Pressure sensitive adhesive-backed labels may be applied to the outside of the package or plain labels slipped onto the bird before overwrapping. A recent development is the use of transparent printed polystyrene film labels. These are used under clear film overwraps. Use of a very soft, highly plasticized, stretchy film such as polyvinyl chloride, can be used to produce a package that resembles a shrink film package. The whole bird is wrapped tightly in the film by rolling the bird end-over-end and sealing the film in place. An absorbent pad is placed on the back of the bird prior to the first wrap. Next the sides of the film are grasped, pulled tight and sealed to the bottom. This yields a very tight and attractive package. For extra durability a band of film about 10 cm wide can be stretched transversely around the bird and sealed in place. Soft stretch films may cause tray distortion or may press blood onto the film surface. Unsightly packages have to be rewrapped to added cost.

Cut-up Birds.—A large proportion of all poultry sold is cut up, ready-to-cook fryers or broilers. These are cut up in the store into 10 or more pieces, put in trays and overwrapped with film. As with red meats, trays used to be paperboard or pulp; now they are made from molded, expanded polystyrene foam or from clear polystyrene. Advantages of pulp trays are low cost, fast film wrapping and moisture absorption. Foam trays offer a clean, bright appearance coupled with increased shelf-life. Plastic trays preserve color and quality for up to 48 hr longer. This is attributed to their low moisture absorption.

In some states, allowable moisture absorption is controlled by regulatory agencies. Plastic trays can mean up to 5% savings in product because no extra meat needs to be added to compensate for moisture loss into the tray. Other plus factors for plastic trays include a fresher product, a more sanitary package and, when frozen, the meat does not stick on the tray. Different sizes are used depending upon the pack; for example a 12.7 × 25.4 × 2.54 cm tray is suitable for a baking hen while a 14 × 20.3 × 2.54 cm tray is recommended for a cut up fryer. Although shrink films make a more attractive package initially, they are not essential for overwrapping fresh poultry. Like soft stretch wraps they can press more fluid out of the bird. Earliest materials were cellophanes, rubber hydrochloride, and nonfogging polyethylene. The introduction of polyester and polyvinylidene chloride films provided better barriers and better puncture resistance; however, their higher costs have held back their use. Stretch PVC and polypropylene films are now used. In addition to the typical tray-pack, other chicken parts such as livers, gizzards, hearts, necks, etc., are often marketed in lined, folded cartons or in thermoformed polystyrene tubs with snap-on transparent covers.

Centrally Prepackaged Whole and Cut-up Poultry

By far the most important development in this field is one which was started about 15 years ago and which is gaining adherents, that is, the prepackaging of chicken at the processing plant. Pioneers in this development were Holly Farms Poultry Industries, Inc. This is a totally integrated operation starting with the purchase of grains for feed and ending with chilled, cut, packaged, weighed and price-marked packages

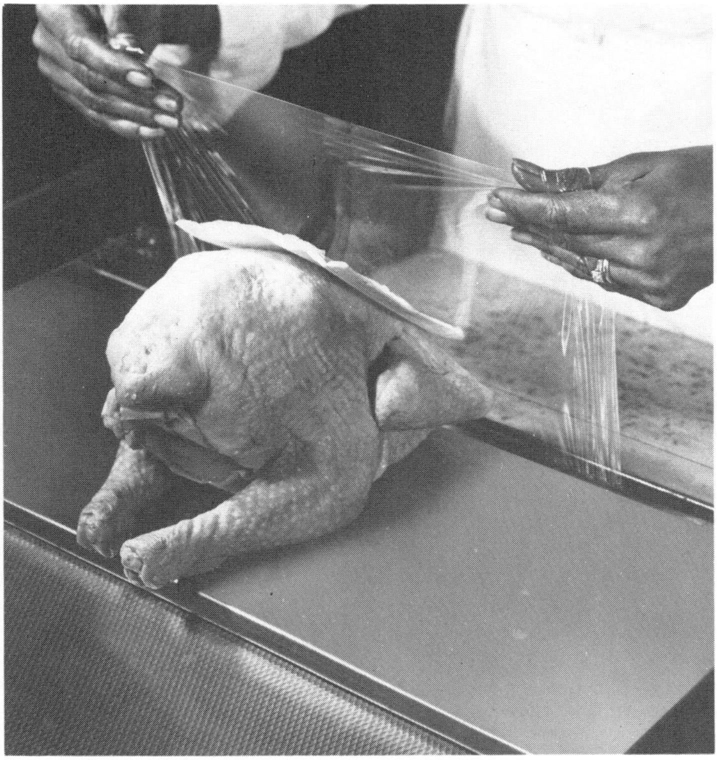

Courtesy of American Viscose Div., FMC Corp.

FIG. 4.1. SOFT FILM POULTRY WRAP
A soft stretchy PVC film is wrapped tightly around a roasting chicken. An absorbent pad is included to soak up juices.

of chicken going direct to the retail store in controlled refrigerated trucks. Over 60,000 broilers per hour are produced by this operation. Whole birds are chilled in 0.5°C water and then drained in a 1.1°C cooler. After insertion into polyethylene bags with a soaker pad over the neck they are further chilled in a −40°C freezer to −2.2°C. After a second holding period at −2.2°C they are weighed, price-labeled, boxed and palletized on returnable 101.6 × 121.9 cm wooden pallets. During warehousing and shipping they are carefully held at −2.2°C. Special nitrogen-cooled trucks are used for transportation. Birds that are to be cut up are packaged in white polystyrene foam trays containing a 15-ply soaker pad with a polyethylene coating on the side which contacts the meat. The trays are put into 2 mil thick stretchable polyethylene bags and heat-sealed on the bottom. These packages also pass through a freezer and into −2.2°C storage. There are several advantages claimed for this system:

Courtesy of American Viscose Div., FMC Corp.

FIG. 4.2. SOFT FILM POULTRY WRAP
An extra band of soft, stretchy PVC film is wrapped tightly around the "waist" of the bird to insure an attractive package.

(1) The meat is not frozen but water and juices are. This avoids drip and promotes ease of handling.
(2) The meat has a lower moisture content (but also loses less in storage). Ice pack chickens contain 12% moisture and frozen chickens have 8%. Chill packs average 6–7%, effectively eliminating "tray weep."
(3) Because of careful control of sanitation and of temperature the processor guarantees a minimum 5 day shelf-life.
(4) Despite the higher cost of the chill-pack chicken, the system saves the retailer money on an overall cost basis. The retailer is getting more meat and less moisture and doesn't have to rewrap. His/her highly paid meat cutters can be assigned work on higher priced meat cuts. There is less waste and more efficient use of space.

FROZEN POULTRY

Cut-up Parts.—Some cut-up chicken parts, when in surplus supply, are frozen by the retailer and sold from the freezer case. This is not prevalent practice, however, because the retailer is not equipped for rapid freezing which is necessary to preserve quality. Cut-up parts of chicken and turkey and some cut-up duckling parts in this country are being prepackaged at central processing plants, frozen and shipped frozen to market. Frozen cut-up poultry suffers from bone discoloration. While overall quality is unaffected, eye appeal is inferior, hence it does not command a high level of buyer acceptance as yet.

As with fresh parts, the product is trayed and overwrapped. Shrink films make a more attractive package and should be resistant to breakage in order to prevent moisture loss and freezer burn. Useful shrink films include irradiated polyethylene, oriented polypropylene and oriented PVC, PVDC and PVDC-coated polyesters. Some work has been done with EVA modified polyethylenes (polyethylene vinyl acetate).

Whole Birds.—Larger chickens, especially those that are stuffed, capons, ducks, geese, turkeys and most small specialty birds such as pigeon, game hen, guinea fowl and the like are sold frozen whole. The once plentiful Long Island Duck has now been displaced from its premier status by both Midwestern and Virginian ducks. Regardless of the type of bird the method of handling is generally the same as for whole chickens through the chillers. Birds are then packaged and frozen. Freezing is usually done in blast freezing tunnels with air at −28.8°C circulating at 396.2−457.3 m per min velocity. Another method is immersion in a supercooled brine for a brief period followed by storage in a noncirculating freezer. Brines may be calcium chloride solutions; other supercooled liquids such as glycols may be substituted. Still another method is immersion in liquid nitrogen, nitrous oxide or carbon dioxide. The freezing by liquid gases takes only about 2½ to 5 min for cut-up fryers packaged in aluminum trays as opposed to 80 min in a brine solution.

Packaging of large whole frozen poultry is usually done by using a special PVC-PVDC copolymer film having a very high puncture resistance and ability to be heat-shrunk by immersion in hot water. Smaller birds that need less puncture resistance or birds which will not be stored long can get by with a cheaper shrinkable polyethylene film. Preformed printed film bags are used. The bird is placed onto a mandrel. The bag is drawn over the bird and gathered around the mandrel. A vacuum is drawn through the mandrel and the neck of the bag is clipped shut. During subsequent heat shrinking of the film the neck is sealed tighter. As with cut-up frozen parts, other shrinkable films used for frozen birds include shrinkable polyesters and cross-linked polyethylenes. Contour wrappers of whole birds have been attempted with ethylene vinyl acetate films. The packaged bird then goes to the freezing line. Depending upon the size of the bird, one or more may be placed in corrugated shippers.The timing of the latter action depends upon the method of freezing. In blast freezers and still-air freezers, birds can be cartoned before freezing. Brine, liquid freezing or liquid nitrogen freezing is done prior to putting the birds in the shippers.

Frozen Turkey.—In preparing turkeys for freezing, live large turkeys are usually electrically stunned and bled. The bleeding process must be complete in order to produce desirable skin color. It takes between 1 and 3 min to bleed the bird. Other methods of killing include the "axe and chopping block" method and the Jewish method whereby birds bleed freely. In the latter method a "Shohet" cuts through the windpipe and gullet with a sharp implement.

After bleeding, the turkeys are put through a scalding tank. This provides for easy removal of the feathers. The water temperature is between 58.8°and 60°C and the bird is immersed for 30 to 75 sec. This process, called subscalding, is widely used on birds that are to be sold frozen.

Defeathering is then conducted by either a machine or manually. The machine method involves rotating rubber fingers which removes all but a few pinfeathers. The latter are commonly removed by hand-picking. After plucking, the bird is eviscerated in a separate cool room. The viscera are inspected by a veterinarian or someone under his/her supervision. The birds passing inspection are thoroughly washed. Difficult to remove organs are dislodged by a suction tube prior to washing of the birds.

The washed birds are then rapidly chilled to 1.6°C with ice slush. They are then drained, sized and graded for quality.

Frozen whole turkeys are usually sold in vacuum-packaged shrink plastic bags. Shrink films commonly used are Cryovac "L," shrinkable polyester (Mylar) and Saran. The bird is inserted into the bag. A vacuum is pulled. Then the bag is secured by twisting and tying a knot or sealing with a metal clip. The sealed, evacuated package is immersed in hot water (90.5°C) in order to shrink the plastic film. The net result is a skin-tight wrap with no air pockets. Also, no voids are present where water vapor can migrate to the package surface and cause freezer burn to the skin. Turkey parts are packaged in film-overwrapped trays.

The birds are placed in telescoping solid-wall shipper cartons and then into freezers where temperatures from −45.5° to −34.4°C with or without fan circulation are used to freeze birds. Obviously the colder the air and the more circulation, the faster the birds will be frozen. Freezing should be accomplished within 2 to 6 hr for optimum quality.

Another method of freezing consists of immersing the film-wrapped birds into a liquid freezing medium such as brine, glycerol or other alcohols, then packing them into shipping cases.

Frozen turkeys are also being packaged in unplasticized polyester and polyamide bags which do not have to be removed for cooking. It is claimed this method permits browning, reduces weight losses, shortens cooking time and improves flavor and succulence.

CURED OR SMOKED POULTRY PRODUCTS

Smoked poultry meat has a unique taste and is treated as a gourmet item usually with a premium price. The cure must be lighter in salt and spices to maintain a delicate and bland flavor and not to make the meat taste identical to ham. Humidity should be controlled during smoking to avoid excessive drying and shrinkage. The light meat turns light pink and the dark meat a mahogany color. As with other meats, hardwood smoke is preferred and the type of wood imparts a characteristic flavor. Birds should be of top grade and quality and should be young to insure tenderness. Poultry sausages made from minced meat and giblets can be smoked also. Chicken and turkey hot dogs are now widely accepted in the market place. Turkey "ham" has more recently been introduced.

Curing solutions are made from salt, sugar, sodium nitrite and spices. Birds are soaked in the pickle at about 1.6° to 4.4°C for about 1½ days per 1 lb (0.45 kg) dressed weight and then soaked in cold water for 1 hr to remove excess salt. Smoking is done for about 5 to 12 hr at temperatures of 43.3° to 54°C. Smoking at higher temperatures (76.6° to 85°C) until the bird reaches 71.1°C internal temperature produces a ready to eat product. Smoked poultry must be kept refrigerated or frozen. Whole birds are wrapped in printed film to keep down desiccation. Polyethylene, polypropylene, polyvinyl chloride and polyvinylidene chloride films are used. Because of the premium price, printed foil laminations can be used as overwraps.

COOKED POULTRY PRODUCTS

Cured or uncured poultry, usually chicken or turkey, can be cooked and then processed into a wide variety of products, such as sausages, patties, sticks, croquettes, frankfurters, bologna, luncheon meats, pies, hashes, stews, soups, potted meats, pates, chow mein, slices and gravy or the familiar "a-la-king." These products may be frozen or canned and are packaged accordingly. Paste or sausage-like products may be packaged in film chubs made from cellophane laminations, polypropylene or polyvinylidene chloride. Canning in tins or jars is done by traditional methods. Canned whole chickens used to be popular but now that frozen chicken is available the canned product is on the wane. The canned product tends to fall off the bones and to contain an excess of gelatin. This makes it suitable for casserole dishes but not for a whole roast. Frozen products are packed in sealed plastic bags (polyethylene-coated polyester film) with printed paperboard cartons for protection. The bag

Courtesy of Armour and Co.

FIG. 4.3. FROZEN WHOLE TURKEY PACKAGE
A heat shrink film-wrapped frozen turkey is labeled "self-basting" because butter has been injected subcutaneously.

can be used for reheating the product in boiling water. Lidded foil containers have also been used for frozen cooked products. Chicken luncheon meat items are packaged like their red meat counterparts.

An increasingly popular item is a boneless roast made by partially precooking the bird, removing the flesh and forming it into a loaf-shaped roast. The meat is usually enclosed in skin and the roast is packaged in a lidded aluminum foil container. Some similar roasts are precooked and need only be thawed to use. Cooking is done in fibrous netting or in plastic film bags (such as polyamide). It takes several hours immersion in 71.1°C water to effect a proper cook. Turkey rolls, roast and parts have been marketed in like manner in paperboard cartons and shrunk films. Foil containers may be replaced by foil-paperboard composites. Cooked,

Courtesy of E.I. du Pont de Nemours and Co., Inc.

FIG. 4.4. FROZEN WHOLE TURKEY PACKAGES
Young turkeys are put in "Mylar" polyester film bags with clip closures, the film is tightly heat shrunk, and then the birds (shown here) are given a 15 min liquid immersion quick freeze.

diced chicken meat has been sold in polyethylene bags (two 5 oz (141 g) bags in a waxed carton) as a frozen convenience item, which can be thawed and added to salads, cream dishes and the like. Frozen, partially cooked poultry meat is found in the familiar TV dinner tray and in chicken pot pies, both packages being formed from aluminum foil.

For fine particle meat users, deboning machines have been developed which debrade the meat from the bone. These accept raw or partly cooked meat, even the poorer cuts such as necks and backs. Brine flotation is used to separate the bone particles from the meat.

DRIED POULTRY PRODUCTS

There are several ways to remove moisture from poultry meat including freeze drying, drum drying, spray drying, oven drying or heating in oil. Each has advantages and produces a somewhat different product. Freeze drying retains flavor and original form while keeping bacterial growth and enzymatic changes to a minimum. Little or no oxidative rancidification can occur because the process involves a vacuum. The products

Courtesy of Reynolds Metals Co.

FIG. 4.5. TAKE-OUT CARTON FOR CHICKEN
Growing use of fast food restaurants and increased market for take-out poultry products have spurred development of foil/board take-out containers.

may be stored without refrigeration and contain less than 1% moisture. They are fragile and porous, somewhat like a solid sponge. They must be packaged to protect against moisture pickup, oxygen and crushing or other mechanical damage. Oxygen must be held under 2% and moisture under 3%. This dictates hermetically sealed cans or pouches either vacuum-packed or inert gas flushed. A suitable pouch material for freeze-dried items would be a cellophane-polyethylene-foil-polyethylene; however, poultry fats pick up or extract off-flavor from polyethylenes. This dictates a different interior component; for example, an ionomer film, polyvinylidene chloride, polyamide or polyvinyl chloride.

The principal market for freeze-dried poultry meat is for military rations, campers and ingredients for prepared dishes such as soups or stews.

Spray drying consists of taking finely comminuted cooked poultry meat and spraying it into a superheated atmosphere. The product is suitable for soup mixes or other preparations where a fine particle can be utilized. Poultry meat in various sizes can be put in trays and oven-heated to dry. Vacuum is usually used to reduce heat and oxygen damage. Heated drums or rollers have been used for the same purpose.

POULTRY BYPRODUCTS

Poultry fat is used as an ingredient in other prepared poultry dishes such as soups. Care must be taken not to use mesenteric fat, which may be contaminated and which is certain to have off-flavor. Heating to remove moisture must be held to below 104.4°C to avoid discoloration and the onset of rancidification. Best results are obtained by centrifuging to remove moisture. Antioxidants, such as BHA and BHT, up to 0.02% by weight may be added for stability. Refrigeration or freezing is highly desirable to avoid deterioration. Since most fat is used commercially, various types of bulk packaging are used. Several consumer packages of chicken fat have been designed employing screw-top glass jars. They are marketed in large cities to the Jewish who use it in cooking.

Finely chopped poultry skin is highly nourishing, has excellent flavor and is added to many commercially prepared recipes. Some work has been done to use deep-fried chicken skin as a snack food; however, it is not marketed on a large scale, if at all.

Frozen duck feet are considered a delicacy in the Far East and are exported from Midwest processors for that purpose.

FUTURE TRENDS

More progress can be expected in genetic development of improved poultry strains. For example work is being done to develop 100 lb (45.36

kg) turkeys. At this weight they could be cut into retail parts like small red meat animals. The all white meat chicken is not far off.

Improvements in automation through machinery or through handling as is done at Holly Farms can be expected to hold prices down and quality up.

Further convenience foods and snack foods are a certainty.

EGGS

Like the flesh of the bird, its egg also varies in character and taste depending upon the species. Duck and goose eggs tend to have a strong flavor, but turkey, pheasant and partridge eggs are almost indistinguishable from chicken eggs.

Eggs are prized for their nutritive value and flavor and have been eaten raw or cooked in many different ways, from egg-nog to omelette. They are an important ingredient of many other foods including baked goods, candy and ice cream.

Although eggs have been traditionally a very popular food and although the total consumption continues to rise each year, the per capita consumption in the United States has declined over the past 25 years from about 390 to about 308. The production of over 65 billion eggs requires a large amount of packaging material, yet until comparatively recent years the egg package has remained singularly unattractive and unimaginative.

Much research has gone into the breeding of poultry for meat or egg production and great strides have been taken in the introduction of automation, integration and amalgamation in order to lower costs. Small operations have been amalgamated into larger organizations with over 1,000,000 laying hens. Farms have been vertically integrated from the production of feed and brood stock to the disposition and sale of the main product as well as byproducts. Production operations have been made more efficient by the use of automatic machinery and new equipment such as more easily cleaned plastic chicken coops. All of this has helped to hold costs down. Even more effort is required, however, to stimulate buying through more imaginative packaging and marketing. Little or no effort has been spent on genetic alteration of the egg itself other than recent work aimed at producing a thicker and more uniform shell. One could ask, why not larger eggs, larger yolks or consistently double or triple yolked eggs?

Characteristics

Nature has given the egg a natural package, the shell, and an inner liner, the membrane.

Despite its relative strength the egg is a fragile object and must be protected against breakage. In addition the egg is alive and must be given a chance to breathe or it will die. Respiration may be slowed by refrigeration or by treating the shell (usually by an oil dip), but if it is completely halted the flavor and texture of the egg will deteriorate more rapidly. Eggs are also susceptible to deterioration from ageing and from attack by bacteria and fungi. Bacterial contamination through microfissures in the shell is minimized by washing the egg. Approved bactericides may be used in the wash water. Careful drying is absolutely esssential to avoid mold. Best conditions for storage of eggs in the shell are 85 to 92% relative humidity and −1.6° to −0.5°C temperature. Under these conditions undamaged eggs can be stored for 4 or 5 months with only slight changes. After 11 months, differences can be detected but whites and yolks still can be separated, a very important factor in some end-uses.

Eggs are susceptible to loss of moisture and particularly sensitive to absorption of foreign odors. Quality loss is evidenced through loss in weight (as much as 5% in 9 months), thinner and runnier white with loss of opalescence and flatter yolk due to absorption of moisture from the white. Very old eggs develop a straw-like taste probably due to enzymatic changes.

Broken-out eggs are sold frozen or dried and are extremely sensitive to bacterial contamination.

PACKAGING PROCEDURES AND MATERIALS

Shell Eggs

Eggs traditionally have been shipped to market in flats molded from paper or pulp. The flats may be stacked in wooden crates, in baskets or in corrugated shippers. Although most shell eggs are sold to the consumer preboxed by the dozen, a few retailers merely display the flats and allow the purchaser to select and box his/her own. Eggs may be sold graded or ungraded, if so identified, but higher grades command premium prices. Grading is usually done at the hatchery or at a central distributing point such as may be maintained by a farmer's cooperative. Grading includes candling, weighing and occasional "breaking out." Top grade eggs have thick opalescent whites, tall-domed yolks and are free of defects such as blood, bits of shell or other foreign matter. In the United States fertilized eggs may not be sold for retail food consumption. In earlier times when eggs were produced in barnyards rather than in modern factories this was more of a problem and candling was the means of detecting the fertilized eggs.

After sizing and grading, eggs are packed into dozen containers and these in turn into larger shippers. They may be held in storage for reasonable periods of time, but most consumers expect and demand fresh eggs. Prompt delivery to retail markets is an inherent part of the system and most egg packages are marked "strictly fresh."

The traditional package for one dozen eggs has been a folded paperboard box with dividers. This was replaced several decades ago by more rigid paperboard cartons with interlocked dividers and by boxes fashioned from molded pulp. The pulpboard box could be shaped to hold three rows of four eggs, as some of the earlier paperboard cartons were shaped, or in the now more popular two rows of six eggs. The interlocked carton is limited by its construction to the latter configuration. A wide variety of closures has been employed including metal clips, wire staples, gummed tapes or labels, glued spots or simple locking-tabs.

Decoration of these older packages was particularly uninspired. Paper or pulp was supplied in off-whites or dull greys. Inks were prosaic reds, blues, greens, blacks and browns, and copy was simple line lettering to name the vendor, brand, if any, and the quality of the contents. They offered almost no visibility and little protection. All too often a cracked or broken egg was found. Breakage in one carton contaminated neighboring cartons with resulting loss. Repacking in the store was an added cost as well.

Over the more recent past years attempts have been made to improve egg carton design. Paperboards have been upgraded to make them whiter, stronger and more printable. Carton designs have been altered to give more support and protection to the egg, particularly during stacking of the boxes. Brighter and more attractive colors have been added. On some folded boxes lithographed process printing has been used to add interest and appeal. Party scenes convey the idea of a deluxe product. Appetizing food dishes suggest tempting uses. Color has also been used to identify grade or size differences.

Recent Trends

Folding Box.—The latest development in the folded paperboard egg carton are designs which enable the buyer to see the egg. One approach is to form the box with an open top and apply a shrink film overwrap such as PVC, PVDC or PE. A 27-point cylinder kraft backboard is used with triple-end panels and honey comb-like construction for added stacking strength.

Molded Pulp.—The molded pulp egg carton is fighting for its life against stiff competition from molded plastic foams. It chief advantage is

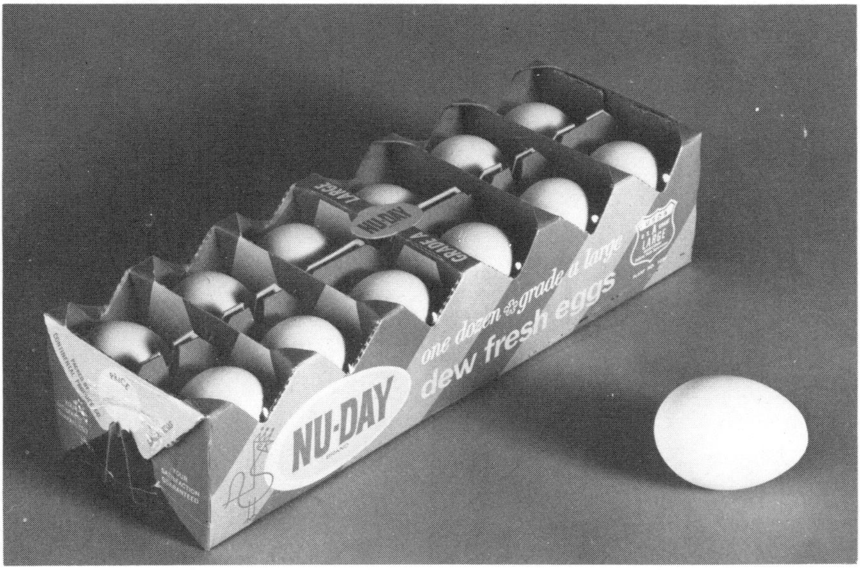

Courtesy of Reynolds Metals Co.

FIG. 4.6. SEE-THROUGH PACKAGING FOR EGGS
New trends in egg packaging stress the see-through advantages of a heat shrink PVC film overwrap. Here the package is formed from folded and interlocked printed paperboard.

its price. Its chief disadvantages are lack of strength and unattractiveness. It is possible to increase strength by using denser pulps or by adding fillers or binders with some added cost.

Attractiveness is limited by the inherent color and surface roughness of the material itself. It is also not conducive to fancy printing. One attempt to solve this has been to combine a deep-celled molded pulp tray with a printed paperboard lid which can be glued or otherwise fastened on. Another attempt substituted a shrinkable polypropylene or polyvinyl chloride film overwrap and a printed paper label for the aforementioned paperboard lid. This gave strength, visibility and decoration. When tried in the Los Angeles area it outsold the conventional packages five to one. Another packager put a pulpboard tray of eggs into a preformed, printed, shrinkable polyethylene bag to achieve the same general purpose. Still greater strength was attained with a transparent "crash helmet" lid made from thermoformed oriented polystyrene. The lid separated and protected each individual egg and was locked by tabs to the pulpboard tray.

Courtesy of The French Institute of Packaging

FIG. 4.7. NOVEL EGG PACKAGE
A thermoformed unit is used to package fresh eggs.

Molded Plastic Foam Egg Cartons.—A radically new approach in the packaging of shell eggs came with the adoption of cartons and trays made from expanded polystyrene foam. Two basic types of material are used. Molded shapes are created by filling polystyrene foam beads into a suitable mold and then applying heat and/or steam to expand the beads into a coalesced mass. The resultant block shape has strength, provides excellent cushioning, is very light weight and is a good insulator. Other shapes are thermoformed from a thin sheet of expanded polystyrene foam. The latter material has a smooth satiny appearance almost as if it had a thin transparent skin. It is very attractive and is presently available in snowy white and pastel colors.

Egg cartons can be molded in almost any shape from these materials. The present chief competitor looks much like the 2 × 6 molded pulpboard container, except for being more attractive. It can be printed with high quality line graphics in simple designs and patterns. It can be filled and closed on fully automatic machinery. Although early versions had some difficulties with variations in smoothness, stiffness, strength and dimensions, most of these have since been overcome. The cushioning against shock and the ability to be formed to any desired shape make this an extremely functional as well as attractive package. Many new ideas are being tested.

One manufacturer produced a block-molded package in a boat shape. After use of the eggs the carton became a child's toy. To achieve a simple closure and better graphics, another company has tested a foam tray with an A-flute printed corrugated sleeve. Others have tried a rigid plastic film sleeve in similar manner.

European designers have experimented with smaller unit packages. One package featured two molded blocks with matched cavities for six eggs. The blocks were designed to close with an adhesive, but male and female locking grooves or tabs would probably serve as well. Openings permitted ventilation and visibility from all sides, yet the package was stackable and the flat top permitted decoration.

Another European package called the "Econopak" also featured a two-part foam box and top. This one held two eggs and two rashers of bacon. A printed plastic outer sleeve completed the package.

Courtesy of Container Corp. of America

FIG. 4.8. FOAMED POLYSTYRENE FOR EGG PACKAGING
Molded foamed polystyrene egg cartons provide superior cushioning and protection against odors and moisture. The package is resistant to fungus and mold growth. Consumers like its attractive luster and the soft pastel colors.

Hermetically Sealed Plastic Packages.—Experimentation has been reported with a two-piece rigid polyvinyl chloride carton fashioned in such a way that it could be flushed with nitrogen and then sealed hermetically by pressure. Grade AA eggs were said to have kept fresh and in good quality in this package for up to 3 weeks. Further research on ideal gas combinations might yield still further advantages.

FROZEN BROKEN-OUT EGGS

Although about 89% of eggs produced in the United States for consumption as food are sold in the shell, about 6–10% are broken out and sold frozen, liquid or dried. Broken-out eggs are processed in different ways depending upon ultimate end-use. They may be frozen whole and later thawed or they may be homogenized and then frozen. They may be separated into whites and yolks and the parts each separately frozen. Eggs which are to be frozen are usually treated to prevent coagulation or rubberiness. Ten percent by weight of salt or sugar or 5% by weight of glycerol may be added for this purpose. Salt is used where the ultimate use will be in salad dressings or nonsweet cooked items, sugar or glycerol when the end-use is in sweet items such as candy or some baked goods. Whole eggs or yolks are sometimes pasteurized before freezing. Eggs for freezing are put in 30 lb (13.6 kg) bulk containers. Freezing at −17.7°C may take up to 44 hr. Use of −28.8°C blast freezers can drastically reduce this time.

Bulk containers can be made of a variety of materials. Large cans, small drums, spiral wound paper canisters and the newer molded plastic pails are all available for the purpose. Flexible polyethylene plastic bags in corrugated shippers would also be satisfactory.

DRIED BROKEN-OUT EGGS

Dried eggs, again, may be processed as whole eggs, yolks or egg whites. Drying is done with steam-heated drum driers or spray driers with or without vacuum. Foam spray-drying is a somewhat more efficient method of spray-drying but properties of final products made from the differently dried eggs may vary. The properties of whole eggs and yolks are sufficiently similar that they may be considered together. Both deteriorate rapidly at room temperature if glucose, which is naturally present in the egg (about 1.1% of egg solids), is not first removed by fermentation or enzymatic oxidation to gluconic acid. Oxygen causes off-flavors to develop and these will vary depending upon whether or not sucrose is added (20–30% on dry solids) prior to drying. Addition of sugar (sucrose) also prevents loss of leavening power of dried egg whole or yolks. This power is what makes a sponge cake spongy. Dried egg white also is susceptible to glucose darkening of color and must therefore be treated in a similar manner. Dried whites can lose their whipping or foaming properties if small amounts of glucose or of egg yolk are retained.

Packaging of dried egg white is principally concerned with protection against moisture absorption, whereas whole egg or yolks must be protected also against oxygen. In the latter case, vacuum or gas packaging is

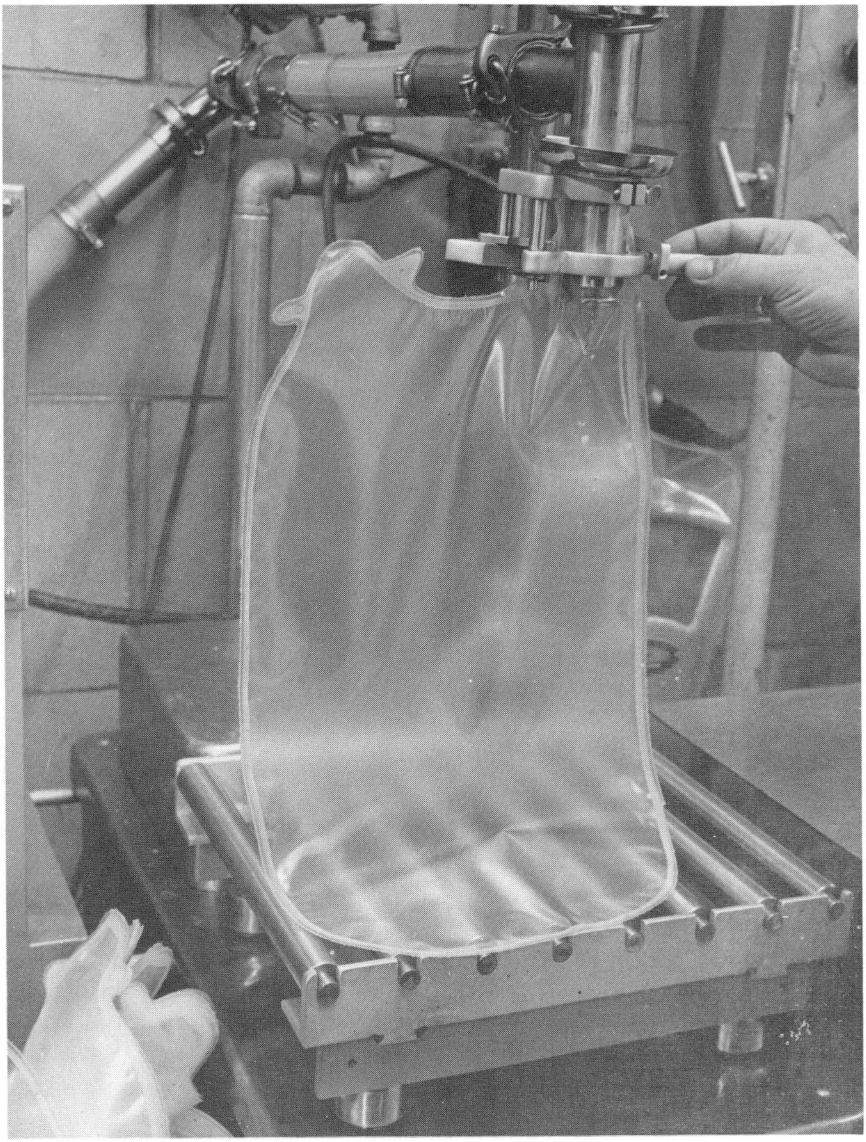

Courtesy of Container Corp. of America

FIG. 4.9. LIQUID EGG PACKAGING
Broken out eggs can be packaged in polyethylene film bags which will hold 10 lb (4.5 kg)—the equivalent of 100 medium sized eggs. The automatic filling device shown here measures the liquid eggs into the bag.

highly desirable. Since very little, if any, dried raw egg is sold at retail, we are again concerned with bulk packaging.

Economics and protection are the prime considerations. Vacuum- or gas-packed cans, cannisters or cartons with hermetically sealed liners or metal drums are suitable for the purpose.

COOKED EGG PRODUCTS

In most cooked egg products the identity of the egg is lost. For example the egg used in a sponge cake is not identifiable. Consequently no attempt is made to cover such products in this chapter. They will be discussed in other appropriate chapters. There are some cooked egg dishes, however, where the egg remains identifiable, for example, an omelette or an egg foo yung. Some egg dishes of this nature have been prepared and then frozen. Packaged in foil laminates or foil containers, they are readily thawed and reconstituted as convenience meals. The U.S. Army has experimented with considerable success in freeze drying many egg dishes.

REFERENCES

ANON. 1966A. Broilers: Ships—without ice. Food Eng. *38* (7) 157.

ANON. 1966B. Showcasing the egg. Mod. Packag. *40* (3) 128–255.

ANON. 1970A. Heat-shrinkable bags. Mod. Packag. *43* (2) 66E.

ANON. 1970B. Hinged-lid moulded plastics egg pack. Env. Packag. Design *86,* 2 (Nov.).

ANON. 1978A. Egg packaging international Poultry Ind. *Oct.,* 8.

ANON. 1978B. Not just a package, more a sales system. Poultry Ind. *Oct.,* 13–15.

CASH, D.B. and CARLIN, A.F. 1968. Quality of frozen boneless turkey roasts pre-cooked to different internal temperatures. Food Technol. *22* (11) 143–146.

DONALDSON, I.S. and LEGGE, D. 1977. The egg: A special product that requires specialised packaging. Packag. Rev. *97* (7) 73–74, 84.

MAY, K.N. *et al.* 1966. A comparison of quality of fresh chicken packed in various containers. Ga. Agric. Exp. Sta. Bull. *168.*

MOUNTNEY, G.J. 1976. Poultry Products Technology, 2nd Ed. AVI, Westport, Conn.

NIRMALA, N. *et al.* 1976. Storage stability of egg powder in flexible pouches. J. Food Sci. Technol. (India) *13* (3) 162–163 (May/June).

RANKEN, M.D. 1967. Plastics in poultry packaging. Proc. Conf. on Advances in Packaging with Plastics, Ppr. *14,* London.

RUSSO, J. 1967. Cartons save 160 man-hours per week. Food Proc. Mrktg. *28* (9) 82–83.

SACHAROW, S. 1967C. Packaging of food. XIII. Prepacks take over poultry market. Paper, Film Foil Converter *41* (4) 102–106.

TAYLOR, M.H. *et al.* 1966. Evacuated packaging of fresh broiler chickens. Poult. Sci. *45* (6) 1207–1211.

WELLS, F.E. *et al.* 1958. Effect of packaging materials and techniques on shelf-life of fresh poultry meat. Food Technol. *12* (8) 425–427.

WILSON, G.D. 1960. Sausage products. *In* The Science of Meat and Meat Products. Amer. Meat Inst. Found. W.H. Freeman and Co., San Francisco and London.

ZEHNER, M.D. 1967. Consumer attitudes toward egg cartons. Michigan State Univ. Agric. Exp. Sta. Q. Bull. *50* (1) 84–92.

5

Milk and Dairy Products

> In May, 1977, a Sevres milk pail made for Queen Marie Antoinette sold at auction for $96,000.
>
> *Sotheby's Mentmore Towers Sale, May 18–27, 1977*

Milk is one of the basic foodstuffs and from it have been derived a variety of processed foods which are known as dairy products. There exists today a frieze that was carved in the third millenium B.C. which shows priests of Sumer milking, straining and presumably buttermaking at a dairy farm sacred to the goddess Ninhursaga. The Brahman Hindu "Rig-Veda" about 1400 B.C. describes a pastoral society which raised cattle, feasted on beef and regarded milk as a highly valued food. The Promised Land of the Israelites was to be "flowing with milk and honey." Because milk soured quickly in the warm climates of early civilizations, it is not surprising that butter and cheesemaking were early discoveries.

The most exotic dairying item to be sold recently has been a pair of Sèvres milk-pails which turned up amoung the curios of the recent Mentmore sale organized by Sotheby's. They were made during the years 1787–88 for Marie Antoinette who used them at Rambouillet.

The ill-fated French Queen's predilection for dressing up as a dairymaid was by no means an unusual hobby for an 18th century great lady. Well into the present century the ability to do well in the dairy was considered as essential to the domestic education of chatelaines as it was to that of cottager's daughters. The care that Queen Victoria lavished on the 1860 renovation of the Royal Dairy at Frogmore, with its gilded vaulting, marble statuary and multicolored tile work, reveals just how important she thought a dairy was.

However, in the 18th century it is possible that Marie Antoinette's preoccupation with milking may have been as much cosmetic as bucolic. Long before Jenner was able to pin down the cause of smallpox and avert the consequences with vaccination, it had been observed that dairymaids, almost uniquely among working-class women, boasted complexions unmarred by this dreadful disease. What more sensible precaution could a famous beauty take than to spend at least some of her spare time "dabbling in the dew" as an insurance for future good looks. (Dairymaids' skins were safe because in the course of their work they contracted the much milder "cow-pox" which provided an immunity against the severer infection.)

MILK

Products

Milk and cream comprise nearly 1/6 of the weight of food consumed annually by the average American family. Fluid milk sales amounted to $4.3 billion in 1977.[1] The per capita consumption of milk is 86 kg. Whole milk has declined in popularity, falling to a production of less than 40 MT in 1977. The reason for the overall increase in milk sales in 1977 vs 1976 is the growing strength of lowfat milk. In this country nearly all milk sold is from cows, although some goats' milk is produced. In other countries, goats' milk is more popular and still elsewhere milk is obtained from other animals, such as camels, donkeys, llamas, reindeer and yaks.

Composition

Whole milk is a complex physicochemical system consisting of a water solution of salts, lactose and lactoalbumin. Proteins are colloidally dispersed in the water solution and milkfat solids are present in a partially emulsified suspension. The proportion of milkfat varies with breed of cow, individual and period of lactation. Fat content according to breed is approximately as follows:

Jersey	5.3%
Guernsey	4.9%
Ayrshire	4.0%
Holstein	3.4%

Protein content remains relatively constant at about 26.5% of total solids and includes casein and the whey proteins. Casein is present as a calcium caseinate: calcium phosphate complex and can be precipitated at

[1] If all the milk produced on United States dairy farms in 1977 were placed in gallon cartons and lined up end to end, they could reach the moon 9 times or reach from the earth to the moon and back 4½ times. From an earthly point of view, the same end-to-end gallon cartons would be enough to form a band around the world 87 times at the equator. Taken out of cartons, the milk produced could create a 4791 km long, 1.2 m deep river.

pH 4.7 by addition of acid. It is used in the manufacture of industrial adhesives and plastics. The whey proteins are albuminous in character. Both they and casein are primarily amino acids high in nitrogen.

Another major constituent of the watery portion is lactose or milk sugar. Lactose is easily fermented by lactobacilli into lactic acid, a principal ingredient of sour milk.

The fatty portion of milk is a mixture of triglycerides of saturated and unsaturated fatty acids. It also contains small amounts of cholesterol and lecithin type phospholipids. The latter have high emulsifying properties.

Characteristics

The glycerides of the lower fatty acids (butyric, caproic, capric, caprylic, lauric and myristic) are mainly responsible for the characteristic flavor of milk fat; the deterioration of these fats causes most off-flavor problems. Rancidity is caused by the lipase enzymes which hydrolyze the glycerides and release the strong flavored and strong smelling fatty acids. Pasteurization inhibits the lipase action and sterilization destroys it. Oxidation of pure milk fat produces a tallowy flavor probably from the formation of hydroperoxides. The reaction is accelerated by heat, acid and metal catalysts, particularly copper.

The distinctive "oxidized" flavor of milk appears to be caused by oxidation of phospholipids. Another fishy off-flavor results from hydrolysis of lecithin to produce trimethylamine.

The vitamin content of milk is affected by exposure to light, heat and oxygen. Ascorbic acid (vitamin C) is deteriorated by heat and oxygen. Thiamin (vitamin B_1) is partially destroyed by heat. Vitamins B_2 and B_6 are destroyed by light. Vitamins A, D and E are fat soluble and heat stable.

Since milk is an animal product and an excellent growing medium for bacteria, it is extremely important that it be produced under sanitary conditions. Prior to the legal enforcement of sanitary codes, diseases such as typhoid fever and tuberculosis were transmitted through distribution and sale of milk. Milk must now be produced from healthy animals under strict sanitary conditions. Packaging of milk, therefore, should be designed to protect the product from contamination by dirt or bacteria and from the effects of light and oxygen.

Milk is produced by the lacteal glands in the udder of the cow. It is drawn from the cow's teat canals by automatic milking machines using a pulsating vacuum principle. The milk goes directly into a bulk-holding tank which is refrigerated at 4.4° to 10°C. After a maximum storage of 2 days, it is transported to the milk processing plant. In some instances, the milk goes to a central receiving station where it is blended with high fat and lowfat milk. It then goes to a processing plant.

TABLE 5.1. MILK PACKAGES (1968–1978)

Type	Market (in percentage) 1968	1978
Glass	28	2
Paperboard	74	66
Plastic	4	30[1]

[1] Nonreturnables: 30%; returnables: 0.5%.

Smaller farms ship the milk in 10 gal. (37.8 liter) milk cans. Most larger farms utilize bulk road tankers. These tankers are made of insulated, stainless steel which can hold up to 5000 gal. (18,927 liters).

At the processing plant, tests are conducted to control the milk quality. Representative tests run are fat and total solids, sediment, bacterial counts, freezing point and evaluation of milk flavor. In some cases, antibiotic and pesticide residues are tested. The milk is held in refrigerated (4.4°C) vats and may be blended to a specified fat content.

The next step consists of classification. A centrifugal classifier is used. The machine removes sediment, body cells from the cows udder and some bacteria. The milk is then ready for pasteurization.

Two methods are used to effect pasteurization: the batch or holding method (62.7°C for 30 min) and the high temperature-short time method (HTST) (71.6°C for 15 sec). Batch pasteurization is conducted in heated, covered vats equipped with an agitator and thermometer. HTST methods employ a more complex system including heating plates, holding tube, flow diversion valve and time-temperature recording charts.

In order to check the effectiveness of the pasteurization method, the phosphatase test is run. Milk is incubated with disodium phenyl phosphate and 2,6 dichloroquinone-chlorimide is added. A blue color indicates improper pasteurization or contamination.

Homogenization generally follows pasteurization. This disperses the fat globules and insures an even-bodied product. Large fat globules are sheared in the homogenizer and reduced in size by about 10 times. They become evenly distributed and no longer rise to the top of the milk. The milk is then cooled and packaged in polyethylene-coated paperboard cartons.

Packaging Materials

Glass.—Until 1950, almost all milk was packaged in glass bottles. Milk was sold to consumers mainly by home delivery or through supermarkets in 1 qt (0.9 liter) bottles. The increase in supermarket shopping and the decrease in home delivered milk have served to lessen the usage of glass milk bottles. In 1965, home delivery accounted for less than 30% of sales.

Glass bottle usage correspondingly decreased from 50 to 40%. Consumers favor a lighter package for the supermarket, while in home delivery the heavier glass bottle reigns supreme. By 1980, the use of glass milk bottles is expected to decline to a negligible amount. In spite of decreasing usage in the United States, glass milk bottles are still the basic package in many other lands.

Glass milk bottles originally were round, tapering to a rather wide mouth with a thick flange. The move to a squared body saved considerable space in the home refrigerator. Quarts, pints and half pints were once the standard sizes, however, there is now an increasing trend toward half gallons (1.9 liters) and gallons (3.78 liters). Glass bottles average about 50 trips and are packed at rates up to 24,000 per hour. In Finland milk is packed in brown glass to filter out harmful light. United States consumers will not accept the brown color despite the added protection it offers.

Bottle closures are formed from aluminum foil, high density polyethylene, polypropylene and paperboard. Most closures are applied by automatic machinery at high speeds. Aluminum foil closures are supplied in reels about 5 cm wide and weighing 5.4 to 7.7 kg. The foil is 2 mils thick and 0.45 kg is enough to cap 1300 bottles. Printing is applied prior to use.

Paperboard.—The introduction of wax-coated paperboard captured the milk market in the late 1940s. Although consumers were attracted to the concept of a disposable milk container, some problems existed. Wax particles were common in the milk. The outer surface had a cloudy, unattractive appearance. Although a few polyvinyl chloride-coated cartons were tried, the introduction of polyethylene-coated paperboard in the early 1960s solved nearly all the problems inherent in milk packaging. It was disposable, clean looking and functional. In the United States about ⅔ of all milk is packaged in polyethylene-coated paperboard cartons. Most retail sizes sold are quarts (0.9 liters) and half-gallons (1.9 liters). Gallon paperboard cartons have a tendency to leak. Cartons of various types account for 85% of the German market, 75% of the Swedish and small percentages of the British and Australian market. Several basic types of cartons are marketed (see below).

Plastics.—An all plastic milk bottle is light weight and tougher than its two competitors. Reclosure is more sanitary than the tear-open carton. Pigmentation can be used to help screen-out light. Disadvantages include difficulties in connection with printing, labeling and various decorating techniques. The basic materials in general use for an all plastic milk bottle are polyethylene, polystyrene and more recently, polycarbonates. Plastics have come from about 7.5% of milk sold in 1970 to nearly ⅓ in 1979, with the greatest amount by far in nonreturnable half gallons (1.9 liters) and gallons (3.78 liters).

Package Forms

Milk Cartons.—There has been a noticeable decline in paperboard usage for milk cartons. It went from 848,800 MT in 1968 to an estimated 823,800 MT in 1978.

Many different cartoning lines exist and varying carton designs are used. A unique system is the Perga carton patented by the Jagenberg Company in Germany. The Perga carton consists of a 2-piece container which is supplied to the dairies preformed, erected and nested in columns. Specially designed machines are used to fill and seal these cartons at rates up to 2400 per hour. Since the filling and sealing operations occupy a small area and the machines are economical, this package offers definite cost advantages especially when there are fluctuations in consumption. The Perga carton is available in ⅓, ½ and 1 pint (0.16, 0.24 and 0.47 liter) sizes. It is widely used in Australia.

In the United States, one of the major packages used is the "Pure-Pak." This is used by the dairies as precut blanks which are formed, filled and sealed on one machine. The package is formed on automatic machinery, milk is filled and the familiar gable top is used as the seal.

Other closures include Sealright's "Sealking" where a foil lined corner is lifted to form a pouring spout and American Can's "Can Co" label which is a flat-topped gable where the user has to tear part of the top off and bend part of the gable out to form the spout.

Paperboard carton sales are in pint (0.47 liter), half-pint (0.24 liter) and half gallon (1.9 liters) sizes. Most attempts to design gallon (3.78 liters) paperboard cartons fail because of manufacturing costs or because the handles don't support the weight. Two successful approaches to the paperboard gallon milk market are:

(1) Two half gallons strapped together with a paperboard handle.
(2) Four quart (0.95 liter) containers shrink-wrapped with a finger-grip carrier for easy transportation.

Preformed.—Preformed cartons are supplied to the dairy in a fully erect form and ready for filling. The "Perga" carton is an example of a preformed carton. Both waxed and polyethylene lined paperboard are used. Paperboard caliper is 0.023 cm for the ⅓ and ½ pint sizes and 0.0279 cm for the 1 pint "Perga" size.

Precut.—In a precut carton system, printed, coated paperboard blanks are supplied in a knocked-down shape. The final carton is setup, formed, filled and sealed on one machine. "Pure-Pak," "Seal-Right," "Blocpak" and "Tetra-Rex" cartons represent variations of precut cartons. Both rectangular and gable tops are used.

Postformed.—Postforming uses rollstock, forms, fills and seals in one continuous operation. The system may use polyethylene or foil-laminated paper. The "Zupack" is a rectangular block while "Tetra-Paks" are tetrahedron-shaped. If an aseptic unit is to be prepared with a "Tetra-Pak" shape, aluminum foil is included in the lamination.

Note: Where paper comes in contact with food its chemical purity and the nontoxicity of its coatings must meet FDA standards. Additionally, the microbiological condition of paper products is rigidly specified by food manufacturers and in certain food ordinances. Thus, the Grade A Pasteurized Milk Ordinance of the U.S. Public Health Service-FDA states that paper for milk cartons and caps be made from sanitary virgin pulp, and contain no more than 250 colonies per gram of disintegrated stock by a standard bacteriological test.

Study will reveal the relative strengths and weaknesses of available packaging materials for specific food applications, and special uses and restrictions will be uncovered. But this does not even begin to convey the many forms that these materials can exist in, depending upon such variables in their manufacture as identity and mixture of polymers, degree of polymerization and molecular weight, spatial polymer orientation, use of plasticizers (softeners) and other chemicals, method of forming such as casting, extrusion or calendering, etc.

Rigid Plastics.—The increase in half gallon (1.9 liters) and gallon (3.78 liters) milk purchases has accelerated the trend toward plastics. The movement started with the country's growing family size and continues in the face of now decreasing family size. Consumers now prefer larger sizes, and the number of quart sizes sold is steadily decreasing. Glass bottles in sizes larger than quarts are cumbersome and heavy and, as previously noted, gallon paperboard cartons tend to leak.

A gallon (3.78 liters) of milk weighs 3.8 kg—a bit ungainly in a paperboard package. Because the all-plastic bottle is more expensive than paperboard or glass bottles, it is impractical in quart (0.9 liters) sizes or smaller. Plastics dominate the gallon milk market with 85% in HDPE and the balance mainly in polycarbonate and paperboard.

When the all-plastic bottle was initially introduced in the United States, closures and design were the basic problems. Caps did not fit satisfactorily and consumers felt that the handles were too small for regular usage. Handles have been improved by the introduction of superior manufacturing techniques. They have been made wider and smoother, permitting easy holding.

Capping and filling procedures must be modified in order to introduce a plastic bottle. The first bottles were closed with simple paperboard plugs. This was the closure method used on glass bottles, however, leakage

Courtesy of Imballaggio, Milan (from an article by Dr. O. Pasquarelli)

FIG. 5.1. MILK CARTONS
A tetrahedral package for milk is shown here beside a rectangular carton. They are named "Tetra-Pak" and "Tetra-Brick" in Europe.

occurred and closures were faulty. Metal screwcaps then appeared and were capable of preventing milk spillage. The cost of the metal cap was high and its use was restricted in several states in the United States. A recent innovation in closure devices is a plastic top made of polyethylene with a diaphragm. The cap is applied to the bottle and the diaphragm is in the center of the cap. When the diaphragm is depressed, the cap expands causing a tight fit. It is tamperproof and may be produced at high speeds. A paper label can be used for brand identification. Aluminum foil screwcaps and reclosable, heat-sealable caps are being developed and will soon be available.

Distribution.—A dairy considering the introduction of a plastic bottle can choose from several methods. The simplest approach is by direct pur-

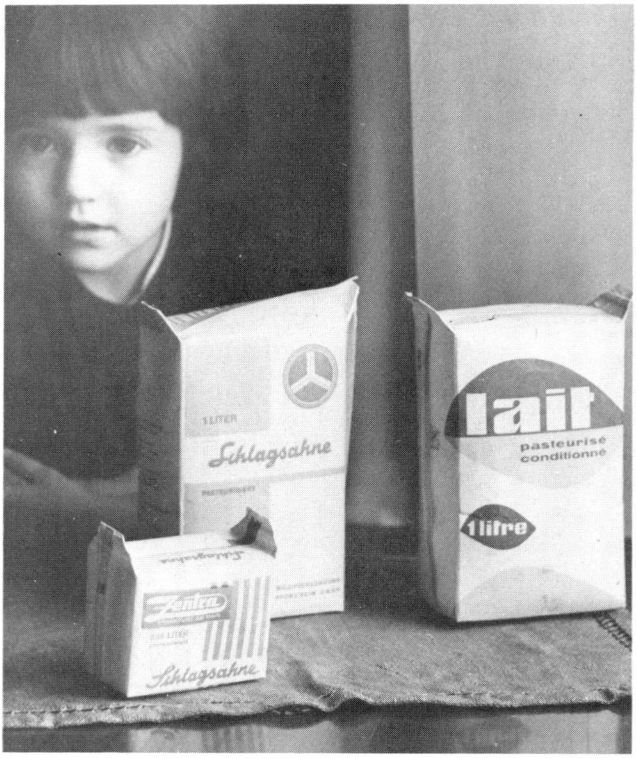

Courtesy of Imballagio, Milan (from an article by Dr. O. Pasquarelli)

FIG. 5.2. ANOTHER EUROPEAN MILK PACKAGE
The "Zu-Pak" in ¼ liter and 1 liter sizes is used commercially in several different European countries. This package is formed from polyethylene-coated paperboard and is capable of standing unsupported.

chase of ready-made bottles from a local blow-molder. No capital investment by the dairy is necessary and deliveries can be fairly rapid. Disadvantages include expense and dependence on the supplier for functional bottles. The dairy may produce bottles as an in-plant operation. Equipment is purchased and the capital investment is rather high. In addition, personnel must be hired to run the machinery. The obvious advantages are low raw material cost and the capacity to supply a sufficient demand for daily milk production. The dairy may contract the services of a bottle manufacturer. An agreement is signed as to specific quantities and the bottle manufacturer installs the machinery in the dairy or in an adjacent plant. The entire program is managed by the supplier, but the total production is captive to the dairy. The dairy may purchase partially

Courtesy of U.S.I. Chemicals Div. Natl. Distillers and Chemicals Corp.

FIG. 5.3. BLOWN RIGID PLASTIC CONTAINERS FOR MILK
Light weight plastic gallon (liter) jugs for milk feature molded-in handles making them easy to grasp.

fabricated bottles and complete the manufacture in its own plant. Not all bottle designs are conducive to this approach. Freight costs are reduced since only partially formed containers are shipped. About 1130 dairies use plastic gallon (3.78 liters) containers. Of that group, 130 do their own blow-molding at a cost of 6.5¢ to 8.5¢ The others use contract blow-molders and pay between 10.5¢ and 12.5¢. Paperboard gallon containers cost between 8.2¢ and 9¢.

Returnable Plastic Bottle.—The higher competitive cost of an all-plastic bottle compared to paperboard and glass has led to the development of a returnable plastic bottle. Polycarbonate returnables are now receiving noticeable interest. Of the 15 million refillable plastic milk bottles and over 20 million refillable glass bottles used today, polycarbonates expect to capture a lion's share of the market. A gallon polycarbonate milk of milk costs about $1.00 to produce but is returnable. This is a definite bonus in view of the current environmental concern and rash of new "bottle bills."

New Trends

All the above approaches are in current use by United States dairies. The final choice depends on a thorough study of economics and production. A dairy must consider the most expedient approach and maintain a steady supply of bottles. A standardized design is also under consideration by many dairies.

Flexible Plastic Pouches.—Liquid milk also may be packaged in plastic film or laminated pouches. The packaging of milk in pouch form was pioneered in France and has subsequently been introduced in several other European countries. Pouches offer economy, compact storage and ease of disposal. Disadvantages include the need for support and an unconventional appearance. Since clear plastic does not offer adequate shelf-life, an opaque laminate is required. Most all-plastic film pouches for milk are prepared from two-ply, low density polyethylene lay-flat tubing. The outerply is white and the innerply black to protect the milk from ultraviolet degradation. The lay-flat tubing is made by extruding two polyethylene resins through a coaxial die and then passing the two films through a second die to produce a 0.001 cm laminate. On the packaging machine the tubing is sterilized by ultraviolet irradiation, cut to the desired length, sealed on the bottom to form a pouch, filled and heat-sealed at the top to close. Many filling systems are available: "Polipack," "Prepac Polymaid," "Pintapac," "Doypac," "Prepak" and "Pitcher-Pak." The pouches are fairly popular in Holland, France, Finland and Canada. A specially designed holder must be used to support the bag upright after opening and during storage.

Paperboard-Plastic Liner.—In recent years, a "Bag-In-Box" concept utilizing a low density polyethylene bag in a corrugated container has become popular. Both the bag and box are completely disposable. The normal size used is 5 gal. (18.9 liters). The innerply consists of either single or doubleply 0.005 cm polyethylene. A spout is heat-sealed into the bag and a plastic valve may be added for dispensing purposes. The bag collapses in use as the level of liquid falls. The corrugated fiberboard container is printed and used as the outer package. Systems available include "Pergall," "Polygal," "Liqui-Box" and "Cubitainer."

Courtesy of Imballagio, Milan (from an article by Dr. O. Pasquarelli)

FIG. 5.4. FLEXIBLE PLASTIC MILK PACKAGES
An example of a flexible polyethylene bag used for milk. The molded rigid plastic pitcher makes it possible for the package to stand up and facilitates pouring.

Aseptic Milk Packaging

Aseptic or long-life milk was originally introduced in Sweden in the early 1960s. Originally called the "Tetra-Pak" system, it utilizes a laminate presterilizer and a filling environment heater. Aluminum foil is an integral part of the flexible laminate in order to provide a barrier against light and gas. Cow's milk is preheated to 73.8°–85°C then rapidly raised to 140.5°–148.8°C. It is held at the latter temperature for 2 to 4 sec and then suddenly cooled by flashing into a vacuum chamber. It must be packaged under completely sterile conditions. No refrigeration is necessary for at least 3 months. If kept under refrigeration, a shelf-life of up to 1 year is possible.

"Brik-Pak" has been used for aseptic milk ever since 1968. It accounts for 50% of the milk sold in Germany and Italy and 35% in Switzerland. Actively attempting to introduce it in the United States, the concept is currently in the hands of the FDA for approval.

Courtesy of Brik-Pak, Inc.

FIG. 5.5. BRIK-PAK ASEPTIC MILK UNITS
A selection of aseptically packed Brik-Paks from around the world.

Canned Milk Packaging

Sterilized canned milk is produced by using lacquered tinplate cans which have a special seam stronger than a conventional can. They must be able to withstand the stresses imposed by high temperatures used to sterilize the contents.

Condensed and Evaporated Milk

Condensed milk is prepared by evaporating milk in a vacuum at 15.5°C. Sugar is added and the total water content is reduced by 40–50%. Evaporated milk is essentially an extension of the same process but contains no sugar.

Both the sweetened and unsweetened types are packed in tinplate cans of varying size. After canning, they are heated like all other canned foods. Sweetened condensed milk holds for a longer time after the can has been opened and has the consistency of thick cream. Both types require refrigeration and may be reconstituted by the addition of water.

CREAM

Cream consists of the intermediate stage between milk and butter. The aqueous phase has been reduced but the emulsion has not been broken.

Since cream contains a high percentage of butterfat, it is very susceptible to spoilage. In addition, it must be protected from water loss. The shelf-life of refrigerated creams ranges between 1 and 2 days without proper protection.

Packaging consists of units similar to those used for milk, i.e., polyethylene-coated paperboard. Since cream is used in lesser amounts, sizes sold are 1 pt (0.47 liters) and smaller.

Newer concepts include portion-control thermoformed packs made from linear polyethylene, polystyrene or polypropylene. These may be closed with a peelable lid or snap-on cover. Tinplate containers have also been used for larger sizes.

Whipped creams and synthetic formulations are sold in aerosol cans and polyethylene tubs with snap-on lids. The recent introduction of many synthetic formulations may spur on the introduction of new packaging concepts. Imitation cream made from soybeans and vegetable oils is often marketed in wax-coated paperboard cartons. These products were originally introduced for the Jewish market and for people suffering from allergies to cow's milk.

Ultrapasteurization has been applied to heavy and light creams. The product then goes into "Pure-Paks." A strong seal is necessary for product protection. A shelf-life of 1 month is obtained by UHT techniques.

ICE CREAM

Wealthy Romans transported snow from the mountains and served it flavored to their guests. Sherbet was an Arab and Persian favorite and may have originated in China. The Chinese are believed to have taught the art of ice cream making to the Indians, Persians and Arabs, and they in turn brought it to Italy where it remained a favorite of the wealthier classes. In 1533, Catherine dé Medici journeyed to France at the age of 14 to marry the heir to the throne. Her Florentine cooks made "iced cream" popular in the French court and by 1666 ice cream making was an industry. It was first served in the United States in 1700 by Governor Bladen of Maryland.

Properties and Characteristics

Ice cream is prepared by mixing milk, cream solids, sugar and a stabilizer. It is carefully pasteurized and homogenized for sanitary reasons as well as for texture. Flavorings are added and then the mix is frozen. The frozen mixture is whipped to incorporate air and develop a smooth texture. The amount of whipping is carefully controlled to achieve a desired overrun (volume increase). Bulk ice cream is then packaged and hardened at a low temperature (−45.5°C). Shaped bars are hardened prior to packaging. Rapid hardening is essential to prevent formation of large ice crystals. Off-flavors or poor textures will occur in ice cream only if improper procedures are followed during its manufacture, storage and handling.

A large amount of ice cream is consumed in the United States. Annual consumption is more than 5 gals. (18.9 liters) per capita. Sherbet ices represent only about 10% of the total.

The widespread growth of supermarkets coupled with larger home freezers has resulted in a steady increase in retail ice cream sales. Many retail packs, however, are deficient in consumer convenience and product protection.

The chief requirements of packages for ice cream are protection against contamination, attractiveness, ease of opening and reclosure, and ease of disposal. Protection against moisture loss and temperature fluctuations is desirable.

Package Form

Most bulk ice cream is packaged in a linerless, bleached sulphite board carton, coated with wax or polyethylene-wax blends for protection from moisture and oxygen. Once the carton is opened it is difficult to reclose and the paperboard tends to warp. Home freezers do not maintain adequate temperature stability for most packs, and condensation and ice accumulation cause deformation. Although economic considerations favor the simple rectangular paperboard carton, improved packaging often leads to higher sales, and a marginal price differential may be offset by higher throughput.

The huge half gallon (1.9 liters) and gallon (3.78 liters) market, on the other hand, demands superior packaging. Consumers complain of poor recloseability, desiccation and oxidative darkening. Shippers are concerned with maintaining constant temperature in order to prevent heat shock.

Aluminum Foil Cartons.—Aluminum foil is a very effective barrier to light and radiant heat. As a result of surface heat reflection, bulk ice cream in foil cartons is kept at lower temperatures than that in paperboard cartons. It has also been reported that the ice cream takes longer to thaw. An additional advantage is the attractive appearance provided by printed foil. Foil cartons are most popular for ice cream on-a-stick packs.

Cylindrical Containers.—Many pint and ½ pint packages are constructed of cylindrical paperboard materials with a variety of covers and coatings and all offering an easily opened and reclosable package with superior storage properties. Cylindrical containers cost more than rectangular cartons, but consumers want a convenient pack.

Examples of cylindrical packs are Sealrights "Convocan" and the round container from the Sweetheart Cup Division or Maryland Cup Corporation. Both overcome the reclosing problem by using a separate lid that fits over the rolled top of the container. Convocan has a PE window allowing for product visibility.

Spirally-wound containers are made by pulling materials over a cylindrical mandrel at an angle to the winding mandrel. A container can be tailor-made for a product by incorporating various different plies. Ice cream packs usually consist of plastic or wax-coated containers with window-lid covers and metal reinforced edges.

Plastic Containers.—The most recent trend in bulk—half-gallon and gallon—ice cream units has been all-plastic cylindrical containers. Most of these are made of polyethylene and have recloseable lids. The inherent properties of polyethylene restrict striking graphical design, and the cloudy appearance makes printing difficult. However, the recent introduction of electrostatic printing may lead to design improvement in the future. Sherbert's bright colors make plastic containers attractive. In contrast, ice creams rely on package graphics for customer appeal.

Other plastics that can be used are polystyrene, PVC and polypropylene. Rigid PVC containers offer clarity and excellent graphical design but lack FDA approval. However, PVC is widely used in Europe and may soon become acceptable in the United States.

Convenience Concepts.—It has been proved that consumers will pay for package convenience. One ice cream manufacturer is offering a pail-type container for a gallon pack, and zip tapes and easy-tear features have been incorporated in half gallon packs. Other ideas may include tack-down lids, folding-top improvements and foil overwraps. The ultimate aim is to prevent product deterioration at a reasonable cost.

BUTTER

The origins of butter date to the earliest civilizations such as Sumer in Mesopotamia and the Aryans of India. It was long deemed a sacred food to be prepared only by priests and consumed by the priesthood and royalty, but inevitably became a staple food for all. Monarchs such as the Persian King Xerxes received it as tribute. Wealthy Egyptians had it buried with them in their tombs.

Over a million kilograms of creamery butter are produced annually in the United States despite inroads made by butter substitutes.

Compositions and Characteristics

Butter consists of the fat which has been separated from the aqueous phase of milk. About 15% by weight of water is trapped in the butter. Because of this high moisture content, butter, unlike solid fats, is susceptible to mold growth. Flavor and odor are easily affected by absorption from other materials or through deterioration of the butter by rancidification. The package must provide protection against these factors. Mold growth is inhibited by exclusion of air and addition of salt during manufacture. Tight intimate wraps are used to help exclude air from the product. Materials and manufacturing areas must be kept sanitized to reduce the population of mold spores.

The package should also protect against moisture loss as too much will result in color darkening as well as short weight.

Flavor and odor preservation is a critical requirement. Refrigeration is imperative in order to avoid rancidification. Light and oxygen promote photochemical oxidation. It is therefore desirable that the package be opaque and preferrably a high barrier against oxygen and foreign odors.

Long-term Storage Conditions

For long-term warehouse storage, butter must be wrapped in impermeable materials in order to eliminate odor pickup and must be stored under refrigerated conditions. In paperboard cartons odor transference may occur. The only satisfactory method for eliminating odor damage is lining the carton with an odor barrier coating or wrapping in an impermeable lamination. Superior materials to parchment are needed. Foil laminations are essential for this purpose.

Packaging Materials

Parchment.—The most commonly used butter-wrap is probably vegetable parchment, which is sterile as produced and free from mold spores. Most butter manufacturers use a plasticized grade in order to meet the high speed packaging machine requirements. If parchment is used, it is essential that it not contain excessive numbers of microscopic pinholes. Also, it must be stored properly. It should contain more than 9% mois-

Courtesy of Chemische Werke Huls Aktiengesellschaft

FIG. 5.6. BUTTER GETS BETTER PACKAGING
Rectangular plastic tubs are filled with butter, hermetically sealed with a foil-laminated membrane and closed with a protective clear plastic lid, all automatically.

ture and must be stored in clean, dust-free places at room temperature and 50–80% humidity. If stored too dry, it becomes brittle. When stored in a damp area, it may develop mold and transfer spores to the butter. Even though parchment paper is a grease barrier, it is not sufficiently impermeable to prevent oxygen penetration, and its transparency allows some light to penetrate which promotes oxidation. For superior product protection and longer term storage, laminates of aluminum foil are particularly useful.

Aluminum Foil Laminates.—Aluminum foil provides opacity and barrier properties. The material generally used to package butter consists of a thin-gauge aluminum foil laminated to paper. The paper is added to effect better machinability and to strengthen the thin-gauge foil. In Europe, a lamination of aluminum foil and parchment is used. The parchment provides wet strength and a sterile surface. Equally effective are laminations of foil and grease-proof paper and foil and tissue paper. The foil can be surface-printed and overcoated with a scuff-resistant material. One advantage in printing aluminum foil in contrast to parchment paper is ink transfer. Parchment paper must be printed with fat-insoluble inks. If the wrong ink formulation is used, ink might migrate through the occasional pinholes in parchment and cause discoloration.

The use of foil-plastic laminates is not recommended because many plastics contribute off-flavor or exhibit insufficent grease resistance. Economics for foil-plastic laminates are not favorable.

Recent developments in smooth-walled rigid aluminum containers offer butter producers a means for obtaining aesthetics as well as convenience. These packages are smooth in shape, colorful and hygienic. Half pound (0.2 kg) or pound (0.45 kg) units with printed or unprinted plastic or heat sealable foil lids can be used at the table. Filling can be done during the peak season when milk is in glut supply and with proper storage and distribution. The packages can be sold throughout the year. Gourmet types of butter flavored with garlic or spices are also being marketed in formed foil containers.

Portion Control

For further convenience in butter packaging the use of small portion control packages in formed foil is popular, particularly for air travel food service and for fine restaurants where a quality image is desired. Butter patties in iced water dishes leave much to be desired as to handling and sanitation. Paper trays with release-paper covers are also poor packages as they do not provide full sanitary protection. Machines are currently available for forming 0.00254 cm to 0.005 cm thick aluminum foil into shallow (0.1524 to 0.4572 cm) trays. The trays can be colored or embossed if desired. Covering lids can be heat-sealable aluminum foil, PVDC-coated cellophane or other suitable barrier materials. Covers are heat-sealed in place automatically after the butter patty is dropped into place, giving complete sanitary protection as well as convenience and quality protection.

GHEE

In India and Pakistan butter is not used in the same way as in western countries. Instead it is produced from cow or buffalo milk which has first been allowed to ripen (partially sour). The curd may be separated at this point and sold as *dahi* or the curdled milk may be churned to produce butter. When a batch of butter has been accumulated, usually with some development of rancidity, it is converted to ghee by heating it at 100° to 112°C until all the water has boiled away and the curd turns a yellowish brown color. The fat is strained and then cooled slowly to 26.6° to 40°C when it is poured into containers and cooled slowly to room temperature. The final product is essentially 100% fat, yellow in color, salt free and rather strong in flavor. (Buffalo ghee is nearly white and has a higher melting point.) A more bland flavored ghee is accomplished if the butter is merely heated to 100°C and then allowed to stand until the water and curd particles settle out. The ghee is decanted and cooled.

Due to the high price and shortage of ghee these countries are now producing a substitute product called *Venaspati* made from hydrogenated fats (see Chapter 8).

Ghee is often made in the home and consumed from home containers of glass, earthenware or tinplate. Commercial packages are usually tinplate.

CHEESE

No one knows who made the first cheese. The ancient Greeks revered cheese so highly that they believed it to be a gift of the gods. According to an ancient legend, it was discovered when an Arab merchant carried some milk in a pouch made from a sheep's stomach across the desert. The bouncing of the camel, the heat and the chemical action of the pouch caused the milk to separate into curd and whey. The whey satisfied the Arab's thirst and the curd was cheese.

Cheese is frequently mentioned in the Bible. It made its way from Asia and Africa to Rome and then carried by the Roman legions, it was introduced to Europe and England. The monks of Europe improved cheesemaking and in the 10th century, Italy became the cheese center of Europe. Gorgonzola was made as early as 879 A.D. in Milan. Roquefort was first mentioned in 1070 in the records of a monastery.

The Pilgrims included cheese in their supplies in 1620. In 1851, the first cheese factory in the United States was built in New York. Modern cheese technology started around 1870 when a commercial rennet preparation was offered for sale in Denmark. Dairy scientists perfected more sophisticated methods for cheesemaking and in recent years, significant advances have been made in mechanization and packaging.

Characteristics

Cheese is defined as the curd of milk (coagulated by rennet) separated from the whey and pressed into a solid mass. Most of it is made from cow's milk; however, smaller quantities are made from the milk of goats, ewes, camels, reindeer and buffalo. Cheese is normally made by coagulation of the milk by rennet and lactic acid. The curd is then broken and most of the whey is removed. The partly dried curd is then ripened and sold as cheese.

Production

Cheddar Cheese.—In making cheddar cheese, the curd is formed under controlled conditions of acidity, temperature and rennin concentration. Pasteurized whole milk is added to a vat together with a starter culture of *Streptococcus lactis*. After 30 min, rennin is added in a dilute acidic solution. The milk is allowed to set and stirring is stopped. Curd forms in about 30 min.

The curd is cut into small cubes (0.6 × 1.27 cm) with special curd knives that are formed from wires strung along a frame. The cubes are then heated with continued stirring to separate the whey. The whey is drained off and the curds fuse together to form a cohesive rubbery mass. Next, the matted curd is cut into blocks. These blocks are turned at 15 min intervals and piled on one another 2 or 3 deep. During this time, additional whey is drained off and acid formation continues. The blocks are then cut into smaller pieces, salted, stirred, drained and pressed at about 20 psi pressure overnight.

After pressing, the cheese is removed and placed in a cool drying room at 15.5°C and 60% rh for 3 to 4 days. In order to prevent mold formation, the cheese wheel or block is dipped in hot paraffin then it is boxed and placed in the curing room for ripening.

Ripening takes place for at least 60 days and for a peak flavor it may be continued for up to several years. Sharp-flavored cheddars are the result of longer ageing.

Stilton Cheese.— Stilton, the only trademarked cheese in Britain, may by law be produced only in the three counties—Leicestershire, Nottinghamshire and Derbyshire—that have historically provided both the environment and the skill required. Produced in farmhouses until the turn of the century, the cheese acquired its name when it was sold at the Bell Inn in the town of Stilton, a stagecoach stop between London and York. Today it is made by 12 dairies in the three counties and the process has been sanitized and shortened, but otherwise unchanged. Stilton requires 3 to 4 months of ripening and an infinite amount of patience.

The process begins with the running of lukewarm, pasteurized milk into large vats, to which are added bacteria that promote acidity and an enzyme to hasten clotting. The whey is drained off, the curd forms and settles, and the following morning is mixed with salt and packed into large, stainless steel hoops.

After draining in the hoops for 6 days, the cheeses are "rubbed" to create an even surface. Each cheese is then wrapped in a diaperlike cloth for 24 hr before it is left to stand on its own. The cheese is then carried into the "white" room where, in 80% humidity and 10°C temperature, it begins to form a yellow coat. After 3 weeks, some of the cheese is packaged and sold as white Stilton.

To become Blue Stilton, however, the cheese must be moved to the dairy's enormous "blue" room. There each cheese—the room can hold 5000—receives some 300 punctures from stainless steel needles, allowing oxygen to enter and blue mold to form. As it has been all through the process, each cheese is turned over by hand daily to guarantee an even texture throughout.

After several weeks, the cheese is moved to a colder room where it forms the distinctive rough brownish crust of traditional Blue Stilton. But before the cheese leaves the dairy, the manager eases a cheese "iron," similar to a hollowed-out pencil, through the crust in two places. The manager sniffs the extracted cheese, examines it for an even spread of veins and inspects the iron for a creamy film, which indicates a high fat content. If the signs are all right, the sample is returned to the cheese, deftly turning the iron to plug the hole.

The result of this labor-intensive process is a cheese with a unique taste, described variously by its adherents as sweet, salty or mellow, but always more subtle than any other blue-veined cheese. In all, the industry produces about 500,000 Stiltons a year, of which some 20,000 are sent to the United States. Each cheese weighs about 6.8 kg and retails in Britain for around $2 per 1 lb (0.45 kg). In New York, Stilton sells for about $4 to $4.50 a pound.

Types of Cheese

Natural cheese is made directly from milk, with or without ageing and ripening by bacterial action or molds. More than 400 different types of natural cheese are produced.

They can be categorized according to hardness and method of manufacture as follows:

(1) Low moisture (30%) very hard, ripened by bacteria, examples include: Sapsago, Romano, Parmesan.
(2) Moderate low moisture (38%) hard, ripened by bacteria, examples include: (no holes) Cheddar, Gouda, Edam, Provolone; (with holes) Swiss.

(3) Medium moisture (45%) semisoft, examples include: (ripened by bacteria and surface microorganisms) Hard, Trappist, Limburger; (ripened by bacteria) Brick; (ripened by interior blue mold) Roquefort, Blue Cheese.

(4) High moisture (50–80%) soft, examples include: (ripened) Camembert; (unripened) Cottage, Cream, Neufchatel, Ricotta, Mysost.

Processed cheeses are produced from natural cheese blended for uniformity of flavor, texture and cooking quality. They are ground together, melted, pasteurized and poured into molds. The first processed cheese was produced by Kraft in 1904. Examples of processed cheese include American Cheese and processed Gruyere. Because of the addition of preservatives, processed cheese is much more stable than natural cheese. Process cheese spreads contain a higher moisture content to produce a more spreadable consistency at room temperature. Both spreads and solid processed cheeses may include other additives such as milk solids, wine flavorings and spices.

Packaging

In 1977 United States cheese sales in supermarkets amounted to $2.3 billion. Over 179 different brands were sold with speciality cheeses leading the list.

Regional packers still handle a share of the cheese business; however, there is a trend toward centralized packing and long distance shipping of cheese products to points of sale. There is a need for durable materials and laminates which will withstand the rigors of rough handling incurred during shipment. Consumers also continue to demand more convenience features in cheese packaging.

Packaging Requirements of Natural Cheeses.—Any material to be used for packaging natural cheeses must

(1) afford general protection,
(2) prevent moisture loss,
(3) improve appearance,
(4) protect against microorganisms, and
(5) prevent oxygen transmission.

Moisture vapor and oxygen barriers are critical in natural cheese packaging applications and as a result, materials used to package natural cheese all involve the use of polyvinylidene chloride (PVDC) to some degree. PVDC gives the final composite extremely low oxygen and moisture vapor transmission and, in combination with polyethylene (PE), the moisture transmission rate becomes even more insignificant.

Most natural cheeses require an oxygen barrier which will pass no more than 77.5 cc oxygen/sq m/24 hr (at 22.7°C and 50% rh).

Packaged chunk cheeses, such as cheddar, are usually marketed in gas or vacuum-packaged pouches. A commonly used machine is the Hayssen RT 118 which makes and fills flat pouches for gas packs. Rates of packaging are 60–100 per min and the machine can also handle sliced cheese. Another machine used for packaging is the Hudson-Sharp. This unit forms and fills flat pouches for vacuum packaging at speeds up to 20–40 per min. It will gas flush at 60–100 per min. Sliced and chunk cheese each require a different machine.

When fabricating pouches for natural cheeses on conventional bag-forming machines, a dust is normally applied to provide slip and easy handling. The dust is generally a derivative of corn starch, but since molds form on the surface of cheese, sorbic acid or its salts may also be applied as a dust up to 2.5–5.0 g per thousand square centimeters of laminate.

Package Forms for Natural Cheese.—The traditional methods of natural cheese packaging were based on bandaging and dressing. Whole cheeses were treated at the farm or factory to protect them for retail distribution. In bandaging, a calico or cheesecloth bandage is applied subsequent to treatment of the cheese with greased muslin. The unit was either sewn or glued with a flour paste. Dressing involved the external treatment of the cheese with oil, fat or wax. Edam and Gouda cheese are famous for their wax dressing. They are washed in hot water, stained red and then waxed or immersed in a red wax.

Although bandaging and waxing are still used on certain cheeses, the trend is toward making most varieties in block form for use with flexible laminates. Traditional methods, such as described above, are expensive and offer limited protection to the cheese. They lead to the formation of a thick rind and are not suitable for modern mass-merchandising methods.

*Rigid Containers.—*Cartons made of waxed paperboard are used to package cheeses having no characteristic form. Cottage cheese is often packed in a rigid paperboard carton since it has a short life, readily loses moisture and must be consumed within a few days after distribution. Vacuum-formed polystyrene tubs are now used almost exclusively for the consumer packaging of cottage cheese. They offer superior protection from fat absorption and breakage. Ricotta cheese is also packaged in either waxed paperboard tubs or polystyrene tubs. Some grated cheeses such as Parmesan and Romano are sold in foil-paperboard composite containers and in glass jars. Since glass is fragile and heavy, it is rarely used for other cheeses. Dutch cheese ripened in glass jars was found to have developed an off-odor.

Plastic-lined metal cans are used for special cheeses such as Camembert. Cans offer excellent protection but are expensive. In addition, the cheese frequently acquires a peculiar taste. Grated cheese has been sold successfully in seamless aluminum cans.

Flexible Packages.—Short-term storage (about 3 days) permits the use of a package made of moisture-proof cellophane. PVDC-coated cellophane and even PVDC film will hold the product for a longer period. These types of packages, however, are poor choices for long-distance shipping which requires more sophisticated composites with superior barrier properties.

Cellophane-PVDC-PE was probably the first major film composite used for long-term storage as a packaging material by the natural cheese industry. The material is moisture-proof and it may be either vacuum-packed or nitrogen-flushed with the PE film serving as a heat-sealing medium. The cellophane may also be reverse-printed and adhesive-laminated to the PE.

The combination of materials is fairly inexpensive and is very useful for packaging sliced Swiss, Muenster and other general types of cheese. However, cellophane composites have the disadvantage of poor abrasion and flexable characteristics which make them particularly unsuited for chunk and brick natural cheeses.

Aluminum foil laminates are widely used for cheese packages. As early as 1930, several European soft cheeses were packed in aluminum foil. Natural cheese appears to ripen more rapidly in foil materials. Port du Salut has been wrapped in aluminum foil and Roquefort has been packaged in an aluminum foil laminate. Cream cheese is now packed in both 3 (85) and 8 oz (226.8 g) sizes in a laminate of foil-paper-foil. This material offers excellent folding characteristics coupled with protection.

Polyester-PVDC-PE materials offer more inherent strength and less moisture sensitivity than most composite structures. There are various grades of polyester film available, varying in their degree of orientation, and these are designed for better abrasion resistance and transportation durability than cellophane.

The use of nylon as part of a lamination offers the cheese maker a film with excellent flex and abrasion, and both barrier and heat-sealing characteristics when used in combination with PVDC and PE. It is important to note, however, that not all nylon films are alike and that there are some which are more suitable for cheese packaging than others. Nylon 6 is superior in machinability and abrasion resistance and is less expensive than other types.

Courtesy of Istituto Italiano Imballaggio

FIG. 5.7. ITALIAN CHEESE PACKAGE
This unit won first prize at a national competition.

Nylon-PVDC-PE is a cheese packaging composite material which is about equal in cost to a polyester composite and yet superior in abrasion and flex. The hydrophilic nature of nylon is altered when PVDC is added to the final structure and may result in difficulties when this material is handled by conventional packaging machinery.

A construction of oriented polypropylene-PE-PVDC-cellophane-PE (Airpolene 200) lately has aroused quite a bit of interest as a packaging material for the cheese industry. It is a rather complex structure and offers excellent abrasion resistance and flexibility. The complexity of the structure has been found to retard its deadfold character. Packages made with this material do not conform tightly to product shape when vacuum-packed. Since it is a rigid film, it has fairly good machinability but, on the other hand, it has a tendency toward delamination between film plies and it produces a loose package. Oriented PP gives the total structure excellent abrasion resistance and durability and the film may be used in the packaging of many chunk cheeses.

Elimination of cellophane will reduce the cost of the total structure to some degree. A material composed of oriented PP-PVDC-PE will

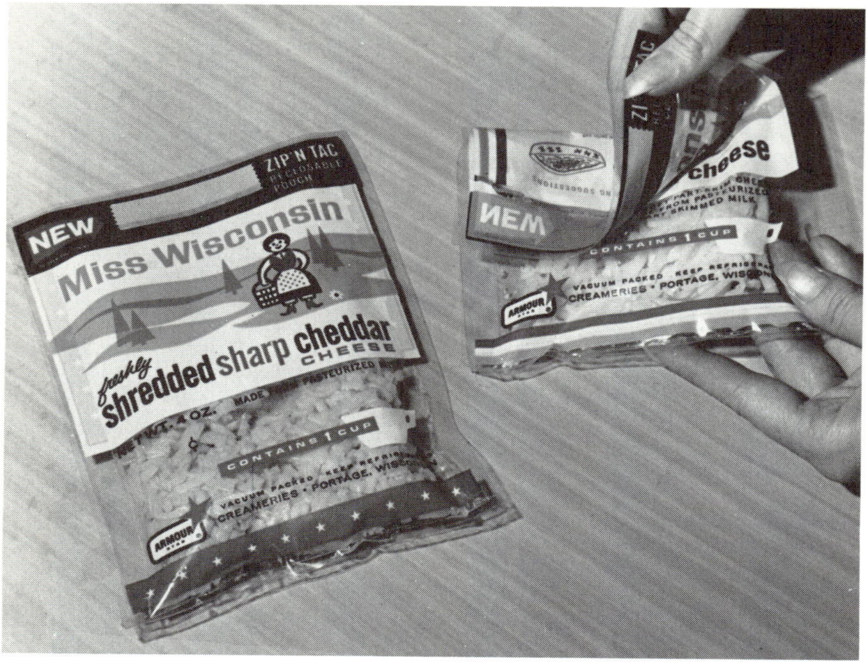

Courtesy of Armour and Co.

FIG. 5.8. FLEXIBLE FILM VACUUM PACKAGE FOR CHEESE
Laminated plastic film pouches provide good protection for vacuum packed shredded cheese. This package features a "Zip'N Tac" reclosure feature.

provide good abrasion resistance, but the narrow heat-sealing range of the oriented PP can be a major difficulty. Recent technological advances by several film suppliers have resulted in improvements. Heat-set and stable films are now available that will yield a nonpuckered seal if machined under the proper conditions.

Rindless Natural Cheese.—When small portions of natural cheese with rind are prepackaged, considerable waste is obtained by discarding the rind. An additional problem is that the variable size and shape of natural cheese leads to additional waste. In order to reduce waste, rindless cheeses of uniform size and shape have been developed. Cheese is ripened in bags made from plastic films similar to the types used for long-term storage. Acetoglycerides and polyvinyl acetate emulsions have been used as coatings to produce rindless varieties of cheese.

Packaging Requirements and Package Forms for Processed Cheese.—As noted previously, processed cheese is a much more stable product than natural varieties. It is sterilized during processing, pack-

Courtesy of Aluminum Foils Ltd.

FIG. 5.9. CREAM CHEESE
This famous package illustrates the attractiveness of embossed aluminum foil-paper laminates. In the past 5 years it had emigrated from the United States to the United Kingdom.

aged in a molten state and is a homogeneous product with superior shelf-life and stability.

Due to the better keeping qualities of processed cheese as well as its artificial conformation, packaging requirements are much less critical. Some of the composite films just discussed are used for processed cheeses. Foil composites with a heat-sealable layer are also used, with cellophane or paper laminated to the outer surface.

A great amount of processed cheese is packed in wax-coated cellophane. Hot cheese is poured into preformed pouches and the sealed package is tight, mold resistant and inexpensive.

The wax-coated cellophane is treated with a mold-preventive agent to retard bacteria growth. Several converters are producing wax-coated cellophane materials for the processed cheese industry in substantial quantities.

Cheese spreads are packed in pry-open glass jars, in paperboard cartons with foil overwraps and in rigid vacuum-formed polyethylene tubs. "Tubbed" cheeses are popular specialty items for local and regional brands.

Packaging Machinery

Automatic packaging machinery has allowed cheese packers to eliminate all handwrapping operations and cheese packages may be formed from either rollstock or prefabricated pouches.

Use of vacuum packing is limited by the soft texture of some cheese slices which tend to cling together under vacuum. Paper inserts, which are placed between each slice of cheese, have done much to eliminate this.

Gas flush has been used fairly extensively and does not cause any deleterious effects on the product. Occasionally, however, a "balloon" pack may result, and if the product releases carbon dioxide, the seals will break and the package will rupture.

Many packers prefer a "cling" package—one under a slight degree of vacuum and not fully tight.

Hayssen Manufacturing Company offers a Model RT 118 machine which is used extensively in the cheese packing industry. It may be used for both chunk and sliced cheese packaging and operates at speeds of 60–100 bags per min. The machine is equipped with a separate chamber for gas flushing which produces a pillow-type package. Most of the composite films just described may be run on the RT machine and its rotary motion enables it to run at the high speeds necessary for profitable production.

The Hudson-Sharp Division of FMCs Packaging Machinery Division offers an automatic high-speed packaging machine capable of vacuum packing 20–40 packages per min. The gas flush operation can run at a more rapid speed (60–100) and wither a tight or "pillow" package can be produced depending on the method used in packing.

European producers of cheese packaging machines include Lerner Machine Company, Masson and Morton, Ltd. and the W. R. Grace machine used for the Cryovac process internationally.

OTHER PRODUCTS

Yoghurt

In recent years, yoghurt has become very popular in many nations. United States sales amounted to $500 million in 1977. Swiss style yoghurts are holding 30% of the market. Excellent advertising coupled with bold package design have made yoghurt a trendy item for the 1970s.

Originally a Balkan food, yoghurt is a specially fermented milk. It was brought from the Balkans to France during the reign of King Francis I. Long associated with health-giving properties, yoghurt is made by reducing partially skimmed milk over slow heat. The milk is then divided into bowls and cooled. A culture is added and the bowls are stored for 28–48 hr before use. By the introduction of a wide variety of flavors, it is rapidly expanding sales in supermarkets.

In the United States, yoghurt is packed in coated paperboard containers with recloseable covers and thermoformed containers. European packaging trends are more imaginative. The product is automatically filled in polystyrene tubs and covered with 0.005 cm aluminum foil. The cover is clamped over the tub and is easily peeled off. The inherent brittleness of polystyrene may become a problem. It has also been reported that the terpene-containing aroma of fruit yoghurt may have an effect on polystyrene. Fruit acids may also cause pitting of unlacquered aluminum lids. A polystyrene tub is nevertheless an excellent package for yoghurt. It is economical, practical and widely used. The unit is effectively displayed and promotes product distribution.

In Germany yoghurt tubs are being sold in expanded polystyrene foam display trays. Each tray holds 20 tubs of yoghurt. Trays are stackable and serve not only as display holders and light weight shippers, but also holders during fermentation of the yoghurt. This saves extra handling and repacking.

Dried Milk

When raw liquid milk is dehydrated, milk powder is produced. The product is highly susceptible to oxidation and moisture. Reconstitution with water yields liquid milk. If the fat is removed as well as moisture, dried skimmed milk (nonfat dried milk) results. This product is not so sensitive to oxidation and does not require vacuum or gas packaging.

Most consumer packages consist of aluminum foil laminates. Paper-polyethylene-foil-polyethylene comprise the vast majority of dried milk portion packs sold. Larger retail units are packaged in cans, jars or lined paperboard cartons. Commercial bulk packaging is either in polyethylene bags inside corrugated cartons, or in 50 (22.7) and 100 lb (45.36 kg) bags made of 2–3 mil polyethylene liners and 4 to 6 ply kraftpaper outers with taped-seam closures.

Fermented Milk

Fermented milk products other than yoghurt are not sold in large quantities in supermarkets. They originate in other lands and have exotic names such as *Dahdi* (India) and *Laban Zebadi* (Egypt). Since they are native and special ethnic drinks, they are packaged in a wide variety of materials from sheepskins to glass bottles.

Courtesy of Aluminum Foils Ltd.

FIG. 5.10. YOGHURT PACKAGING
Flavored yoghurts are very popular. This package is made from a molded plastic tub with printed foil lids.

Courtesy of The Metal Box Co., Ltd.

FIG. 5.11. DRIED MILK POWDERS
This dried nonfat skimmed milk package comprises an inner bag and an attractively printed sealed paperboard carton called "Hermetet."

REFERENCES

ANON. 1968A. Bag-in-box packaging cuts indirect costs. Packag. Eng. *13* (3) 79–80.

ANON. 1968B. Pourer for milk powder pack scarcely cuts protection. Packag. News *15* (6) 12.

ANON. 1969. Milk-in-sachet system. Only one customer says "no" to pack. Packag. News *16* (10) 3–4.

ANON. 1970A. Dutch dairy installs first Tetra Brik aseptic unit. Packag. News *17* (1) 6–8.

ANON. 1970B. The poly-trip plastic bottle. Good Packag. *31* (23).

ANON. 1970C. Pouch packs milk at 1 cent/quart. Food Eng. *22* (3) 117–120.

ANON. 1970D. Foil seal keeps cheese fresh in reclosable PVC tub. Packag. News *17* (9) 14.

ANON. 1970E. The milk industry looks at plastics. Packag. Rev. *90* (10) 32–39.

ANON. 1970F. In-dairy bottle welding cuts milk packaging costs. Packag. News *17* (10) 11–12.

ANON. 1971. Whipped cream in aerosols. Aust. Packag. *19* (3) 25–27.

ANON. 1976. Cheese film. Food Drug Packag. *34* (6) 21 (March).

ANON. 1977. Say milk jug's success hinges on marketing. Food Drug Packag. *37* (9) 4, 11 (Nov.).

ANON. 1979. Selling products at the door. Packaging *50* (587) 10–11 (February).

BRISTON, J.H. 1968. Progress in milk packaging depends on consumers. Inst. Packag. J. *14* (98) 20–22.

DAVIS, J.G. 1965. Cheese. Vol. 1, Basic Technology. American Elsevier Publishing Co., New York.

DAVIS, J.G. 1971. The milk container of the future. Food Process Ind. *40* (477) 27–30 (July).

DE MAN, J.M. 1978. Possibilities of prevention of light induced quality loss of milk. Can. Inst. Food Sci. Technol. *11* (3) 152–154.

DOWNEY, W.K. and MURPHY, M.F. 1968. Light-barrier properties of various butter wrapping materials. J. Soc. Dairy Technol. *21* (2) 104–106.

HANRAHAN, E.P. *et al.* 1967. Experimental equipment for cooling and packaging foam spray-dried milk in the absence of oxygen. J. Dairy Sci. *50.*

HORNER, E.C.A. 1969. Plastics materials for packaging milk and milk products. J. Soc. Dairy Technol. *22* (1) 3, 15.

KNOX, J. *et al.* 1977. The benefits of aluminum foil in butter packing presealed at Symp. or Alum Foil in Pkging. Nov. 16–17, London.

LISIECKI, R.E. 1971. Aseptic carton system. Mod. Packag. *44* (1) 77–79.

MANN, E.J. 1969. Aseptic packaging of milk and milk products. Dairy Ind. *34* (1) 30–31.

PEARSON, S.D. and SCOTT, C.R. 1978. Three major cheese packaging methods, "which is best?" Food Eng. *50* (3) 105–108.

PLEETH, M.J. 1970. Single-service containers for milk. The case for a rectangular carton. Dairy Ind. *35* (2) 85–89.

RANGER, H.O. 1969. Films for cheese. Paper Film Foil Convert. *43* (11) 73–76.

RIPPEN, A.L. 1970. Aseptic packaging of grade A dairy products. J. Dairy Sci. *53* (1) 111–115.

RUTHERFORD, R.E. 1967. Returnable plastics bottles in an American dairy. Dairy Ind. *32* (5) 371–375.

SERCHUK, A. 1978. Milk packaging: Still in transition. Mod. Packag. *51* (4) 37–40.

TOMLINSON, N. and DIXON, B.D. 1977. A comparison of wrapping materials for export butter. Aust. J. Dairy Technol. *32* (2) 70–71.

6

Fish and Shellfish

> The aboundance of Sea-Fish are almost beyond beleeving, and sure I whould scarce have baleeved it except I had seene it with mine owne eyes.
>
> *Francis Higginson, 1630*

FISH

General Marketing Concepts

Although the sale of all seafood items in the United States in 1978 amounted to about $3.5 billion, the packaging of fish has been one of the most neglected areas of packaging technology in the country. Most packaging advances have originated in Europe where fishing has remained a much more important part of the diet and where the fishing industry has received greater governmental support. Despite the rise of supermarkets in the United States with mass merchandising to the consumer, Americans ate only 5.80 kg of fish per person in 1977 (as compared to 5.89 kg in 1976) in contrast to 29.48 kg for Japan and 20.4 kg for Sweden. The nature of the product, the relatively small level of consumption, legal requirements, and the methods of processing and packaging of fish are all interrelated—all being the cause and result of the other. United States law forbids the landing of fish at American ports in other than vessels built in the United States. In effect, U.S. fishermen are barred from buying their boats at competitive prices in foreign countries. In March 1977 the 200-mile limit was imposed on foreign fishing fleets. People who watched Atlantic fishing shrivel up under foreign competition soon rushed in to cash in on the new found exclusiveness of America's offshore domain. But problems still remain for domestic fishing fleets. Many boats are at least 20 years old and are comparatively small and inefficient. An additional problem are the quotas placed upon cod, haddock and flounder catches. These species are tightly regulated by the Federal Government in order that they can replenish in supply.

There are over 750 species of fish commonly consumed by humans. If combined with lesser known varieties the number would be well over 2000. Added to the complexity is that the fish industry uses many different names for the same fish. Examples are many, i.e., swellfish, blowfish, globefish, puffer and swell food are all the same fish (they may also appear as sea squab). The leading variety of fish caught is not a food fish; it is menhaden which is used for oil and fertilizer. To cover the subject adequately one must divide this foood source into two major categories: fresh fish and processed fish. Each category requires special packaging materials and different handling procedures.

Fresh Fish

Fresh fish are one of the most perishable of all foods. The rate of spoilage doubles for every 5.5°C rise in temperature. They must be refrigerated or frozen immediately after harvest and kept refrigerated until eaten. Any departure from immediate and continued refrigeration will lead to flavor and texture losses. If the fish are of poor quality to start with, the quality loss will be even more pronounced. The best package cannot improve the quality of the contents. The fish must be of high quality prior to packaging.

Packaging Requirements.—The rapid decline of retail seafood stores has been coupled with a change in consumer buying habits. More and more supermarkets now maintain fresh seafood counters as well as offering prepackaged "wet" fish. The maintenance of proper shelf-life by packaging becomes an essential prerequisite for marketing success. Close attention must be paid to the display cabinets used and to careful and rapid handling of the product. What might be insignificant departure limits for other products could lead to a 3 day shelf-life being reduced to a few hours.

A suitable fresh fish package must:
(1) reduce fat oxidation,
(2) reduce dehydration,
(3) provide for less bacterial and chemical spoilage,
(4) eliminate drip, and
(5) prevent odor permeation.

Fat Oxidation.—The highly unsaturated nature of fish oils and fats makes them very susceptible to rancidification. Other spoilage occurs also from chemical, enzymatic or bacterial action. At room temperature the lipid (fatty) components rancidify rapidly in the presence of oxygen. The use of a good oxygen barrier package and cold storage drastically reduce this action.

Dehydration.—Since fresh fish are wet, excess moisture loss leads to texture, flavor and color changes. Poor storage conditions or a poor barrier package will allow dehydration.

Bacterial and Chemical Spoilage.—Enzymes and bacteria present in fish cause rapid spoilage. Powerful proteases in the digestive tract may, after death, perforate the intestinal wall and attack the surrounding flesh. This is why fish should be gutted and washed promptly or immediately frozen. The characteristic fishy odor of trimethylamine is not present in fresh fish. It is formed by enzyme action and by secondary enzymes produced by bacterial growth. The flesh of healthy fish likewise is free of bacteria, but the slime and digestive tracts contain many bacterial flora. Immediately after death these bacteria begin to grow and spoil the fish with development of off-flavors and odors and pronounced textural changes. As with enzymes, immediate cleaning is the best means for preventing spoilage. Refrigeration alone is ineffective in preventing spoilage as many bacteria asssociated with fish (of the genera *Pseudomonas, Achromobacter, Flavo-bacteria* and others) are psychrophylic—that is, they will grow at 0°C and below, some as low as −7.2°C. Freezing will retard spoilage provided the temperature is held low enough to prevent further bacterial growth.

Drip.—When a piece of fish is packaged, some juice may be lost and trapped in the package. The result is messy and unsightly. An absorbent pad may be placed in the package to soak up the drip. Another approach is to dip the fish in a polyphosphate solution long enough to swell the outer surface layer of cells. The rupturing of cell walls helps to prevent the escape of internal fluids.

Odor Permeation.—Odors from fish become stronger in permeable packages stored at 1.6°C within several hours. Although this odor may dissipate, a considerable amount of flavor volatiles will be lost. Foreign odors are easily picked up by fish. A good odor barrier is an essential part of a good package.

Bulk Shipment.—Formerly, iced fresh fish were shipped in wooden boxes by rail. In recent years, the decline in rail service coupled with a rise in consumer affluency has spurred the development of lighter and more economical shipping units. In order to ship fresh fish to markets with greater speed, air transport is required. Here a lightweight and protective container becomes a necessity. Air shipment is an emerging trend in fresh seafood distribution.

The rigid-when-wet corrugated carton is acquiring a share of the wooden box market. Although several variations exist, corrugated boxes may be printed, are cheaper than wooden boxes, are better insulators, and are significantly lighter in weight. Most cartons used are wax-coated and leak-proof. For further protection, inner insulation may be used. Slab polysytrene foam is effective in providing increased insulation. The choice between an insulated wax carton and one that is not is dependent on the season the shipment is made. In extremely cold weather, added foam insulation may not be necessary. The Bureau of Commercial Fisheries Technological Laboratory in Gloucester, Massachusetts, has developed a corrugated carton that will carry fresh fish successfully in all forms of transport with substantial savings. Insulation is used and very little ice is required.

Other types of cartons used for fresh seafood shipment include containers made of expanded polystyrene. They are very lightweight and offer excellent insulation. One French design consists of a container with a sliding lid. It offers a secure lid for the package and aids in stacking. Many other designs exist in European markets.

Courtesy of I.C.D. Chemical Industries, Ltd.

FIG. 6.1. NOVEL FISH PACKAGE
A polyester/PE laminate is used as a "boil-in-bag" for smoked kipper fillets.

Retail Packaging.—The most popular package for fresh refrigerated fish consists of a shallow tray and transparent film overwrap. If the product is large or irregular in shape, a direct film overwrap may be used. Tray packs are usually overwrapped with polyvinyl chloride. The tray may be fabricated from molded pulp, foam polystyrene or clear polystyrene. Foam and clear polystyrene require the use of absorbent blotters. Product identification may be achieved by means of a pressure sensitive label or a printed overwrap. In order to obtain improved rigidity for direct film overwrapped packages, a plastic or wax-coated paperboard insert is used. Occasionally, a vacuum pack is used to improve shelf-life. Higher barrier films are required such as PVDC-coated cellophanes, polyethylenes, polypropylenes or polyesters.

Courtesy of White Fish Authority

FIG. 6.2. RETAIL PACKAGING OF FRESH FISH
Film overwrapped trays and sealed film bags are used for prepackaged fresh fish fillets.

Processed Fish—Frozen

A great deal of fish is sold in the frozen food section of the supermarket in this country. In recent years, frozen batter fish portion sales have been on the increase. Wherever fish-and-chippers expand, business increases for breaded and battered items. About $660 million of frozen seafood was sold in 1977. Unprepared seafood amounted to $330 million, fish was $167 million and shellfish amounted to $163 million.

One problem with frozen fish, but one that does not produce deterioration or spoilage, is "drip." This is liquid that escapes from the fish during thawing. Drip can be serious, however, since it represents a loss in the natural fish juices. Less drip occurs when frozen fish is thawed if the fish is fresh. In tests comparing rate of thawing, it was found that very rapid or very slow thawing gave the poorest product. Moderately slow or moderately fast thawing were recommended for highest quality frozen fish.

Courtesy of Nixon-Baldwin Div., Tenneco Chemicals, Inc.

FIG. 6.3. RIGID PLASTIC CONTAINERS
A formed rigid vinyl tray is used to package frozen breaded fish sticks. A clear flexible film lid is heat sealed to the flange of the container.

The factors affecting the quality of frozen fish are: (1) moisture loss, (2) oxidation, (3) rancidity, (4) change in odor and flavor, (5) loss of volatile flavors, (6) enzymatic activity, and (7) loss of vitamins.

Glazes and Antioxidants.—For many years an ice glaze was used on frozen fish. This consisted of dipping the frozen fish in cold water and withdrawing them into air in a cold room. Thus, the adhering film of water froze and glazed the seafood. Actually, glazing was not too satisfactory since the glaze was quite brittle and cracked or chipped easily. In addition, it didn't last long, as it soon sublimed in dry atmospheres. Glazing also increased the weight of the fish thereby increasing transportation costs. When glazed fish thawed, the amount of drip increased and the fish flesh absorbed some of this moisture. This made frying of the fillet more difficult because excess moisture caused the grease to spatter. Ascorbic acid solution and ethyl gallate solution glazes have proved to be more beneficial than water glazes. However, these chemical glazes have not found wide acceptance commercially. At the present time, most fisheries are not using glazes on fillets as they believe this is not necessary.

Packaging Requirements.—Proper packaging for frozen fish is essential. Efficient packaging will help to offset the detrimental effects of oxygen and of desiccation by barring the entry of air.

The package in most widespread use is a polyethylene, wax or hot-melt coated carton with or without a waxed paper overwrap. There is a definite trend towards printed cartons throughout the frozen seafood industry. Although the industry has lagged behind the times in both marketing and package design, changes are now appearing.

Wax-coated cartons were the first packages used. Problems were encountered with wax flaking off and with dehydration through the score lines on the package. The arrival of polyethylene-coated paperboard offered good flexibility and better moisture protection. Although this was an improvement over wax cartons, problems still remained in heat sealing, ink adhesion and delamination when in contact with fish juices.

At present, most cartons are coated with petroleum wax-resin blends (hot melts). If the hot melt is used on the inside of the carton, a high gloss is left on the product surface. This offers pleasing aesthetics. The hot melts are easily heat-sealed. Cartons may incorporate tear strip openings or other convenience features.

Overwrapped cartons for seafood have substantially decreased in use. One large remaining area lies with foreign copackers. Over ⅔ of the United States fish supply is imported. Copackers produce many varieties of frozen fish in plain cartons. They then wrap the cartons in paper overwraps supplied by their customers. The overwrap may be coated with wax, hot-melts or polyethylene.

Frozen fish may also be found in overwrapped trays displayed in the frozen fish cabinet. Individual fillets are overwrapped with cellophane or PVC, with or without paperboard inserts.

Vacuum packaging is not generally done because of the added expenses of operation and handling. In some cases, frozen fish is vacuum-packed at a low temperature and frozen. The most suitable material must maintain an excellent oxygen barrier. Laminated structures such as cellophane-aluminum foil-polyethylene or polyester-PVDC-polyethylene are desirable materials. As determined by peroxide values, rancidity formation decreases drastically in vacuum packs employing low oxygen permeable laminates. Overall shelf-life is extended by almost 100%.

Since vacuum packaging does not remove all the oxygen from the fish, lipid oxidation still occurs, but at a much slower rate. Proper freezing storage is essential for all packaged fish, whether it be vacuum-packed or overwrapped with a plastic film.

An interesting frozen fish item is smoked salmon (lox) in tinplate or easy-open aluminum cans. Each layer of the sliced fish is separated by a parchment paper insert.

Courtesy of The Metal Box Co., Ltd.

FIG. 6.4. CARTONED FROZEN FISH
Cartons are used for frozen whiting fillets. These cartons feature an easy-opening end.

Courtesy of Vita Food Products, Inc.

FIG. 6.5. CANNED FROZEN FISH
Frozen "Lox" smoked salmon in metal cans are available in both tinplate and aluminum cans. The latter feature top tear-out panels in the lids.

Processed Fish—Canned

Salmon, tuna and sardines form the main fish products currently canned. Tuna is canned in an oil or water pack. Salmon is usually canned in oil. Sardines are either smoked prior to canning or put up in brine and with tomato sauce.

The final quality of canned fish is dependent on the condition of the raw fish, the methods used during canning and changes which may occur during storage after canning. Fish to be canned must be as fresh as possible. In tuna processing, a proper refrigeration cycle is mandatory prior to canning. Fresh fish does not necessarily imply high quality. If salmon is caught late in the season, the fish may be soft and have a very low oil content. Tuna sales were \$837 million in 1977. The catch was down from 269.4 million kg against 298.4 million in 1976. The United States relies on imported catch for 45–60% of its tuna.

Canned Tuna.—The choicest tuna species for canning is the albacore. Other species canned are the bluefin and yellowfin tuna. After the fish are caught they are bled and gutted, then washed with a water spray. Considerable care must be exercised to assure proper refrigeration.

In the factory, the fish are hung up and allowed to drain prior to cooking. The cooking process involves placing the tuna in pans which are in turn placed on portable racks. The racks are inserted into a steam chamber and exposed to 100°–102.2°C for 2½ to 4 hr. Longer times are needed for larger fish. This allows all the natural oils to drain and permits thorough cooking of the flesh. After standing for 1 day, the skin, head and bones are removed and dark meat is separated from the white meat.

The white meat is put on trays and cut to proper length for cans. The trays are then sent to the filling tables and the cans are filled. The main part of the fill consists of a thick piece of meat while smaller ends or bits are used to reach final weight. Olive oil, salt, soybean oil or water is added either before or after the meat is filled into the cans. There has been an increased acceptance of water-pack tunas recently due to dietary reasons. Also, a good upswing in albacore and white meat tunas are evident despite the high prices they carry in comparison to the industry's best selling chunk light.

The canned tuna is given a 10–12 min exhaust and sealed. Since the cans may become greasy through handling, they are subjected to a weak alkali bath prior to being processed. After processing, cans are cooled, labeled and packed in shippers.

The process time and temperature employed for canned tuna is dependent on the size of the can. A typical No. 1 tuna can process when initial temperature is 21.1°C would be 95 min at 115.5°C and 80 min at 121.1°C. The type of can used may vary; however, enameled tinplate cans are usually employed. A commonly used coating is called "seafood enamel," which is a phenolic type material. Aluminum cans are particularly good for tuna as they maintain fresh color, odor and flavor.

Canned tunafish is inspected by the voluntary inspection service of the industry as well as U.S. Food and Drug Administration (FDA) officials. Some states also regulate both quality of fish and canning conditions.

Errors in time or temperature are very rare in commercial canneries. However, constant vigilance and careful control are imperative. In the early 1960s, three women in Detroit ate tunafish-salad sandwiches. All became ill from botulism and two died. Investigation by health authorities at muncipal, state and federal levels traced the source to one particular cannery and the use of cans having defective seams, which permitted *Clostridium botulinum* organisms to enter the cans after sterilization, possibly from the cooling water. The cannery was quarantined and all products were located and withdrawn from the market. Despite the fact that the problem was identified and localized, the entire tuna canning industry was injured due to public loss of confidence. It is estimated the loss in sales was about $40 million.

Processed Fish—Smoked, Salted and Marinated

Smoking or smoke-curing is a method of preservation effected by a combination of drying and the deposition of naturally produced chemicals resulting from the thermal breakdown of wood. Salting often accompanies smoking or it may be used as a separate method of preservation. Marinated fish are treated with vinegar and salt. Spices and herbs are also often added. Most marinated varieties are sold in metal cans and glass jars. Examples include "Schmaltz herring," "Rollmops," "Lunch herring" and herring in cream sauces. Other types include herrings in dill, lemon and wine sauces. They are refrigerated during storage.

Curing is used to preserve the shelf-life of fish as well as to provide improved flavor. A suitable packaging material must prevent moisture loss or gain and be impermeable to organic volatile flavoring constituents. The odor and flavor of cured fish may be lost through package permeability. In addition, there is a risk of the intermingling of flavors between different products in a refrigerated cabinet.

Smoked Salmon.—The two basic types of salmon are the Atlantic and the Pacific, with five varieties of the latter found in west coast waters and a sixth in Asia. Atlantic salmon, whether from waters around Ireland, Scotland, Scandinavia, Nova Scotia, Newfoundland or the Gaspé Peninsula, is the more highly prized fish because of the more delicate flavor of its meat and the firmer, more satiny texture that enables it to be sliced to a translucent thinness. The best Atlantic salmon is taken off Ireland and Scotland, where fish feed in the plankton-rich waters, making their meat even firmer and more flavorful than those on this side of the Atlantic. Pacific salmon is used to make lox, salmon cured in a highly saturated salt brine. This soft, salty salmon is much favored on bagels with cream cheese, or soaked and then mixed into scrambled eggs with sauteed onions. Pacific salmon can be delicious and satisfactory as smoked fish, but it simply does not have the oily richness that gives the flavor and delicate texture of the Atlantic fish.

Courtesy of Vita Food Products, Inc.

FIG. 6.6. PICKLED FISH PRODUCTS IN GLASS
Marinated herring in various sauces is shown packed in glass jars with metal screw on caps.

A good deal of Pacific salmon is smoked by the prestigious London smokehouse, Forman & Sons, who smoke Chinook and coho salmon so well that it takes on some of the texture and flavor of the Atlantic product. In Mount Kisco, New York, Peter Heineman has also begun to smoke Alaskan salmon in the Scotch manner. The flavor is excellent, although the texture is heavier and stickier than the Scotch salmon formerly served there.

If the product is a Pacific fish, in Europe it is known as smoked salmon. It may be called Novy, Novie or Nova, but it may not be called Nova Scotia, according to the New York City Department of Consumer Affairs.

The terms "smoked Scotch," "Scottish salmon" and "Scotch smoked salmon" are all understood to mean salmon caught and smoked in Scotland.

Because Pacific salmon is more abundant and cheaper, it is often passed off as Atlantic or Eastern fish. There is very little authentic Eastern salmon being sold in New York City. Since many of the differences disappear or are obscured in the trimming and smoking of the fish, the most reliable guide is the shape and size of the side. Atlantic salmon are long and narrow and smaller and thinner than Pacific. The body is shaped much like that of a large trout, with a top that is almost straight and tubular, tapering only toward the tail. The meat tends to be paler than the Pacific salmon and has a much finer grain.

The varieties of Pacific salmon used for smoking are on the average larger than the Atlantic, and all are blimp-shaped, widening greatly toward the middle. They are very thick and meaty, and the meat will have a coarser, wider grain. It will be leaner and generally closer to deep coral color than the Atlantic fish. Salmon are best caught in salt water before they have spawned because their meat is then at its firmest, fattiest and most flavorful. Because bruises on the fish sometimes affect flavor and keeping qualities, care must be taken in how they are caught. The best methods are trolling with huge hook lines or trapping, but most Atlantic fish are caught in nets. Once caught, fish must be bled thoroughly before they are frozen or cured. Otherwise, unpleasant bloody veins remain around the belly and these also can impair flavor or keeping qualities and must be trimmed off.

Pacific fish are almost always given a wet cure, meaning they are placed in a salt brine (which may also contain sugar and other seasonings) to soak. Atlantic fish, are generally given a dry cure, meaning salt and sugar are rubbed into the flesh of the fish. All Nova Scotia, Gaspé and Eastern salmon found in New York markets are cured in this country, mostly in Brooklyn and Queens; no salmon smoked in or around Nova Scotia was found in retail stores in the city. Commercially smoked salmon, here and in Ireland and Scotland, is cold smoked and cured, as compared to hot smoked and cooked. The cold method results in a firmer fish that lasts much longer. In Ireland, peat is used for smoking, alone or in combination with hardwood, and the fish has a rougher burr to its flavor than does the Scotch, traditionally smoked over oak alone.

Smokehouses in the United States use hickory primarily, along with other hardwoods. The type of wood also flavors the salmon and so must be nonresinous. In Scotland and Ireland it is traditional to hang fish for smoking with a cord running through the shoulder of the side. This method was also used in the United States until about 25 years ago when the added labor cost of stringing the fish became prohibitive. Nova Scotia fish is therefore smoked on a rack, a practice now being followed to some extent in Ireland and Scotland. Fish smoked on a rack will have grid marks on the skin side instead of the string or the hole left by it. Hanging is considered superior because it enables the fish to dry out more thoroughly, thereby rendering its flesh that much firmer. But there is also more of a weight loss to the producer, as there is also between wet and dry cures. All such factors affect the price.

Once smoked and cooled and, preferably, allowed to rest for a few days, the fish is ready for shipping. Nova Scotia salmon is almost always sold fresh. Scotch and Irish salmon are almost all shipped vacuum-packed and sometimes frozen as well. The first condition is far less deleterious than the second. The vacuum packing used to prevent bacterial deterioration does make the meat ever so slightly softer, the degree depending on the length of time the fish is in the pack.

Frozen fish loses much more flavor and texture, again depending on how carefully it is handled. It suffers less if it is thawed slowly in the refrigerator after having been taken out of its vacuum pack; that way ice crystals melt slowly and the moisture evaporates instead of going back into the fish. Frozen fish will almost never slice as smoothly or as thinly as nonfrozen.

The taste of smoked fish varies according to fattiness or juiciness and leanness or dryness. The amount of oil varies not only from one fish to another but from the tail (dry) to the top (moist). The European taste is for the drier fish that allows more of the flavor to come through. Because it is thinner, the tail portion will also be saltier than the thicker part of the fish, especially if it has been hung.

The way the retailer trims the fish is also of considerable importance. The closer the trim, the more loss incurred; varying retail prices often reflect varying degrees of trimming. All fins and oily edges must be trimmed and then, with a thin-bladed sharp knife, the smoked crust should be pared off. Bones should then be removed with tweezers. The best trimmed sides will have no bits of vertebrae. This white beaded line prevents the fish from being sliced in crosswise slices. Larger sides of Atlantic salmon, i.e., 5½ to 7 lb (2.5 to 3.17 kg), are premium fish. Smaller sides cost the dealer and the customer less per lb or kg because they have not developed the same degree of oily richness, firmness and flavor.

Although most vacuum-packed, smoked salmon fillets are presently packed in clear plastic pouches, the use of aluminum foil pouches is exhibiting moderate market growth. Specialty Seafoods, Inc., uses a thin profile, three-ply laminate of aluminum foil, polyester film and polypropylene film (Reynolds Metals Company's "Flex-Can") for its products.

The packager begins with frozen salmon which have been cleaned and beheaded. Filleting, trimming and deboning are done at the plant before the fish is smoked over alder and maplewood. After smoking, the fish are trimmed again, checked for remaining bones and placed into 18.4 × 45.7 cm foil pouches, which are then vacuum-sealed. A heat process is used which renders the fish shelf stable. No refrigeration is required until the package is opened.

Courtesy of British Cellophane, Ltd.

FIG. 6.7. FLEXIBLE FILM POUCHES FOR SMOKED FISH
Smoked salmon packaged in a "Cellophane" laminate can be frozen or refrigerated.

Packaging.—Vacuum packaging can also be used for cured fish. It is extremely important to maintain proper storage conditions during refrigeration or freezing. In a fermented or lightly cured product, the shelf-life is extended; however, the process has little effect on *Cl. botulinum* formation.

Most cured fish, i.e., kippers, smoked whitefish, smoked haddock, are drier and more stable than fresh fish. They can be kept for several days completely unwrapped. For wholesale storage, they are packed in wooden boxes. Cured kippers are packed for consumer sale in overwraps of cellophane. Hygienic protection is obtained as well as proper brand identification. In addition, condensation within the package is reduced.

Smoked fish can be frozen and cold-stored, but the shelf-life at a specific temperature is not as good as unsmoked fish. In smoked fatty fish rancidity may develop in the frozen state. Proper storage conditions must be maintained at all times or there can be tragic consequences.

In 1963, a man and wife died from type E botulism caused by eating smoked whitefish bought in Michigan and a few weeks later four more people died in Knoxville, Tennessee, also after eating Michigan packed-smoked whitefish. Authorities traced the contaminated fish to inadequate smoking and subsequent mishandling. The fish were shipped in an unrefrigerated truck in summer and were routed first to Chicago where they sat for several days before going on to Knoxville. At no time was the shipment treated as perishable.

As a result several regulations have been adopted by the government agencies in the United States and Canada regarding the smoking process and inspection of plants. Canada prohibits the sale of smoked fish unless frozen. This incident all but destroyed the smoked fish industry, however, it did recover in time under new regulations.

Recent developments in cured fish packaging include the use of polyamide-polyethylene laminated film and polyester-polyethylene laminated film. Kippers in buttersauce or smoked salmon may be inserted in a boilable pouch, evacuated and frozen. Vacuum packaging prevents the pouch from floating to the surface during heating. High-density polyethylene film and polyester-polyethylene laminated film are being used. Smoked ciscoes have also been sold in polyester-polyethylene pouches.

Processed Fish—Sardines

The canning of sardines is an industry in itself which began in 1834 in Nantes, France, and spread to other countries including Spain, Portugal, Norway and Sweden. It was begun in this country at Eastport, Maine, in 1875 and with varying success, due to supply of fish, in California. The sardine is by definition a small herring and originally received its name from the island of Sardinia. In the United States the eastern sardine is smaller and is packed in flat, oblong ¼ (0.1), ½ (0.2) and 1 lb (0.45 kg) cans. The longer western (pilchard) sardine is packed in oval and round cans in ¼ and 1 lb units. In 1977, sardine sales amounted to $60 million. Imports accounted for ⅔ of the United States supply with Maine, Norway, Portugal and more recently Peru being the major source of both domestic and foreign sardines.

TABLE 6.1. CANNED FISH—1977 SUPERMARKET SALES

Fish	Volume (in Million $)
Tuna	837
Salmon	104
Sardines	60
Anchovies	6
Herring and kippers	5
Mackerel	6
Crab	26
Shrimp	40
Clams	31
Oysters	15
All others	12

Care is exerted in bulk handling of sardines in the fishing boats as damaged skins or broken fish are not acceptable for canning.

Preprocessing.—Fish are scaled, beheaded, cleaned and detailed. (Some very small varieties packed in other countries are left with heads and tails.) All wastes and rejects are saved and processed into fish meal. After rinsing, the fish are brined to remove slime, blood and water, to toughen the skin, to whiten the meat and to salt the fish. Larger fish require a more concentrated brine and a longer period of immersion.

The fish are dried in warm air (37.7°C) for up to 90 min in order to dehydrate them and toughen the skins further.

Processing.—*Oil Frying.*—Fish are placed in wire baskets and immersed in hot (110°C) cottonseed or sardine oil for several minutes, depending on size. Frying is complete if the backbone can be removed easily and shows no redness. After steaming or hot-water washing to remove excess oil, they are cooled, air-dried and packed. Quality depends on the quality and freshness of the frying oil.

Brine Frying.—Boiling saturated brine is used and the temperature is 104.4°–108.3°C. After 6 to 10 min fish are removed, sprayed with brine, cooled and dried. This method is seldom used.

Precooking.—Fish are placed in cans and subjected to steam, superheated steam, gas heat or a combination in order to drive out liquid and oil, which are drained away.

Packing.—Fish are placed in cans according to sizes and weights and the desired sauces or oils are added. These may consist of tomato sauce, mustard sauce, olive oil, cottonseed oil or soybean oil. Oils are usually added hot (104.4°C), but tomato sauce must be added at 65.5°C and mustard sauce cold.

TABLE 6.2. COMMONLY USED FISH NAMES

Fish	Common Names
Cod	Atlantic cod, grouper, black snapper, common cod, true cod, gray cod, sea bass
Sea bass	Common sea bass, rock bass, bluefish, blackwill, hannahill, humpback, giant sea bass, black sea bass, grouper bass
Bluefish	Tailor, fatback, snapping mackerel
Butterfish	Harvestfish, hardhead, tomcod
Red lake	Mud lake
Striped bass	Striper, greenhead, squidhound, rock bass, rockfish
Pollock	Boston bluefish, pollack, silver cod, green cod, green salmon, queddy salmon, sea salmon
Flounder	Gray sole, lemon sole, fluke, dab, sand dab, black-back (the correct name)

Vacuum Packing.—One other type of pack does not precook or fry the fish but instead after saucing, the can is closed in a vacuum chamber to an average can vacuum of 7.6 to 10 cm. The result is cheap and simple but a rather watery product.

Sterilization.—After closure, cans are retorted for various times and at various temperatures depending upon can size and shape, the size of the fish, sauce and initial product, and the temperature.

Processed Fish—Fish Meal

Fish meal is a product which has increased in importance annually. It is made by removing the liquid substances from fish and grinding the remainder. Present uses are feeds for poultry, pigs and cattle. Current legislation is pending in several nations for its use as a potential source of protein in human diets. Under these circumstances the product is thoroughly sterilized and ground to a dry powder known as fish flour. Fish flour does not spontaneously overheat (see below) because sterilization has destroyed bacteria.

Whole fish meal is a fatty and hygroscopic material and must be protected from oxygen and moisture. It is packed in air-tight containers such as multiwall paper sacks with an inner liner of polyethylene. Antioxidants such as BHT aned Ethoxyquin may be added in amounts of 0.45 and 0.11 k per MT, respectively. The liner serves as a moisture barrier and retards oxidation. It also reduces the possibility of spontaneous heating in transit. The problem of spontaneous heating is a serious one and often necessitates the use of metal-lined, airtight containers. Laminates of aluminum foil as an inner liner would appear to be extremely useful for bulk packaging. In the United Kingdom, the British Ministry of Transport requires that fish meal containing less than 6% or more than 12% moisture be packed in containers devoid of paper or burlap.

Overheating of fish meal does not occur when its moisture content is less than 6%, but when moisture rises to greater than 12%, bacterial action occurs and this gives rise to overheating. In packing fish meal, the bags must be cooled and properly handled prior to shipment.

SHELLFISH

General Marketing Concepts

Shellfish are very similar to fish in respect to their flesh and how it is processed, packaged and consumed. There are two groups of shellfish: the mollusks, which include oysters, clams, scallops, mussels and abalone; and the crustaceans which include shrimps, prawns, lobsters, crayfish and many species of crabs.

Most mollusks are taken alive and kept alive until just prior to their use. One exception is the scallop which has successfully been shucked and frozen.

A few oysters have been sold frozen but this is not prevalent practice. Crustaceans can be sold live or frozen and both methods are used extensively depending mostly on how close the market is to the fishing port. A few typical shellfish will be discussed in detail. Their characteristics are representive of what can be expected of similar seafoods.

Shrimp.—Some fresh shrimp are sold live at dockside. Most of the catch which is to be shipped to market is beheaded and iced at sea. A better quality product can be expected if the shrimp are frozen at sea. In 1977, $40 million of canned shrimp was sold in United States supermarkets. The U.S. set a record for shrimp catch in 1977, netting 216.4 million kg.

When shrimp have been shucked, graded and deveined, they are usually placed in waxed cartons and frozen. Some of these cartons have an inner liner to aid in protecting the product against oxidation and loss of moisture. Other cartons have holes in the bottom and after the shrimp are frozen, the cartons are immersed in cold water to glaze the shrimp. Or the carton may be opened and the shrimp glazed by spraying with water.

Glazing presents a problem of extra handling and also adds about 1½ lb (0.68 kg) extra weight to a 5 lb (2.268 kg) carton, thus increasing transportation costs. If a package with the advantage of easy filling, puncture proofness and high moisture-vapor proofness were used, it would eliminate the loss due to dehydration and would eliminate any need for the glazing process.

TABLE 6.3. GENERAL TYPES OF CAN COATINGS

Coating	Typical Uses	Type
Fruit enamel	Dark colored berries, cherries and other fruits requiring protection from metallic salts.	Oleoresinous
C-enamel	Corn, peas, and other sulfur-bearing products, including some sea foods.	Oleoresinous with suspended zinc oxide pigment.
Citrus enamel	Citrus products and concentrates.	Modified oleoresinous
Seafood enamel	Fish products and meat spreads.	Phenolic
Meat enamels	Meat and various specialty products.	Modified epons with aluminum pigment.
Milk enamel	Milk, eggs, and other dairy products.	Epons
Beverage can enamel (noncarbonated beverages)	Vegetable juices; red fruit juices; highly corrosive fruits; noncarbonated beverages.	Two-coat system with oleoresinous type base coat and vinyl top coat.
Beer can enamel	Beer and carbonated beverages.	Two-coat system with oleoresinous or polybutadiene type base coat and vinyl top coat.

Source: Joslyn and Heid (1963).

Experiments show that unglazed shrimp can be packaged successfully for 12 months storage at −17.7°C in a waxed carton with a satisfactory overwrap. Both glazed and unglazed shrimp from packages with no overwrap were slightly tough but had no off-flavor. Elimination of the glaze reduced the defrosting time of the 5 lb (2.268 kg) block of frozen shrimp from 36 to 55% (under running water) and also reduced the shipping weight of the box by 21%. Glazed shrimp packaged without an overwrap showed high weight losses, disappearance of the ice glaze and desiccation of the surface shrimp. The principal adverse factor resulting from dehydration appears to be texture change.

The development of individually quick-frozen shrimp has expanded sales at the retail level. IQF shrimp are packed in polyethylene bags of various sizes. An interesting development has been thermoformed linear polythylene containers covered with a heat-sealable lid. The use of the covering material allows for the printing and product identification. In order to prevent crushing and other types of damage, an outer paperboard carton is used.

Canning of Shrimp.—The large shrimp caught off the east coast, south of Savannah, and in the Gulf of Mexico are not only sold fresh and frozen but also are canned. The largest sizes are marketed as prawns. The shrimp are graded for size and are sold by the barrel which contains 200 lb (90.7 kg). First they are iced to make peeling easier. Peeling consists of

Courtesy of Nixon-Baldwin Div., Tenneco Chemicals, Inc.

FIG. 6.8. COMBINATION PACK FOR COOKED SHRIMP
Cooked individually quick-frozen shrimp packed in a thermoformed PVC/plastic container featuring an auxiliary inner package containing cocktail sauce. The inner pack is hermetically-sealed with a coated-film membrane lid.

removal of the head, thorax and shell leaving the muscular portion of the abdomen and tail. After a thorough washing they are placed in wire baskets and parboiled in hot brine for about 4 min. This changes the flesh to a firm white color and the surface to a bright pink. After cooling, inspection and removal of bits and fragments they are prepared for filling. Wet pack shrimp can be filled immediately. Dry pack shrimp are allowed to drain. A 1% brine is added to the wet pack. Care must be taken not to overfill, as this can mat the shrimp and injure their appearance. Weights of fill for standard cans are specified by the FDA. After exhausting and closing, cans are retorted. Retort processing varies from 12 to 30 min on wet packs depending on can size, initial temperature and process temperature (115.5° to 121.1°C.) Dry packs vary from 50 to 80 min. C-enamel cans are recommended. This enamel is an oleoresinous type containing zinc oxide pigment which counteracts the action of the sulfur in the food and prevents blackening.

Scallops

Scallops are shucked at sea and the scallop meats are washed and packed in ice. It is usually 7–8 days before the meat is unloaded and rapid processing is extremely important. If excessive dehydration occurs, shelf-life decreases and flavor and nutrient loss occurs. A moisture-proof durable package is mandatory for scallop packaging.

Most commonly used packages include hot-melt or polyethylene-coated cartons in both small retail sizes as well as the larger 5 lb (2.268 kg) institutional carton. In spite of the increasing use of packaging, most fresh scallops are still sold iced and are individually weighed into containers by the retailer.

Crabs

Because of the seasonal nature of crab production, the fresh product does not appear on the market in uniform quantities throughout the year. During the months when the crab catch is large, some method of preservation must be utilized. Since the fresh characteristics of crab meat are lost in freezing, this method of preservation is used for short storage periods. However, Blue Crab meat does not retain its original quality for any length of time when in a frozen storage; it loses its flaky texture, becomes rather tough and loses some of its flavor and color.

Crab meat wrapped in MSAT cellophane and stored at −17.7°C remains palatable for 9 months before becoming undesirably tough. Cooked crab legs that have been frozen, ice-glazed, packed in fiberboard cartons and stored at −17.7°C also remain palatable for 9 months. Crab meat packaged in hermetically sealed tin cans and stored at −17.7°C is edible after 12 months. Covering the meat with 1 or 3% salt solutions delays the onset of toughening but does not extend storage life beyond 12 months due to off-flavors developing. The two most important factors which lower the quality of frozen crab meat are changes in color and texture. Crabmeat that has been carelessly washed will show bluish discoloration after storage. Blue Crab meat that has been packaged loosely with many air spaces in the package is likely to turn yellowish. In order to minimize toughening and discoloration, the meat should be packaged tightly in a moisture-vapor proof container. The freezing and storage of raw crab legs is not recommended. After only 3 months the product becomes discolored and off-flavored.

Cooked Alaskan King Crab legs are marketed frozen in paperboard cartons sealed shut with hot melts and cellophane overwrapped.

Canning of Crabs.—The canning of crabmeat from Blue Crabs (and lobstermeat) encountered difficulties in the early days, as plain tinplate cans reacted with the hydrogen sulfide released by the fish to produce a black discoloration. Parchment liners helped the problem and, later, enameled cans all but eliminated it. Acid dipping of the shellfish prior to canning also is helpful. Canned King Crab amounted to 1977 sales of $12 million and all other crab products amounted to $14 million.

Very little crabmeat is canned in the United States because the crabs are smaller than Japanese crabs and the meat is so tornup in removal as to be much less attractive. Crabs must be processed rapidly. Live crabs are steamed for 20 to 25 min then the meat is picked out and washed in a weak salt solution. C-enameled cans are used. The Japanese use double-enameled cans and sometimes a parchment inner-liner in addition. Dipping in 1% citric acid solution and packing in acidified brine give best results in avoiding discoloration of the meat. A tight fill is necessary. After filling, cans are exhausted 10 to 15 min before closing. Retorting varies from 20 to 95 min depending upon can size and processing temperature (110°–121.1°C). Standard packs are 5¼ (148.8) and 7¾ oz (219.7 g) drained weight.

Oysters

Oysters are harvested by hand-tonging or dredging all along the East Coast, the Gulf Coast and the West Coast of the United States. The West Coast fields are rather new having been started when it was discovered Japanese oysters could be cultivated in those waters. Oyster sales were $15 million in 1977 (canned).

Freezing of oysters should offer wide opportunities for expanding the market for these shellfish. Although oysters are highly perishable and are produced in quantity only during the colder months of the year, only a small amount is frozen. Bags made of cellophane, pliofilm, nylon, polyethylene or other films, and papers give comparable results as far as retention of moisture-vapor. A sealed moisture-vapor proof bag within a waxed carton that is then overwrapped, would be an ideal package for this product. An overwrap would provide added protection against leakage and normal handling.

Fresh oysters are shucked into paperboard or thermoformed plastic (polystyrene or polyethylene) containers with friction or snap-on lids and sold refrigerated. The product has a limited shelf-life; once shucked, the oysters should be used promptly.

Shucked oysters may show a pink or red discolored liquor when they are thawed. "Pink Yeast" organisms are responsible and they are capable of growing at temperatures of −17.7°C or lower. Thorough washing and sanitary handling during preparation will prevent this occurrence.

Canning of Oysters.—Oysters for canning are thoroughly washed; dead oysters (open-shelled) or damaged oysters are discarded. The oysters are then loaded into a wheeled "steamer-car," which will hold 225 kg, and thoroughly washed again. The cars are passed through a steamer where the oysters are steamed 5 to 10 min at .7 kg/cm^2 pressure. This kills the oyster and opens the shell enough to permit removal of the meat. Oysters are cooled, shucked and inspected, discarding all that are discolored. The oysters are then conveyed to a washing tank where they again are washed to remove shell and grit. After draining for 15 to 30 min the meats are filled and weighed into cans. Slack weight must be avoided. Cans are topped with boiling brine, double-seamed and processed. Plain coke cans are used for oysters, although C-enameled cans may be used if desired. Retorting times vary with can size and processing temperature.

Some oysters are smoked and then packed in a sauce. These come packaged in cans or jars and are usually sold as a gourmet item.

Lobsters

Most lobsters are sold live from chilled tanks of seawater, and connoisseurs demand it that way. Only a small quantity are frozen due to their cost and the toughening of the meat during storage. In order to facilitate removal of the meat from the shell, the lobster is immersed in boiling water for an interval of 15 sec to 5 min depending upon the thickness of the shell. A heating period of 1½ min is sufficient for a 1 lb (0.45 kg) lobster. After this treatment, the lobster is cooled and frozen. Frozen, cooked lobster meat in sealed, waxed packages keeps less satisfactorily than whole precooked lobster. The storage life of the latter is reported to be 3½ months at −28.8°C.

The spiny lobster from Africa is used for frozen lobster tails. The tails are removed from the lobster, wrapped in nitrocellulose coated cellophane and packed in a waxed-board carton. The cartons are frozen 2½ hr in a plate freezer and stored.

Canning of Lobster.—Small lobsters are boiled in salt water for 20−30 min and then cooled. The meat is removed from the shell and packed in enameled cans using parchment paper liners. Three can sizes are standard 3–4 oz (85–113.4), 6–8 oz (170–226.8), and 12–16 oz (340.2–453.6 g).

Brine is added and the cans are exhausted to 60°C before closing. Retorting at 118.3°C varies from 35 to 60 min depending upon the can size.

An interesting package for lobster consists of an air transported unit sold directly by the lobster catcher to the consumer. A round metal container is used and seaweed forms the cushioning material for the lobsters. Steamer clams are added to the unit for use in a clambake. The lobsters are packed live by hand into the container with claws pegged so that they are not in a position to attack each other. Packaged ice is placed below and on top. The container serves as the refrigerated shipper and also as the cooking vessel. The outer package is double walled in order to insulate the contents. A shelf-life of 48 hr is obtained, which is enough for air transportation.

Lobster has also been vacuum packed in a polyamide-polyethylene laminate for ease in marketing.

Clams

Clams are treated essentially the same as oysters. United States sales of canned clams were $31 million in 1977.

In canning only C-enamel cans may be used as black discoloration is a common problem. Clam juice, which is collected during steaming of the clams, is also canned as a beverage, either straight or blended with tomato juice.

Other Fish Products

Fish Roe.—Shad or herring row are byproducts and can be sold fresh, frozen or canned. Fresh roe is packed in sealed polyethylene film bags and kept refrigerated. Frozen roe may be vacuum-packed before freezing. Polyethylene-coated polyester film or polyethylene-coated nylon film may be used for this purpose.

Some fresh roe is canned. East Coast roe are usually minced and canned with brine. West Coast canners preserve the shape of the roe and can it with brine, oil or sauce.

In the USSR, black caviar (beluga) is obtained while the fish is still alive. If the fish dies, the membrane of each roe sac breaks and all is lost. The fisherman slits the fish and carefully extracts the roe. As much as one and a half million eggs are wrapped in the thin membrane of the sac. It is also possible for one roe to weigh as much as 100 kg.

The roe is then given a cold shower and carefully put into a large wooden sieve. A caviar taster grades the lot for taste. Caviar from different fish are never mixed. Grades are based on color, taste, grain size and hardness. Each lot is packed in separate batches. In the sieve, the roe is mixed by hand. The membrane floats to the surface, while the eggs slip through into a wide container which holds exactly 12 kg. The caviar is spread out evenly since the eggs may be crushed by their own weight. The final step involves sprinkling with very fine dry salt (3% by weight).

Courtesy of Romanoff Caviar Co.

FIG. 6.9. CAVIAR BULK PACKAGES FOR EXPORT
Iranian workers at the Iranian fisheries examining 2-kilo tins that will be used to pack precious sturgeon roe.

Granular caviar is tightly packed in flat jars with pry-open tops. It is also packed in wooden barrels. Barrelled caviar is given an additional salt treatment (7 to 10 % by weight). The exact amount of salt depends on three factors: geographical position of the packing plant, season of the year and quality of the roe. Oak barrels are used and they hold about 13 gal. (42.2 liters). They are given a thin coating of parafin wax on the inside and brushed with drying oil on the outside. Granular caviar is also usually pasteurized, which eliminates the need for antiseptics and increases shelf-life.

Pressed caviar is prepared by immersing the roe in a hot saturated brine for 2 min. It is then scooped out with a sieve into linen sacks and squeezed under a press. The caviar turns into a dough-like mass. After this operation, it is packed into oak barrels lined inside with white fabric and wax paper. It is also packed in 1.6 kg cans.

Roe caviar is prepared from roe with immature or weak eggs unsuitable for granular or pressed caviar. It is inferior to the others in taste. Packaging is usually done in flat glass jars.

Whale Meat.—Whales, dolphins and porpoises are the only mammals that live entirely in the water. Most whales are captured for their oil, but whale meat is consumed abroad particularly in Japan, the USSR and Scandinavia. Whale meat is very similar to that of other red-meated animals except it contains tiny amounts of trimethylamine oxide and methylamines, which give it a slightly fishy flavor. It tends also to be somewhat oily. The fat is very high in unsaturates and hence readily subject to rancidification. The meat can be chilled, canned or frozen, and packaging which would be suitable for fish will be suitable for whale. Whale meat is ground into a meal rather than being processed as steaks or roasts. Whale meat consists of approximately 70% moisture, 4% fat and the remainder carbohydrate and protein.

A comparison was made of the protection against dehydration and oxidation of products frozen in various sheetings. It was concluded that PVDC film was superior to polyethylene film, foil bonded to Kraft paper, and greaseproof paper. After 6 months, rancidity had developed even in the PVDC wrapped meat.

PREPARED FISH FOODS

A wide variety of fish and shellfish foods have been prepared, cooked and are usually sold frozen. Examples include breaded fish sticks, scallops, clams or oysters; lobster, crab and shrimp newburg; lobster thermidor; cod fish cakes; tunafish pot pies; the fish fillet TV dinner and others. Breaded items are packaged in waxed-paper overwrapped cartons with parchment liners or separations. Those which contain sauces are packaged in sealed plastic bags which can be put in boiling water to reconsititute. The pouches are usually polyethylene-coated polyester film and are overpackaged in a printed paperboard carton. Pot pies and TV

Courtesy of Vita Food Products, Inc.

FIG. 6.10. HERRING SALAD IN ALUMINUM CONTAINERS
Herring salad spreads are marketed in formed aluminum tubs with pull tab easy-open feature.

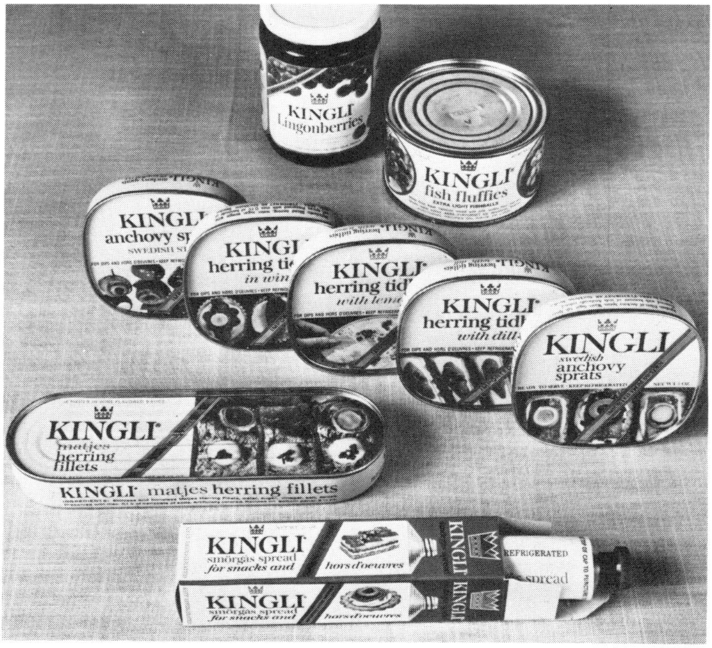

Courtesy of Vita Food Products, Inc.

FIG. 6.11. FISH HORS D'OEUVRES
This picture shows several types of packages used for hos d'oeuvres fish products. Note particularly the collapsible metal tube used for the fish spread.

dinners are packaged in foil pans and trays, which are also enclosed in printed paperboard cartons.

The collapsible metal tube has been used for anchovy spreads as well as salmon and other types of patés. It is an excellent package useful for hors'd'oeuvres preparation.

Gefilte fish is packed in glass jars and metal cans. It is a combination of two or three kinds of freshwater fish minced together with onions, carrots, matzo meal and spices. Combinations of fishes used include haddock, cod, whiting and halibut.

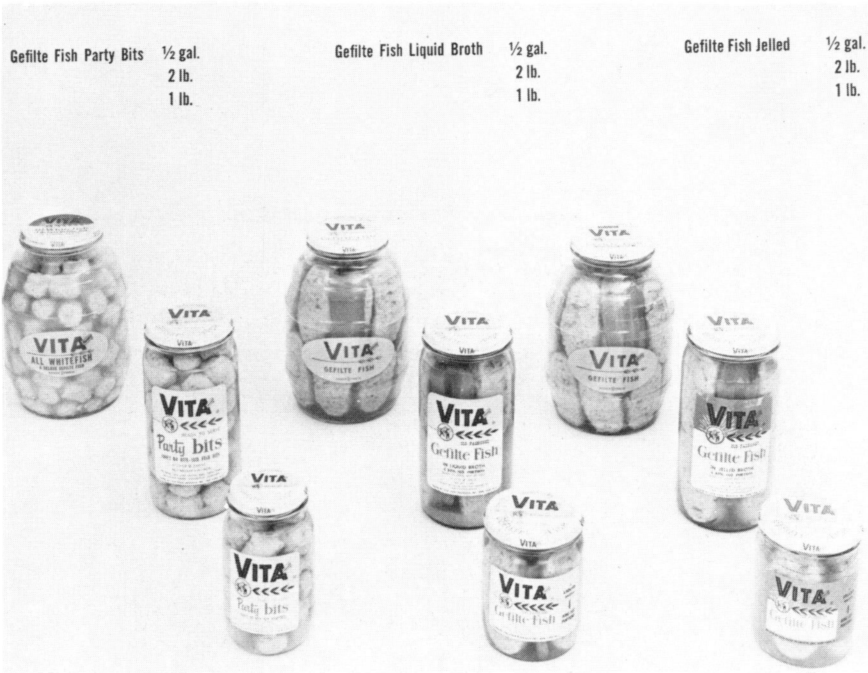

Courtesy of Vita Food Products, Inc.

FIG. 6.12. COOKED FISH PRODUCT IN JARS
Different shaped glass jars used for packing several varieties of gefilte fish.

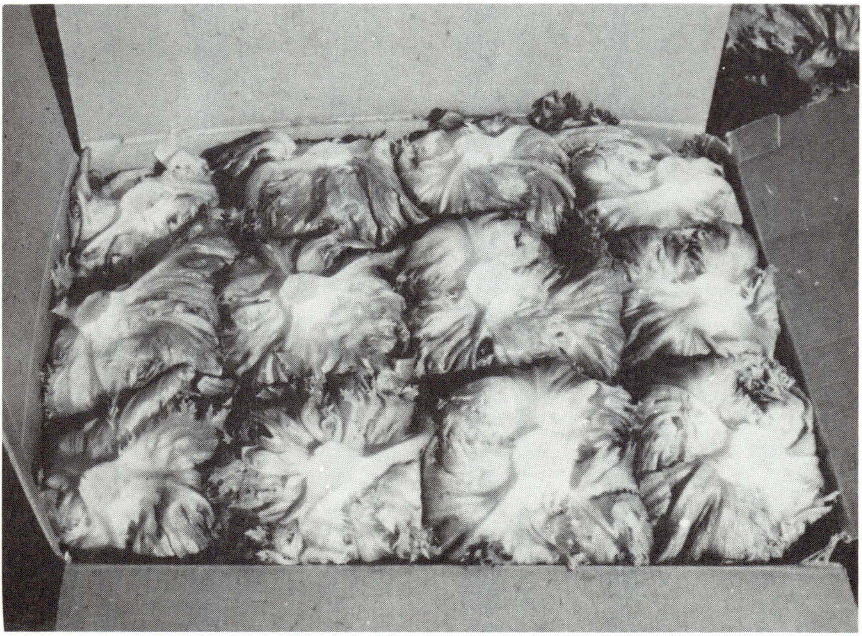

7

Fruits and Vegetables

The apple in the Garden of Eden was not an apple, it was a banana.

The Koran

Fruits and vegetables originated in many different parts of the globe and for thousands of years accompanied man's travels. The earliest record of the grape and olive show them to have been cultivated in Transcaucasia about 6000 B.C. Lentils, cucumbers and figs were Stone Age foods from the Himalayas and Mesopotamia. Cabbage, broad beans and spinach were raised in the Bronze Age in the Mediterranean area, Asia and Afghanistan.

The Myceneans, Greeks and Romans spoke highly of the quality of Persian melons and peaches, Russian plums and cherries, Arabian asparagus and Syrian eggplant.

An interesting 19th century lithograph by Bouchot is composed of two figures madeup entirely of vegetables and set in glowing colors.

FRESH FRUITS AND VEGETABLES

The market for fresh produce is large and varied. Even the most popular items number well over 100. It is therefore impossible to discuss all types in this book. Rather, it will be necessary to discuss general categories with a few examples, seeking, where possible, a common denominator.

In 1977, $9.51 million worth of fresh produce was sold for direct consumption in the United States. This does not include exported goods, which were processed into other foods, or those which were preserved by freezing, canning or the like. Of this total about ⅔ were vegetables and ⅓ were fruits. Yet, per capita consumption has shown a downward trend. Less than 50% of all fruit sold in the United States is fresh. This is down from over 87% prior to World War I.

This is attributed to the industry's failure to make fresh produce competitive with canned and frozen products in terms of convenience and to find ways of preserving fresh quality long enough to satisfy the consumer. Other experts say the decline in consumption of fresh fruits and vegetables is due to the increase in working women who do not have as much time for food preparation and therefore demand convenience foods. During this same period advertising for canned and frozen foods was tremendously increased, and the money spent on research and development by food processors more than doubled. Fresh produce companies could not compete either with advertising or research.

Product Characteristics

Because fruits and vegetables are living organisms even after harvesting, they can remain fresh only as long as normal metabolism continues. Metabolism involves the absorption of oxygen which breaks down the carbohydrates in the product to water and carbon dioxide. If the availability of oxygen is restricted, the chemical reaction changes and small quantities of alcohol are produced. This results in off-odors and flavors and a breakdown of plant cells, a series of events called "anaerobic decay" which can spoil fruits or vegetables within a few hours.

Fruits and vegetables are very high in moisture, ranging from 75–95%. Their equilibrium humidities are as high as 98%. Under any normal atmospheric condition they will dry rapidly. This causes wilting and shriveling due to loss of rigidity and shrinkage of the cells. Proper packaging is able to prolong the storage life of fresh fruits and vegetables by preventing wilting. Serious wilting occurs if the loss of moisture exceeds 10% of the product weight. The rate of moisture loss varies with the product and water vapor permeability of the packaging film. The use of small perforations for oxygen permeation has an insignificant effect upon moisture loss.

A very prevalent type of spoilage of fruits and vegetables is that caused by microorganisms such as yeasts, molds and bacteria. These organisms can cause destruction by growing on the exterior of the product or they may invade the interior through a surface bruise or cut and cause internal decay. This is why careful handling and packaging is so important in the preservation of freshness and quality.

Normal ripening of fruits and vegetables causes alterations in color, texture, odor and flavor. At some point, for each type, ideal ripeness is achieved. Beyond that point the product becomes overripe and its quality deteriorates. The primary goal of fresh-produce merchandising is to deliver the product to the consumer at such a point in the ripening scale that it will achieve perfect ripeness at time of eating. This, of course, is extremely difficult to do. In practice the produce is delivered somewhat underripe at time of purchase and the consumer delays consumption until ripening is complete.

Since all of these processes are highly sensitive to temperature, they can be slowed by storing the produce under refrigeration. Each fruit or vegetable has an ideal temperature for storage. If this temperature is not used, deleterious results occur; for example, tomatoes will not ripen if chilled below 4.4°C, bananas will turn black below 10°C and potatoes develop a sweet flavor below 5°C.

An additional reason for the necessity of refrigeration for fresh produce is the heat generated due to metabolism or respiration. For example, green beans, sweet corn, broccoli, green peas, spinach and strawberries generate from 15,000 to 50,000 Btu per MT per 24 hr of storage at 15.5°C. Even when chilled to 0° to 4.4°C they still evolve from 2500–17,000 Btu per MT per 24 hr. This heat must be taken into consideration in the design of refrigeration equipment.

Some types of fresh produce give off volatile compounds during ripening which will impart unacceptable odor and flavor if not allowed to escape, or they may prematurely ripen the fruit.

Packaging Requirements

For the most part, packaging cannot delay or prevent fresh fruits and vegetables from spoiling. Incorrect packaging can accelerate spoilage. However, packaging can serve to protect against contamination, damage and excess moisture loss. Too much of a moisture barrier will cause an excessively high relative humidity in the package and result in accelerated spoilage due to microorganisms or in skin splitting on some fruits.

Prepackaging of Produce

The packaging of produce is not new, but there is little recorded about it. In very early times most communities raised and consumed their own and had no need to package it. Later in times of plenty, surpluses were taken to markets where they were vended. Transportation was by boat, horse-drawn cart or by similar means. Little or no protection was given, perhaps only a covering of sorts such as a goatskin in early times, a mat or cloth or tarpaulin in later periods. Any convenient box, barrel, basket or sack served as a container for transporting smaller loads. Fresh produce market scenes were featured in many engravings of the 15–18th centuries. The Dutch engravers, Bloemaert, Vander Borcht and Van De Velde published vivid scenes of their contemporary markets. Even later, an etching by Whistler, "T.A. Nash's Fruit Shop," illustrated produce being sold in Chelsea, London, *c* 1885. Obviously, none of the items were prepackaged for sale. During this period, Van Gogh painted his "The Potato Peeler" (now in the Metropolitan Museum of Art). This oil illustrates how important the potato was to the diet of the day. The contemporary Korean artist, Po Kim, is known for his studies of fruit and vegetables featuring form and substance.

Courtesy of French Government Tourist Office

FIG. 7.1. A SCENE FROM LES HALLES PRODUCE MARKET IN PARIS
Fresh produce traditionally was shipped to market in wooden crates and baskets.

Not until 1856 was any attempt made to give additional protection to a fresh fruit. In that year paper was used to wrap and package fruit. In

1879 oranges that were shipped from Australia to England were wrapped in paper; by 1880 the practice was commonplace. Yet, wrappers were not used for vegetables until about 1900.

In 1895 United States apple growers began to ship boxes of paper-wrapped apples. Over the years they built a reputation for quality fruit, free of scald and bruises. In 1907 Samuel Frazer sold apples in New York City in paperboard boxes containing 6 or 12 units. In 1910, a New York cooperative tried prepackaging potatoes in 10-lb bags and half-peck containers, but the public was not ready. Not until the late 1920s and early 1930s did some prepackaging of potatoes, onions and oranges by shippers and retailers meet with any success. In the late 1930s and early 1940s special terminal prepackaging plants began to prepackage tomatoes, kale, salad mixes, spinach and soup mixes, and growers began to package mushrooms and cranberries. During the 1950s, the trend continued with apples, celery and other citrus fruits joining the group. But general acceptance by produce vendors was slow. They could see no good reason for investing in equipment in the face of slow consumer acceptance. They wondered about the real cost of the operation.

With considerable foresight, in 1944 a group of companies pooled their interests and sponsored a project to evaluate prepackaging of all items in a produce station. This became known as the Columbus Experiment. Merchandise was brought to the warehouse in conventional containers and there it was washed, trimmed and repackaged in consumer units. Moisture-proof cellophane was used as wrapping material and each unit was labeled, coded and price-marked. All produce was kept refrigerated from start to finish. Results were conclusive. Prepackaging together with refrigeration was proven to reduce distribution losses by as much as half, to save labor costs, to lengthen shelf-life and to achieve good consumer acceptance.

A typical market was found to lose as much as 30% in conventional in-store handling, trimming and reconditioning of produce, whereas prepackaged produce had negligible losses. Very little markdown or discarding was needed.

Shelf-life of some products was greatly improved. For example, prepackaged spinach was 90% saleable when bulk spinach had totally spoiled. Prepackaged cauliflower was 100% saleable after 5 days, whereas ordinary cauliflower without refrigeration was 50% unsaleable in the same period.

Carrots, radishes, parsnips, beets, turnips and all root crops were extensively prepackaged in the 1950s, some at shipping points and some at specialized packaging plants (terminals). In the early 1960s bananas, pears, asparagus, grapes, cauliflower, more apples, more citrus, snap beans, celery and early potatoes were prepackaged at the shipping point.

Cut-up squash and parsley were packaged at terminals, and soft fruits in the retail stores.

From 1960 to 1970 great strides were made in developing new and improved packaging materials and machinery for their application. Film manufacturers began to produce different kinds of films aimed at providing necessary combinations of strength, heat sealability, heat shrinkability, moisture retention, gas permeability, clarity, breathability, stiffness, durability, fogging resistance, water resistance and printability. The goal was to extend the shelf-life of fruits and vegetables and also to offer the consumer much more effective and attractive packages. By selection of materials and by redesign, containers have been developed which reduce injury and bruising during shipment and display.

Much research has been conducted on the use of controlled atmospheres in storage and transportation of produce to delay deterioration on long shipments in specially equipped trucks, railroad cars and even ships. One method used was storage of apples and pears in boxes sealed within polyethylene liners. When held at low temperature the respiration of the fruit reduced the oxygen and increased the carbon dioxide in the container thereby slowing the ripening of the fruit by as much as several weeks. On removal from storage, bags must be perforated to avoid anaerobic decay and off-flavor development at ambient temperatures.

Recent developments include the use of hypobaric storage (packaging at reduced pressure or partial vacuum) to prolong freshness between harvest and retail display, and the process that won General Mills the 1980 IFT industrial achievement award. General Mills grows leafy produce hydroponically under artificial light. The produce—such as lettuce—is packaged root and all in a sealed and air inflated polyethylene bag. Moisture lost by the respiration of the leaves is taken up and reabsorbed by the roots. Losses from wilt, crushing, and decay are dramatically reduced as also costs of cleaning and trimming.

General Advantages of Prepackaging

Permits brand identification, advertising, price and product information, and recipes.
Reduces waste throughout the marketing system.
Reduces transportation and handling charges.
Reduces retail labor costs.
Increases quantity sold per customer.
Speeds up sales in stores.
Increases maintenance quality and shelf-life of produce.
Fits in with trend toward unitization and self-service retailing.
Makes a cleaner, neater and more attractive display.

More economical and convenient handling.
Provides a service and becomes more of a convenience item.
Decreases possibility of accidents in produce area, thus lowering insurance rates.
Decreases spoilage losses and markdowns.
Impulse buying is stimulated because packaged produce is attractive.
Product sanitation is improved because contact with dirt, insects and consumer handling is prevented by the package.
Pilfering of small produce is minimized.
Fresh appearance is maintained longer.
Reduces possibilities of errors in weighing and pricing at retail level.
Permits customers, regardless of the time of day, to select produce which has not been "picked over" and handled.
Permits an increase in variety of produce displayed in smaller areas being allotted to the produce department.
More convenient storage of consumer packages in the home.
Provides for easier inventory control and proper stock rotation.

General Disadvantages of Prepackaging

Cost of prepackaging in relation to cost of product may be high.
A general lack of knowledge as to what, how and where to prepackage.
Resistance to change by wholesalers, retailers, growers, shippers and consumers.
Requires high standards and uniformity.
Quality control difficult on perishables.
Not all produce is adaptable to prepackaging.
Requires considerable investment.
Consumers like to pick out own merchandise.
Difficult to package in sufficient sizes for all families.
Deterioration of product causes loss of merchandise as well as packaging costs.
Increased refrigeration requirements throughout market system.

Packaging of product is currently being done at three primary levels: shipping point or source level in the production region, terminal packagers or central warehouses at the distribution level and in-store or retail level. Each level at which packaging is done has some advantages and disadvantages and it is the general consensus that packaging will continue to be done at all levels for the foreseeable future.

Source Level Packaging Advantages

Saves transportation charges on waste discarded before shipping.

Allows brand identification and national advertising.

Cheaper labor and property costs.

Can standardize and grade for specific standards.

Better control over product and how it reaches consumer.

Large scale permits mechanization and improved skills in product packaging.

Producer gets larger share of consumer's dollar.

Producer-shipper becomes more familiar with consumer demands and can better adjust to the market.

Less bruising of product in transit.

Number of times that produce is handled is reduced.

Fewer commodities handled should permit greater specialization leading to more efficient operations at source level.

Facilitates the adjustment of supplies to demand because of greater choice of alternate markets.

Permits packer to grade and sort to various standards and to take advantage of area preferences for particular grades of merchandise and size and type of packages.

TABLE 7.1. MOST POPULAR CANNED FRUITS

Fruit	Number of Cases[1] (000 omitted)	Cumulative %
Peaches	32,764	32.8
Apples and applesauce	16,737	49.5
Pineapples	14,982	64.5
Mixed and fruit cocktail	13,726	78.2
Pears	5,633	83.8
Apricots	4,051	87.8
Cranberry sauce	3,307	91.1
Olives	2,451	93.5
Citrus fruit	2,103	95.6
Plums	1,314	96.9
Cherries (sour)	946	97.8
Cherries (sweet)	503	98.3
Spiced peaches	436	98.7
Figs	395	99.1
Blueberries	296	99.4
Blackberries	239	99.6
Raspberries	107	99.7
Other berries	129	99.8
All others	—	—
	100,119	

[1]Case of 24 size 2½ cans (holds 26 oz approx).

Disadvantages

Seasonal marketing may preclude use of machinery and facilities or seriously limit the efficiency.

Some deterioration will occur in the marketing channels due to the perishability of the produce.

Additional investment and costs which prepacker must pass on to purchaser.

Distribution Level Packaging Advantages

Greater efficiency, thus lower costs due to larger volume, mechanization, specialization, skilled personnel and year round operation.

Can maintain a more complete line and achieve a satisfactory selection of quality and price by buying from different parts of the country for various commodities.

Closer to retail market than grower-shipper. Can adjust to changes.

Reduction in investment for equipment and materials at retail and source level.

Quality control improved over source packaging since merchandise is packaged hours rather than days before reaching retail outlets.

Permits more careful grading for quality and ripening than source packaging, and package sizes can be varied to meet local consumer demands.

Can justify considerable investment in machinery and labor saving devices due to large volumes handled, and relatively long packaging seasons.

Disadvantages

Higher rent than at source and space usually at a premium. Union labor costs are higher, also fringe benefits, than at source level.

Transportation costs are higher since nonusable parts of produce are shipped from source.

Cost of waste disposal is relatively high.

Retail Level Advantages

Can package better for store's clientele since preferences can be determined easier.

Operations can be changed to meet demands.

Can price commodities when packaged. This is difficult to accomplish consistently at other points.

Make more efficient use of labor by smoothing out labor peaks during the week.

Quality control can be better established and maintained since inspection and trimming is done just prior to sale.
Less chance of spoiled merchandise being put on shelves.

Disadvantages

Small-scale operation cannot justify expenditure for full mechanization.
Labor costs are very high—additional personnel required.
Space is at a premium.
Personnel training and knowledge insufficient for wide variety of products needed.
Additional investment for equipment, materials and space.
Time required for packaging to permit adequate stockage may not be available.
Packaging operations are basically a hand operation, on a small scale, so packaging costs are relatively high.

A recent trend in produce marketing has been the use of more opened shipper cases in supermarket displays. This is to stimulate the concept of fresh items being sold in a nonpackaged state.

Packages and Packaging Materials

Bulk Shippers.—In the early days of shipping produce barrels were used and the fruit at "the bottom of the barrel" left much to be desired. Also due to the close packing and heavy shaking occasioned during transport, unless the fruit was resorted on delivery, rapid spoilage could be expected, for everyone knew it took only "one rotten apple to spoil the barrel." Careful studies over the years have proven the virtue of specially designed bulk shippers. For example, potatoes are shipped in corrugated boxes, and apples and similar fruit are shipped in molded pulp trays. New deep-pocket trays are proving to be even more effective than the former shallow cups. Easily bruised fruits are protected by using foamed polystyrene trays, shrink films or paper wrappers to prevent their jostling against one another. Round baskets were found to be wasteful of shipping space and have given way to rectangular boxes. Jumble packing has been studied and found to reduce bruising by as much as 50% over place-packing or careful facing of fruit. Jumble packing has also proven to save money due to reduced labor costs, product losses or more efficient usage of shipping space.

Pretrimming of produce in the field also saves on shipping costs. Trimming carrot tops and cauliflower jackets in the fields saved almost half

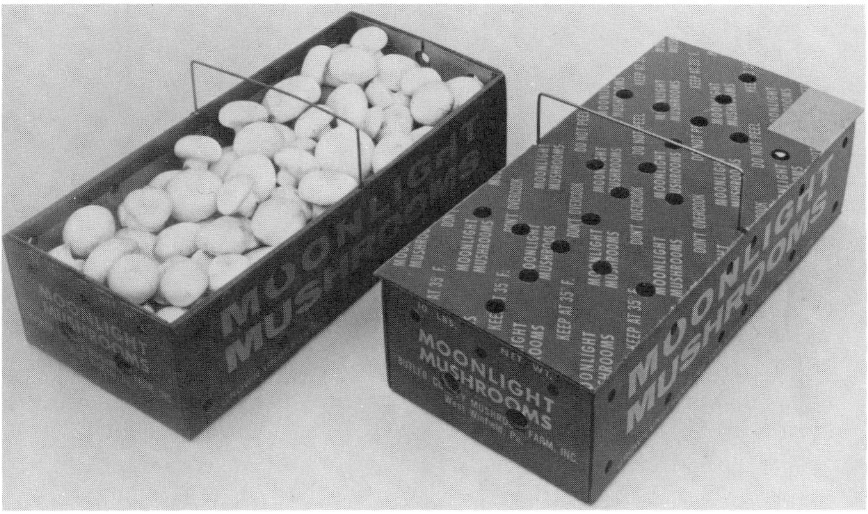

Courtesy of St. Regis Paper Co.

FIG. 7.2. SPECIALLY DESIGNED MUSHROOM SHIPPER
This corrugated shipper eliminates splintering and puncturing of the product—a frequent occurrence with previously used wooden shippers. Product brand name can be attractively displayed, and the handle pushes down to facilitate stacking.

the space and required far less icing to deliver these products to the market. The savings amounts to $4.5 million per year. There is a great potential for similar savings in other products such as celery and lettuce. Over $6.5 million could be saved on these two products alone.

Air shipment of produce is becoming more and more frequent. In 1977, 27.2 million kg of California produce was air-shipped. It is not unreasonable to expect air-shipped produce will amount to 30% of the total shipments by 1981. Special light-weight shippers are being developed which utilize fiberboard and foamed polystyrene for strength, weight-saving, protection and insulation, thereby saving on amount of ice required.

With the increased use of mechanical harvesting equipment, more and more bulk packing will be jumble-packed in large capacity bins which can be palletized and shipped to terminals for prepackaging. Whether these bins will be throwaway shippers, reusable metal or plastic containers remains to be established. Standardization of sizes may ultimately be expected, but lack of standardization on dimensions of transport equipment is delaying this desirable goal. There is no question that it will result in substantial savings in warehousing, inventories and handling costs.

Courtesy of E.I. du Pont de Nemours and Co.

FIG. 7.3. CORRUGATED SHIPPERS COATED WITH PLASTIC FOR MOISTURE RESISTANCE

Containerization may also lead to the elimination of damage due to extra handling. Ideally, produce will be loaded into shipping containers and these into van containers and then not unloaded until arrival at the receiver's warehouse. One eastern apple-shipper made three van container shipments to Great Britain and one to Panama with reports that the apples arrived in excellent condition. Again, more standardization will be required before this can be adopted on a wide-scale.

Retail Packages.—The materials used for retail packaging of produce are extremely varied, however, they can be categorized as films, backings, boxes, bags, bands, closures and labels.

Films.—The first study involving the use of plastic films for fresh produce was conducted by Platenius. He based his ideas on the observation that the changeover from normal respiration to anaerobic fermentation occurs when the oxygen concentration in the atmosphere has fallen below 3%. He believed 5% to be the minimum safe permissable concentration for oxygen in the atmosphere and then determined the rates at which oxygen was utilized by other commodities at this level of concentration. A wide variation existed and he found that no existing film possessed enough oxygen permeability in a completely sealed package. Thus, he formulated the concept of "vents." He demonstrated that two small holes, each 0.38 cm in diameter, in a package were sufficient to prevent anaerobic fermentation under all conditions.

Films are purchased in the form of sheets, rolls, bags and sleeves, and may be nonshrink or shrinkable types. In the nonshrink category coated cellophanes were probably the most widely used until the early 1970s but have given way to plasticized PVC at the in-store level. They come in a wide assortment of sizes and may have a coating on one or both sides of nitrocellulose or polyvinylidene chloride to provide moisture-proofness, heat sealability and water resistance. Because cellophane is only semipermeable to oxygen, it must be perforated in order to allow normal respiration. A few small holes the size of a pencil point are usually sufficient. Some stores have a roller studded with sharp needles which can be pushed across a stack of film sheets to perforate. Of course if the package uses the film only as a cover or a sleeve, leaving openings elsewhere for respiration, perforating is not necessary. There are two major United States suppliers of cellophanes each offering many types. It is also possible to purchase imported cellophanes and to purchase special laminations from converters.

Polyvinyl chloride stretch films have excellent clarity, are nonfogging and when punctures do occur, the film will not rip, run or tear. It is an excellent film for in-store wrapping and is the most widely used material today at the in-store level.

Polyethylene is available as a nonshrink film from many suppliers. It is usually low density and easily heat-sealed. It is not as transparent as cellophane, however, this property has been improved and research continues. Polyethylene is the most widely used film for bagging applications and excellent for firm products such as oranges, potatoes, onions and carrots. It must be perforated for those products requiring moderate to high respiration. Polyethylene sleeves are used for celery and small bags are used for lettuce. It is readily heat-sealable, quite strong and tear-resistant.

Cellulose acetate film is highly transparent and sparkly, making an attractive package. It is relatively high in permeability to oxygen and carbon dioxide. Consequently perforations are not needed on produce having low respiratory rates. It has been used as an overwrap for tomatoes in cartons and may also be used as the window in window cartons.

Polystyrene film is similar to cellulose acetate in that it is clear, crisp and has high permeability. It is also used extensively for lettuce and tomatoes, and is also available as a heat shrinkable film.

Perforated polypropylene is being used by a number of growers as contour wrap on produce such as lettuce, cabbage and cauliflower.

Other heat-shrinkable films are used which have been mentioned before in other chapters. These include irradiated polyethylene, polyvinyl chloride, polyvinylidene chloride and polypropylene. All find considerable general use in shrink wrapping of fresh produce and preferably are used as sleeves or complete overwraps. In the latter instance, perforations are needed.

Backings.—This term is used to embrace a wide variety of tray type containers used to hold produce in conjunction with film overwraps or sleeves. They are made from molded pulp, formed paperboard, formed plastic, molded plastic foam, laminations and aluminum foil. Each is designed for a particular item and quantity. One may hold four round fruits such as tomatoes or oranges. Another may hold two ears of corn; still another six ears of corn or five or six large baking potatoes. Some provide maximum view of the contents, others provide maximum protection against handling damage.

Boxes.—In this category are included wooden, pulp and plastic "tills" for small fruits like berries and cherry tomatoes, window cartons, lidded boxes and cups. Window cartons offer more opportunity for brand identification and decoration, but the traditional till still hangs on. Pulpboard has taken over from wood, and injection-molded plastic tills are becoming popular.

Bags.—This category includes all bags (other than plastic film) and includes regular kraft paper sacks, double-walled paper bags, window bags, fiber net bags, plastic net bags, paper mesh and plastic mesh bags. The mesh and net bags have long been used in packaging potatoes, onions and some citrus fruits. A plastic mesh netting made from high density polyethylene has found an excellent application in field-bagging of grapes and citrus fruits. Prepackaged grapes confined by the netting showed reduced losses of 1½ (0.68) to 2½ lb (1.134 kg) per standard bag. Grapes are easily visible and can be washed or stored right in the bag.

Bands.—A number of different materials are used for banding produce items, including simple elastic rubber bands, printed paper bands that can be heat-sealed or otherwise locked together.

Closures.—Twist ties are the most popular method of closing bagged goods, but cellophane adhesive and stapling are also widely used.

Specific Applications

The particular type of package used depends upon the shape and perishability of the product. There are five main classifications: soft

Courtesy of American Viscose Div., FMC Corp.

FIG. 7.4. SOFT PVC FILM STRETCH WRAP
Another type of PVC film packaging for produce is illustrated by American Viscose's PVC P-21, a soft vinyl film which is stretched tight around the product.

fruit, hard fruit, stem products, root vegetables and green vegetables. Soft fruits are highly perishable and easily subject to anaerobic spoilage. They bruise and squash easily which leads to rotting. They are best packaged in semirigid containers with a cover of cellophane, cellulose acetate, polystyrene or other suitable film cover. Adequate ventilation is a must to avoid fogging. Handling must be careful and avoided as much as possible. Shelf-life is limited due to individual damage and decay.

GCMI ITEM NO.	CAP. O'FLOW FL.OZS.	WT. MAX. OZS.	A	B MAX.	D	E	F	SPECIMEN FINISH
15-46	14 3/4	6 5/8	4 1/2	3 3/64	2 1/4	2 23/64	3/4	63-1710
15-47	15	6 5/8	4 1/2	3 1/8	2 1/4	2 15/32	3/4	66-1710
15-53	17	7 1/4	4 11/16	3 1/4	2 3/8	2 15/32	13/16	66-1710
15-71	28 3/8	11 1/4	4 7/8	4 3/32	2 1/4	3 9/64	1 3/64	83-1710

NOTES:—

1. WHEN OTHER FINISHES ARE USED, CAPACITY, WEIGHT AND HEIGHT SPECIFICATIONS ARE ADJUSTABLE WITHIN THE REQUIREMENTS OF THE FINISH USED. SEE NOTE 2.
2. HEIGHT (DIMENSION 'A') IS BASED UPON USE OF SPECIMEN FINISH SHOWN.
3. THE SPECIFICATIONS SHOWN MAY VARY MODERATELY ACCORDING TO COMMERCIAL TOLERANCES AND INDIVIDUAL MANUFACTURER'S PRACTICE.
4. 'B' DIMENSION IS VARIED TO MAINTAIN CAPACITY.

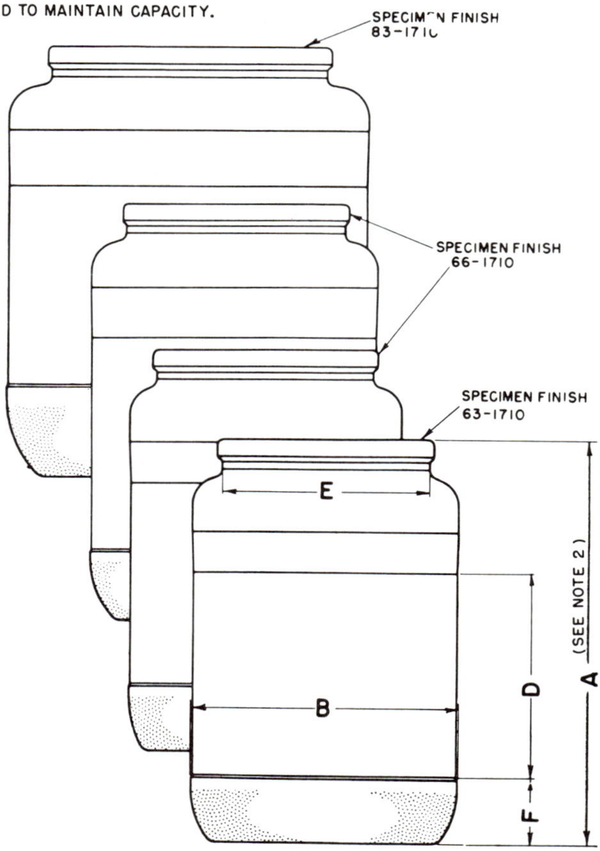

From Joslyn and Heid (1963)

FIG. 7.5. STANDARD DIMENSIONS OF SHOULDER TYPE FRUIT AND VEGETABLE JARS

Courtesy of Reynolds Metals Co.

FIG. 7.6. SHRINKABLE REYNOLON PVC FILM FOR PINEAPPLES
A printed heat shrink PVC film provides brand identification as well as helping to preseve freshness of the product. Tear resistant Reynolon films can survive the sharp contours of this spiny fruit.

Some berries under ideal conditions only remain top quality for 2 or 3 days. Typical soft fruits are currants, cherries, grapes, blueberries, strawberries, raspberries, plums, etc.

Hard fruits are better able to resist damage from handling. They are also less perishable and have lower respiratory rates. Shelf-life is weeks rather than days. The most common package is an open tray or boat with a plastic film overwrap or sleeve. Hard fruits may also be bagged in

Courtesy of Instituto Italiano Imballagio

FIG. 7.7. ITALIAN INGENUITY CREATES AN APPLE BASKET FROM A CARTON
This container is die cut and folded from a two side hot-melt coated white kraft paperboard. The carton is easily assembled, compact, stackable and provides its own carrying handle. This was a winner in the "Oscar for Packaging 1968" contest sponsored by the Italian Institute of Packaging.

perforated polyethylene film or in nets. Examples of hard fruits are apples, avocados, bananas, citrus fruits, peaches, pears and tomatoes.

Stem products are highly perishable as they rapidly lose moisture. They should be bagged or wrapped in moisture-proof cellophane or polyethylene with ventilation, or they should be banded or sleeved with shrink film. Typical stem products are celery, rhubarb and asparagus.

Root vegetables are not highly perishable. They can be stored for long periods; however, it is desirable to protect them against moisture losses. They are washed, graded and sized prior to packaging, which is usually in durable polyethylene bags. Typical root vegetables include carrots, parsnips, turnips, radishes, rutabagas, onions, beets, yams and potatoes. Potatoes are susceptible to light which causes greening. This can be avoided by printing the film or tinting it an amber color.

Green vegetables tend to dry out readily and wilt. Moisture-proof packaging helps to keep them crisp. However, they have high respiratory rates and are very sensitive to anaerobic spoilage. Ventilation is extremely important. This group includes brussels sprouts, cabbage, lettuce, broccoli and cauliflower. (See prepackaging above).

Courtesy of E.I. du Pont de Nemours and Co.

FIG. 7.8. "VEXAR" HIGH DENSITY POLYETHYLENE NETTING
This netting helps reduce damage and losses in soft products such as grapes.

Equipment

In order to package fresh produce either in-store or at warehouses, special equipment has been developed which wrap packages, shrink film, heat-seal film, fill bags, close bags, apply bands, weigh, compute price, price mark and print labels. Other machines make ice, insert staples, tie bags, convey product, chop, slice, sort and grade, and dispose of waste. Each represents an investment but overall the savings to the stores more than justifies the expense.

Recent Trends

There are several areas where improvements can be expected and these have already been mentioned. They include improved standardization of

Courtesy of American Viscose Div., FMC Corp.

FIG. 7.9. CELLOPHANE BAGGING HELPS SELL PREPACKAGED MIXES

Prepackaged combinations of vegetables are already marketed. Future trends will probably include paring, dicing and other types of preparation now performed by the consumer.

bulk shippers, more air shipment, controlled atmosphere storage, and of course improved design packages and new packaging materials. The most likely area for the future development of fresh produce sales lies in convenience packaging. Imaginative efforts can help fresh produce compete with canned, frozen and dried products.

Some examples are as follows: prepared vegetables for soups or stews, tossed salad, washed and bagged spinach, collards, kale and other greens, washed and bagged beet greens, chopped endive, precut green beans, peeled winter squash, peeled (sliced) parsnips, peeled and diced turnips, sliced melons and melon balls, sliced tomatoes, citrus salad, chop suey mix, washed and wrapped baking potatoes, prepared cole slaw and prepared relish tray.

It is also quite likely that more effort will be placed on attractive merchandise displays, advertising, suggested recipes and instructions on how to prepare them. With such an effort perhaps the per capita consumption of fresh produce can be put on the increase again.

There will also be a greater demand for produce to be packaged at the grower level or distribution level. This concept is based on the promise

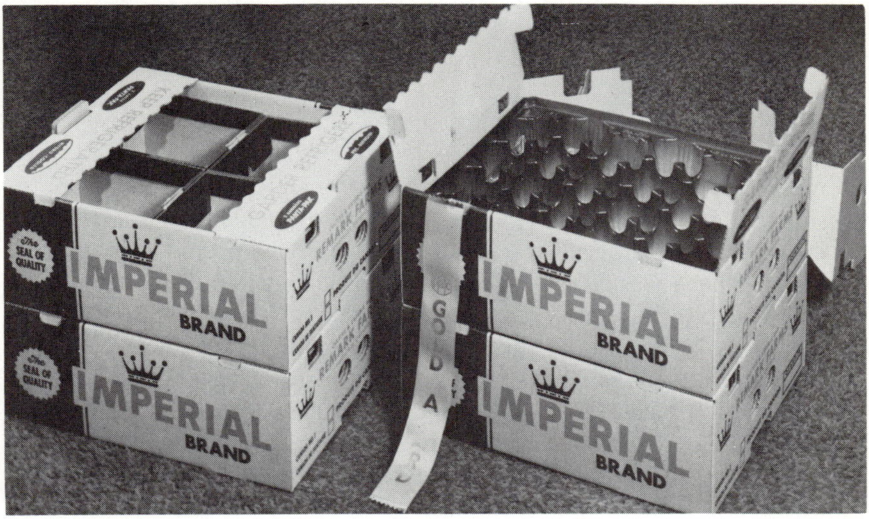

Courtesy of Packaging Association of Canada

FIG. 7.10. CANADIAN WINNER
A novel corrugated produce shipper from Canada.

that high-volume packaging increases production and thereby reduces unit cost. It also enables better supervision of the packaging and a greater degree of quality control. The industry foresees more mechanization of the produce packaging operation at the grower and distribution levels. In-store packaging will reduce proportionally, since this level will no longer be able to compete economically with the packagers at the grower and distribution levels.

FROZEN FRUITS AND VEGETABLES

In 1977, over $102 million worth of fruit and $1176 million worth of vegetables were frozen. The freezing of fruits in this country for later use in making preserves, jellies, pies, etc., began in 1905 on the east coast. Pacific Northwest berries were frozen in 1910. The commercial freezing of vegetables started in Oregon in 1929. Although it was known in the Great Lakes Region that fish left on the ice by winter fishermen froze rapidly and commanded a premium price due to higher quality, few people related this fact to the freezing of fruits and vegetables. Slow and unreliable methods of freezing resulted in such poor qualities that many states and municipalities passed laws requiring merchants to inform their customers when they were selling thawed frozen foods. Cold storage was not a connotation of quality. Today frozen food products have widespread acceptance and are one of the fastest growing segments of the food industry.

Product Characteristic

Frozen fruits and vegetables have a few properties that differ from the fresh products.

First is the fact that most frozen foods have been prepared in order to eliminate the cost of freezing unnecessary components. Stems, skins, tops, hulls, pits, etc., are removed and discarded, and the remaining product may be treated whole or sliced, diced, shredded, chopped or even pureed. During this preprocessing some alterations can be encountered unless proper precautions are taken. Vegetables lose flavor if there is a delay between harvesting and blanching. Corn loses up to ½ of its sugar if held 24 hr at 30°C and peas up to ⅓ if held 6 hr at 25°C after picking. Spinach, lima beans, asparagus and mushrooms lose sugar rapidly. Corn, peas, asparagus, mushrooms and potatoes lose appreciable amounts of glutamates on standing. These again represent losses in flavor. Rapid handling and cooling is the only way to minimize those losses.

Bananas, peaches, apples and pears turn brown when exposed to air after peeling. This seems to be related to the presence of catecholtannins, which are readily oxidized by oxygen in the presence of polyphenoloxidase enzymes. Freezing with nitrogen avoids this discoloration as it displaces oxygen and slows enzymatic action. Ascorbic acid also helps prevent browning. Hulled strawberries deteriorate rapidly on exposure to air and to drying and acquire a stale taste.

Blanching of vegetables in steam or hot water is used to inactivate certain enzymes, such as catalase, which cause off-flavors. Blanching also preserves the color of the vegetable so that in later cooking it will not wash out. Blanching should be kept to a minimum, however, as it leaches out valuable nutrients and vitamins. It is not desirable to blanch to the point the vegetable is cooked, as frozen cooked vegetables don't keep as well.

The freezing process, unless very rapid, forms large ice crystals within tissue cells. These cause structural breakdown of the product and limpness or mushiness. Fluctuations in temperature of storage can cause further growth of internal ice crystals. Even after freezing some fruits are still sensitive to oxygen and will darken, discolor or develop substandard flavors.

Freezing and subsequent thawing have a combined effect of shortening the cooking time needed for fruits and vegetables. The chemical actions and enzymatic actions that occur rapidly at room temperature in the fresh products also take place at freezing temperatures, although much slower. At −17.7°C and below the changes are not serious for reasonable periods of time (several months to a year). Upon thawing, however, the deterioration is much faster than before freezing. Freezing and thawing can cause some fruits such as plums to become very sour, hence sugar is usually added. Other fruits may "weep" or leak fluid.

Because of their acid, pH fruits and berries will not support the growth of dangerous pathogenic organisms such as *Clostridium botulinum*, but they are very likely to ferment or mold.

Packaging Requirements, Materials and Packages

Frozen produce packaging is principally aimed at protection against moisture loss and convenience in handling. Some protection against light is desirable but not always bothered with. Protection against oxygen is not an essential, but may be employed on particularly sensitive fruits.

Earliest frozen products were packaged in waxed cartons with moisture-proof cellophane liners. Clarence Birdseye indicated that E.I. du-Pont developed coated cellophanes to meet his requirements.

Frozen berries were shipped from the west coast in paraffin-coated wooden barrels of 5 (18.9), 10 (37.8), 30 (113.5) and 50 gal. (189.3 liter) capacity. Lacquered tins with slip covered lids were used for cold pack fruits and bulk quantities (30 lb (13.6 kg) units) of IQF (individually quick frozen) vegetables such as lima beans and peas were packaged in tins with friction lids. In the early days of frozen foods, processors were afraid to use tin cans, even though they would have made excellent packages, because they were afraid the consumer would get confused and leave frozen tins on the shelf where the product would spoil. By the time the public was familiar with proper handling of frozen foods the composite can made from wax-impregnated paperboard and tinplate ends had proved to be as good in performance and less costly than the conventional can. Glass containers have been used for frozen products but they have not been popular.

By far the most commonly used package for frozen foods has been paperboard cartons and containers. Paper cups, tubs and cylinders formed from waxed solid sulfite or sulfite-backed manila papers have provided rigid, water-tight sanitary containers that stacked easily when full and nested conveniently when empty. No setup equipment is required. The container is merely filled and closed. Paper spiral-wound containers have been made in a variety of shapes and in sizes up to 5 gal. (18.9 liters).

For a limited time a few duplex paper and plastic film bags were used for frozen foods.

The folded paperboard carton or box has been the work horse of frozen produce packages until very recent years. Several styles have been provided—end opening, top opening and two piece, tuck, glued and locked flaps—in sizes ranging from 8 oz (226.8 g) capacities up to 10 lb (4.5 kg). The boxes were made of good grade bleached sulfite on white patent-coated manila waxed so that the package is moisture-proof. Board thicknesses ranged from 14 to 16 points in small sizes to 17 and 24 points for institutional sizes. The boxes were used with liners, outerwrappers or both. End-filled cartons were convenient for loose-frozen smaller fruits and vegetables, while the top-opening cartons were best for bulky items like spinach and broccoli. One carton was made in which the liner material was laminated to the inside of the carton. Liner materials were moisture-proof cellophanes, laminated-coated wet strength glassines and parchment paper. Outer wrappers were cellophane or waxed paper, and later foil-wax-paper.

Courtesy of Reynolds Metals Co.

FIG. 7.11. ALUMINUM FOIL LAMINATED OVERWRAP
Foil overwraps provide attractive packages and added protection against freezer burn and cavity ice for traditional frozen food cartons.

Newer Packages

Several materials had very strong influences on the packaging of frozen produce. By far the most important was the introduction of polyethylene films, coatings and laminations. Polyethylene has much better moisture barrier properties and low temperature properties than wax and wax-coated papers or any of the previously used materials, except aluminum foil laminates. Its transparency, low cost, ease of opening and ease of reclosure quickly brought polyethylene to the forefront in packaging of IQF vegetables. In fact, ⅓ of the frozen vegetable packages sold at retail are now packaged in polyethylene film bags.

The polyethylene-coated polyester film bag became the work horse of the "boil-in-bags," prepared vegetable package. Vegetable preparations with cream and butter sauces in different combinations such as peas and mushrooms or buttered corn with red peppers were put into boil-in-bags and two bags were placed in a folded outer carton. The consumer needs only to place the unopened bag in a pot of boiling water for a few minutes in order to bring the food to serving temperature. Whatever extra cooking may be needed is accomplished during the reheating cycle. Another use for the boil-in-bag is the "thaw pack." This is the same package but the contents are frozen fruits usually in a sugar syrup. Immersion in boiling water is used only to thaw the ice crystals. The package is opened and the fruit served when it is still thoroughly chilled. Rapid thaw is accomplished because the frozen slab of fruit is relatively thin.

One of the objections to the small oblong-shaped, metal-end composite can was the difficulty in opening it. New opening devices such as the aluminum pull-tab and the plastic film tear-out seaming tapes have made this package a competitor again.

The folding carton is getting a new lease on life as it now comes with a hot-melt coating (polyvinylidene chloride). This makes the package a much better barrier and also permits the ends to be sealed shut. Opening is done by tearing out a strip along perforations. For heat sensitive products, cold-sealing adhesives can be spotted on the end flaps. The improved barrier characteristics eliminate the need for liners or for overwraps which more than offsets the cost of the coating. Production line speeds in filling and closing are also increased. Similar coatings on the interiors of corrugated shippers give added product protection in several ways. The coatings prevent the shipper from weakening by absorption of internal moisture. They are smoother and therefore do not abrade the packages packed inside. They also give added moisture and gas barrier protection to the food.

Another innovation has been the ability of converters using new printing presses with better web tension controls to produce high quality

TABLE 7.2. MOST POPULAR CANNED VEGETABLES

Vegetable	Number of Cases[1] (000 omitted)	Cumulative %
Tomato products	81,413	30.0
Corn	44,152	46.3
Green and wax beans	37,667	60.2
Green peas	33,588	72.6
Beets	12,665	77.3
Sauerkraut	9,400	80.8
Asparagus	9,263	84.2
Sweet potato	8,756	87.4
Spinach	8,031	90.4
Carrots	5,100	92.3
Pumpkin and squash	4,015	93.8
White potato	3,508	95.1
Lima beans	3,089	96.2
Leafy greens	2,757	97.2
Mixed vegetables	2,706	98.2
Field peas	2,083	99.0
Carrots and peas	1,893	99.7
Others	803	100.0
	271,407	

[1]Case of 24–303 cans (holds 15 oz approx).

process printing on extensible films both by flexography and by gravure. The better body and stiffness of polypropylene and its excellent clarity may find it competing with low density polyethylene for the bagged frozen vegetable market. The future will also certainly bring new frozen foods into the marketplace.

Insulation through the use of foams may be the innovation of the future. Excessive temperature fluctuation is the chief cause of the formation of cavity ice in frozen foods. The use of foamed polystyrene sheeting or molded polystyrene foam as an outer container for a multipack of pouches of frozen vegetables or fruits would result in both an attractive and a functional package.

Boil-In-Bag Corn

Whole-grain corn used for freezing is harvested from 1–2 days less mature than the variety to be canned. The corn is harvested by mechanical harvesters, and large sloping bins with slatted bottoms hold the raw corn.

The storage bins are connected to a conveyor which transfers the corn to the husking operation. Between 30 and 50 MT of raw unhusked corn is held in these storage bins.

A double husker is used to husk the corn and it is then inspected, washed and the ears conveyed to cutting machines where ends are trimmed and kernels are then cut from the ear. The kernels are rinsed

Courtesy of The Metal Box Co., Ltd.

FIG. 7.12. MODERN FROZEN FOOD PACKAGES FEATURE PRINTED CARTONS WITHOUT OVERWRAPS
The use of hot-melt coatings or other moisture barrier coatings provide sufficient protection to allow elimination of carton overwraps.

and then conveyed to steam blanchers where blanching is done in about 2–4 min.

After the blanchers the corn is screened, flumed and washed to remove bits of husk, silk, etc. It is then conveyed to picking tables where workers pick-out all foreign matter. Boil-in-bag corn is IQF frozen in trays which are placed on racks.

A commonly used packaging machine for boil-in-bags is the Bartelt form-fill-seal type. The packaging operation consists of:

(1) An operator controls the level of incoming corn on an overhead conveyor, and inspects it for discoloration and damage. A control handle is used to reduce the product flow.
(2) On the machine, a roll of polyester/PE passes over guide rails and is heat-sealed at bottom. The film then passes through two side sealers and cooling bars. A preprinted eyemark controls the position of the side seal.
(3) The pouches are notched or slit for easy-tear opening and severed from the web. Individual pouches are grasped by clamps mounted on an endless chain and thus brought to the filling head.
(4) The Bartelt hopper receives the frozen corn and feeds 4 oz (113.4 g) into 6 volumetric cups around the outer edge of a turret. The turret rotates 60° for each cycle.
(5) The pouch is opened and each pouch receives a charge about half full. It moves along the line where it is then completely filled by a second filler turret.

(6) A timed interval valve operates a nozzle which injects butter sauce into each pouch. Then a vacuum draw snorkel removes the air.
(7) Just after the snorkel is withdrawn, a flat bar seals the top seal and then a round seal bar applies a second seal on the top.
(8) Each filled pouch then drops down a slide and into individual cartoner conveyor pockets.
(9) A cam-operated arm pushes the pouch into a carton, the end is closed and the filled carton moves to the tray assembly line and thence into freezer storage.

DRIED FRUITS AND VEGETABLES

Because dried fruits and vegetables do not spoil and tremendous savings in weight and volume are realized, drying has a very strong economic reason for being a preferred method of preservation. In wartime when shipping space is vitally important dehydration of foods is imperative. In peace time some foods are not saleable when dried because they have undesirable color, flavor, odor and appearance, or because they are difficult to prepare. Nevertheless, over $207 million of dried fruits and vegetables were sold in the United States in 1977.

The most popular of dried produce items are fruits which have a high level of sugar. These include apricots, cherries, dates, figs, grapes, loganberries, peaches, pears, plums, and some raspberries and strawberries. Most fruits when dried to a moisture content of 23% or less will not mold. This is especially true if they have been treated with sulfur dioxide. Packaging is therefore aimed at keeping moisture out and preventing volatilization of flavor and odor constituents.

Vegetables and fruits may be sun, oven or freeze-dried (see Chapter 2).

Sun Drying

Apricots, figs, peaches, prunes (plums), raisins (grapes), pears and dates are commonly sundried in California. Sun-dried fruits are also imported from countries surrounding the Mediterranean Sea.

Pears, peaches and apricots are exposed to sulfur dioxide fumes for 3 to 6 hr before being spread out on trays to dry. Some prunes and grapes are dipped in lye to crack the skins slightly. This makes them dry faster. Care is taken not to dry too much as this would be an economic loss (low weight), while underdried fruit will spoil. Usually the last stages of drying are done in the shade away from the full sun.

Cherries, berries and walnuts are also sun-dried in small amounts.

Artificial Dehydration

Dehydration through proper use of heated air ovens produces a dried product which, when cooked, more closely resembles the cooked fresh fruit in color and flavor. There is less chance for spoilage from lack of moisture control and fewer losses to insects, rodents, birds or the weather. The principal vegetables that have been oven-dried are potatoes, white and sweet, onions, carrots, pumpkin (marrow squash), tomatoes, peppers, pimientos, sweet corn, string beans, peas, spinach, cabbage, cauliflower, brussel sprouts, okra, celery, rhubarb, garlic, horseradish and various beans. In addition to the fruits that are sun-dried other fruits are dehydrated by artificial heat. These include apples, bananas and on a limited scale pineapple and citrus fruits.

Unlike dehydrated fruits, many dehydrated vegetables have not been a popular product. This is probably because earlier processors didn't bother to blanch. This is essential. If peroxidase and catalase enzymes are not destroyed by blanching, flavor, color, odor and texture will deteriorate during drying and storage. Best results with vegetables are attained by drying to 5% moisture content or less.

Some vegetables are sulfite-treated in addition to blanching to further preserve color. Some fruits have been advantageously soaked in fruit syrup before drying.

Dried fruits and vegetables are pulverized and sold as small flakes or fine powders. Powders can be reconstituted as juices or used as flavorings. Flakes are exclusively used as the latter. Dried onion, parsley, garlic, etc., are used in soups and sauces. Dried potato granules or flakes have been used extensively for instant mashed potato recipes.

Dehydrated Potatoes

Freeze-dried fruits and vegetables are being used to a limited extent as whole or chunk additives to cereals and soup mixes, and as flakes and powders to a variety of prepared dishes, sauces, gravies, etc. Because they more rapidly resorb water than oven-dried products, they are used extensively as condiments, flavorings and "instant" beverage powders.

Potato granules consist of precooked dehydrated potatoes reduced to about 70-mesh. When reconstituted with hot water or milk, they resemble mashed potatoes.

The processing method generally used consists of the "add-back" method. Potatoes of high solids content are washed, lye-peeled, trimmed, sliced and steamed until well done. They are then gently mashed and mixed with dried potato granules in the ratio of 15:85 (by weight). This is done to obtain an overall moisture content of 35–40%. After standing for a tempering period, further agitation in another mixer induces granulation. The mixture is then dried first by flotation in a stream of heated air to 12% moisture and then in a secondary drier to 7%.

After this the dried product is put through a coarse 70-mesh screen to remove any lumps or dried skin. Of the finished dried product 85% is recycled and 15% goes to packaging.

Care must be exerted not to rupture individual cells during agitation and mixing lest the product become sticky.

Packaging is done in unsupported opaque polyethylene film bags made on form-fill-seal machinery. Some packages are made from paper-polyethylene-foil-polyethylene laminates. During filling, control of moisture is critical. The flakes are filled from a hopper and through an inner tube inserted into the bag. Bags are filled and sealed automatically. High moisture-barrier laminates are used as the product must be kept dry during normal storage. Although many packers merchandise the product in bags, some are also placed in sealed liners or pouches and merchandised in folded cartons.

Package Requirements

In the packing of dried fruits or vegetables one of the greatest problems is insect infestation. Products must be thoroughly fumigated to destroy eggs, larvae, pupae or adult insects, or if practical they may be heat-treated to deinfest. Packages should be treated to prevent infestation, also, and should be made from material that insects will not penetrate. It is also necessary that the package protect the dried product from moist environments otherwise it may cake or develop mold.

Package Forms and Materials

Whole dried fruits used to be packed in large 10 (4.5), 25 (11.3) and 50 lb (22.6 kg) wooden boxes for shipment to bakers who used them in pies, cookies and the like, or used in candies and confections. Other bulk containers have been used including corrugated cartons, spiral-wound paper canisters and large tins. For retail consumption, packages for dried fruits include paperboard-folded cartons with foil-laminated liners and /or overwraps. Coated cellophane bags and polyethylene or polypropylene bags are now also being used. Dried fruit flakes or powders are marketed in friction-lidded tins, plastic screw-capped glass bottles or foil-laminated flexible pouches. Desiccants can be included in hermetically sealed containers or the packages can be evacuated prior to sealing. Cellophane-polyethylene-foil-polyethylene laminates have been the most commonly used flexible materials and paper-polyethylene-foil-polyeth-

Courtesy of Reynolds Metals Co.

FIG. 7.13. FOIL PROVIDES PROTECTION FOR POTATO GRANULES
Instant potato granules require the high level of protection provided by vacuum or nitrogen gas packaging. The use of aluminum foil laminates in inexpensive flexible pouches provides the requisite barrier properties.

vegetables are produced annually in the United States. While individual crops vary from year to year, it is fairly easy to establish an over-all listing of the individual products which are sold in the largest volumes. Table 7.1 lists the most popular canned fruits and Table 7.2 lists the most popular canned vegetables in the United States.

Canned Peaches

Peaches for canning should be picked 4 to 6 days from maturity. Upon arrival at the cannery, they are graded according to size and held until optimum ripeness is reached (firm flesh, orange yellow color). Most canned peaches are packed as halves or slices; spiced peaches are canned whole. Halving is done by machines which also pit the fruit.

For Clingstone peaches a lye peeler is used. A hot lye solution is poured over the skin side of the halves at 101.6°–104.4°C for 45–60 sec. They are then washed with cold water to remove lye and peels.

ylene has also been used. For small packs of dried fruits used as confections, Saran-coated polypropylenes are very effective. Dried legumes such as beans, peas, lentils, etc., require very little protection. Simple plastic bags are now the most popular package. Here again cellophanes served well for many years but were supplanted by low density polyethylene film.

CANNED FRUITS AND VEGETABLES

Canned fruits and vegetables account for the second largest proportion of the total crop, second only to fresh produce. Over $2 billion of canned

Courtesy of British Cellophane Ltd.

FIG. 7.14. DRIED PRUNE PACKAGE
"Cellophane"-polyethylene lamination is used to provide a convenient package of dried fruit which requires good moisture barrier protection.

Freestone peaches are peeled by steaming. The halves are placed pit cavity down on a belt which passes through a steam scalder. They are then cooled with water sprays and the skins are removed. Other methods used for Freestone peaches are scalding in hot water followed by hand peeling. Some varieties are lye-peeled.

After being peeled, the peaches are inspected. Good halves go directly to filling lines, damaged fruit is diverted to cutting for pie fillings, good halves that are to be sliced are diverted to slicing machines. The cans are filled as rapidly as possible to avoid discoloration caused by exposure to air. Peach halves are filled in-line automatically while slices are hand-filled in order to remove defective pieces.

The cans used are tinplate with enameled electrotinplate ends. If colored spiced peaches are packed, inside enameled electrotinplate cans are used.

Syrup is then added by rotary and straight line syrupers or prevacuumizing syrupers. If hot syrup or decorative syrup is used, cans may be closed without exhausting. Otherwise the cans are usually thermally exhausted by a steam exhaust for approximately 6 min at 87.7°–90.5°C and then closed. A positive headspace of 0.762 cm is necessary.

Processing is conducted immediately after exhausting so that the heat of the exhaust is not lost before processing begins. The conditions employed vary according to use of water or syrup pack, type of peach, size of can and initial temperature prior to processing. For Freestone peaches with continuous reel type cookers, the following is recommended:

Can size	*Exhaust*	*Time and temperature in cooker*
301 × 411	6 min at 93.3°C	14–16 min at 98.8°C

After processing, the cans are water-cooled until the temperature of the contents reaches 35°–40.5°C. Cans are then sent to labelers and then to packers and warehousing.

Canned Green and Wax Beans

Beans used for canning are harvested when average maturity and size of the crop appears to be optimum. If they are to be held for any length of time prior to canning, the beans should be spread out to permit ventilation.

Mechanically harvested beans are cleaned by shakers, declumpers and air cleaners. This removes dirt, leaves and vines. They are then washed and snipped. Mechanical snippers are used. A pregrader separates the beans according to thickness and length to minimize snipper loss. After snipping, the beans are inspected and defective beans are removed manually.

The beans are then put through size graders for separation based on size. Small sizes are usually packed whole. Larger sizes are cross cut into various lengths or Frenched (longitudinally sliced). They are then washed and blanched (76.6°–82.2°C) for 2–3 min. Cooling follows by sprays of cold water.

The filling operation varies according to the type of bean. Whole beans are filled with a semiautomatic hand-pack filler. The cans are placed in a tray which is covered with a metal plate pierced with specified apertures. The tray is filled with beans and agitated until most of the beans fall into the cans. Hand operators complete the fill.

For cut beans, circular disc filling machines are commonly used. The vertical pack involves packing whole beans parallel for insertion into the can. After the cans are filled, brine is added at 93.3°C. An exhaust is normally employed for larger size cans and is also recommended for smaller sizes. It is important that the cans be well filled with brine in order to prevent discoloration of the top layer of beans. The fill-in weights are then added and the cans are sealed.

Beans must be processed in order to cook them and sterilize them. They are cooled quickly after sterilizing and processing. Recommended process time and temperature depends on the type of bean canned and can size. For whole or cut green or wax beans, the following is used:

Initial temperature of food	*Cooker Temperature*[1]		
	115.5°C	118.3°C	121.1°C
21.1°C	21 min	16 min	12 min
48.8°C	20 min	15 min	11 min

Product Characteristics

Although much has been done to improve the techniques of processing and to develop better varieties of fruits and vegetables, the very nature of the canning process sacrifices some of the desirable characteristics of the fresh product in order to insure prolonged preservation without fear of spoilage and with complete safety from food poisoning. Since much

[1]For No. 2307 × 409, and smaller cans.

trouble has been undergone to destroy spoilage bacteria, it is of course extremely essential that the package prevent later reentry of other spoilage organisms. Consequently, packages for canned fruits and vegetables have been the sealed tin can and the sealed glass jar or bottle.

So much has been written on the can and the glass jar that it would be redundant to repeat it here. Sizes, shapes and closures have been improved upon as have the basic materials. Today we have better glass, better tinplate, better enamels and seaming compounds. We even have better aluminum cans. But the basic package is the same today as it was 100 years ago.

Recent Trends

Much work has been conducted for a number of years to develop a flexible aluminum foil-laminated package which can be used to replace the can or jar, the retort pouch. Due to the rectangular rather than cylindrical geometry of this package, it can be brought to sterilization temperature more rapidly. A shorter cook-time in the sterilizer means less degradation of the food. Results are foods having odor, flavor, color and texture more like that of fresh or of frozen produce rather than like the typical canned product. The convenience of the flexible package, its economics and the superior quality of the food should make it a success in the marketplace. Most retortable laminations contain a polyester film as the outer component, aluminum foil and an inner plastic component. Differences lie in the thicknesses selected, in the choice of laminating adhesives and in the inner food contacting resin. Polyamide films, polycarbonates, polyvinyl chlorides, polypropylenes, and modified high density polyethylenes have all been evaluated for this purpose. Modified polypropylenes are now the most commonly used.

REFERENCES

ANON. 1967A. A "standard" way to tight fill pears. Produce Mktg. *10* (2) 20.

ANON. 1967B. Container switch pays off in pears. Produce Mktg. *10* (2) 28–29.

ANON. 1967C. Berries by air, latest results. Produce Mktg. *10* (3) 15–16.

ANON. 1967D. Swedish cartoning systems cover full range of Italian frozen vegetables. Packag. News *14* (3) 17–18.

ANON. 1967E. Fibreboard streamlines fruit production. Austr. Packag. *15* (4) 22–24.

ANON. 1967F. Experiment in apple packing. Austr. Packag. News *8* (4) 2.

ANON. 1967G. Better bananas, faster with new packaging methods. Packag. News *14* (5) 65–67.

ANON. 1967H. Moisture-repelling fibre tray ousts wood for lettuces. Packag. News *14* (5) 67.

ANON. 1967I. Fruit packaging breakthrough. Paperboard Packag. *52* (9) 50–57.

ANON. 1967J. Canned fruit and vegetables. Canning Packing *37* (440) 17.

ANON. 1967K. Keeping moisture where it belongs. Mod. Packag. *40* (13) 133.

ANON. 1967L. Uses PVC tubs, foil lids for pre-cooked potatoes. Food Drug Packag. *17* (9) 1–6.

ANON. 1968A. New tomato packs a great success. N. Z. Packag. *5* (2) 11.

ANON. 1968B. Golden dates in new style pack. Food Trade Rev. *38* (3) 111.

ANON. 1968C. Still more scope for plastics in produce packaging. Packag. News *15* (5) 28–40.

ANON. 1968D. Strength of moulded tray brings cut in cost of apple cases. Packag. News *15* (5) 41–42.

ANON. 1968E. The right pallet box for soft fruits. Mech. Handling *55* (6) 936–937.

ANON. 1969A. Ready-to-use vegetable packs. Packaging *40* (475) 86, 88.

ANON. 1969B. Form/Fill/Seal line for citrus fruits. Mod. Packag. *42* (10) 60.

ANON. 1970A. Delicate fruit comes in shrink-wrapped trays with carry handle. Packag. Eng. *15* (1) 50–52.

ANON. 1970B. How to package a pack of peppers. Can. Packag. *23* (6) 44.

ANON. 1970C. Waxed produce shipping containers: Where the action is. Paperbd. Packag. *55* (10) 25–28, 90, 93.

ANON. 1977. Vacuum packed shredded lettuce stays fresh for 12 full days. Food Eng. *49* (4) 81–83.

ANON. 1978. Study shows overwrap protects produce. Mod. Packag. *51* (9) 12, 14.

AHMORE, M. 1979. Celery grower making, filling HDPE shippers. Food Drug Packag. *40* (3) 3, 12 (Feb.).

BENCHLEY, B. 1978. Stretch netting ups efficiency of produce firm. Food Drug Packag. *38* (11) 1, 4. (June).

BRISTON, J. 1968. Fiberboard box growth stunted for horticultural produce? Converter *5* (6) 44–45.

BRISTON, J.H. 1969. Plastics for horticultural packaging. Rubb. Plast. Age *50* (10) 784.

DEULLIN, R. 1967. Trends in the use of refrigerated containers for the transport of tropical fruit by sea. Fruits *22* (8) 368–370.

FORSYTH, F.R. *et al.* 1967. Controlling ethylene levels in the atmosphere of small containers of apples. Can. J. Plant Sci. *47* (6) 717–718.

JOSLYN, M.A. and HEID, J.L. 1963. Food Processing Operations, Vol. 2. AVI Publishing Co., Westport, Conn.

MELVILLE, F. 1969. Packaging fruit for export and home markets Austr. Packag. *17* (9) 31–32.

NICHOLS, R. and HAMMOND, J.B.W. 1975. The relationship between respiration, atmosphere and quality in intact and perforated mushroom prepacks. J. Food Technol. *10* (4) 427–435.

NIEBOER, S.F.T. 1970. Flexible vacuum packs for processed vegetables. Food Mfg. *45* (2) 60–63.

PLATENIUS, H. 1943. Effect of oxygen concentration on the respiration of some vegetables. Plant Physiol. *18*, 671.

PLATENIUS, H. 1946. Films for produce . . . physical characteristics and requirements. Mod. Packag. *20* (2) 139.

PRESTON, L.N. 1967. Flexible packaging. VII. Fresh fruit and vegetables. Food Packag. Design Technol. *May,* 36–41.

SACHAROW, S. 1967. Fresh produce . . . growing market for flex packs. Paper, Film, Foil Converter *41* (7) 41–42.

SACHAROW, S. 1971. How flexible packaging can meet the challenge of downward sales in produce. Flexography *16* (1) 14–17, 24.

SUGIYAMA, H. and RUTLEDGE, K.S. 1978. Failure of clostridium botulinum to grow in fresh mushrooms packaged in plastic film overwraps with holes. J. Food Prot. *41* (5) 348–350.

VEERAJU, P. and KAREL, M. 1966. Controlling atmosphere in a fresh fruit package. Mod. Packag. *40* (2) 168–254.

8

Fats and Oils

. . . food foul and stinking. . . .soused in oil like a drunkard's mess and sprinkled with some horrible fishy liquid.

Liutprand, Bishop of Cremona (10th century)

Olive oil pressed from the fruit of the olive tree was a primary cooking oil for many civilizations bordering the Mediterranean Sea. The Minoans had a prosperous trade with Phoenecia and Egypt through their export of olive oil. The ancient peoples used olive oil for salads, cooking, lamp light and cosmetic purposes. Olive plantations existed in Spain and Portugal in 1700 B.C. Since it took 30 years to reach full production, the destruction of an olive grove by war or nature was a great tragedy. In Greece, oil was also extracted from the walnut.

Sesame oil pressed from seeds was of oriental origin. It was used by Portuguese explorers in the 1500s. The Chinese also discovered the oil of the soybean. The peanut which yields peanut oil was grown first in Brazil. It was transported to Africa by slave traders and from thence to Virginia. Palm oils were found in various parts of the globe. The coconut palm being the most important, perhaps of all.

Modern methods for extraction of oils and for refining or hydrogenating them have led to many new and valuable products used both in foods and in other industries as well. Cottonseed oil, corn oil and, more recently, safflower seed oil have found great acceptance in table oils and margarines. Useful oils have also been pressed from peach, prune, apricot, cherry and almond pits as well as tomato and grape seeds.

The ability to convert unsaturated oils to solid fats (margarines or shortenings) stems from the discovery by a German worker, Normann, in 1901 that these products could be reacted with hydrogen in the presence of a nickel catalyst.

Hydrogenated vegetable shortenings almost replaced lard in American kitchens, but they themselves came to be challenged by refined cooking oils.

World production of animal fats and oils probably exceeds 36.3 billion kg. Of this amount about 20–25% are nonedible and used as lubricants or drying oils. Some of the edible varieties are used for other purposes, for example, peanut oil is used as a lubricant and coconut oil is used in soaps. Many of the edible fats and oils are further processed before reaching the consumer. Great quantities are consumed by the baking industry and the candy and confectionery industry. As the per capita consumption of salad dressings increases, Americans use over 11½ pints (24.15 liters) of these oil-rich dressings annually. The retail sale of cooking fats and oils in grocery markets in the United States amounts to over $2 billion.

OLEOMARGARINE

Oleomargarine or margarine is a fatty food which is made to look like butter and to be used as a butter substitute. Different countries have different regulations but principally they all agree margarine must not derive its fat content except to a minor extent from milk fat. The product is based on the invention of a French chemist in 1869, who converted fresh beef tallow into two fractions. The higher melting fraction was a hard white fat and was called oleostearine. The lower melting fraction was a soft yellow fat which looked, smelled and tasted like butter and was called oleomargarine. After separating the two fractions he mixed the oleomargarine with skim milk, water, some bicarbonate of soda and some macerated cows' udder. He then agitated the mixture until he had an emulsion looking like cream. Upon churning the "cream" he obtained a product which he named "Butterine" and later "Margarine."

As prices and availability of beef tallow fluctuated, other readily available fats and oils were substituted. These included peanut oil, olive oil, sesame oil, coconut oil and palm kernel oil. With the introduction of hydrogenation processes other cheaper vegetable and animal oils became usable, such as cottonseed oil, soybean oil, palm oil and sunflower oil. In more recent times corn oil and safflower seed oil have come into general use in the United States.

Modern production plants for margarine are now operating with continuous processes including emulsification, cooling, plasticizing (kneading) and final packing—all out of contact with air to reduce oxidative degradation.

The production of margarine is carried on in many countries. In most countries the production of margarine is rivaling that of butter and in many of them it exceeds butter production substantially. In 1977, $1.1 billion of margarine was sold in contrast to $600 million of butter. Private label margarine amounted to 29%, branded premium, 17% and intermediate, 54%.

Product Characteristics and Packaging Requirements

The different countries making margarine use whatever natural or hydrogenated oil is most economically available to them. Russia and Hungary, for example, use sunflower seed oil. Poppyseed oil and tea seed oil have been used in the Orient and pumpkin seed oil and tobacco seed oil have been used in the Balkan countries. Lard, tallow and whale oils are used extensively in Norway and Russia.

In the production of margarine much care is devoted to the selection of blends of fats and oils which will give the most desired properties in the final product: (a) stability, (b) consistency, (c) plasticity (or spreadability), (d) minimum variation in softness with temperature, (e) flavor, (f) odor, and (g) color.

Care must be taken during manufacture to avoid excessive exposure of the fats to air as oxidation will occur in a matter of a few days at temperatures of 40°–50°C.

Manufacture

Margarine is made largely from vegetable oils that have been hydrogenated or crystallized to achieve proper spreading texture.

Two mixtures are usually made. All of the oil and fat soluble ingredients are mixed together while the other mixture consists of water and all water soluble ingredients. The melted oil mixture and the water mixture are then pumped into a high-speed refrigerated cylindrical mixing chamber. This causes a uniform distribution of the water phase throughout the oil phase. The oils crystallize and solidify with the water droplets trapped in the mass. In order to develop optimum-sized fat crystals, proper temperature control is essential. Additional agitation then further subdivides the water droplets and promotes crystallization.

The resultant semisolid margarine is continuously extruded and packaged in-line. Margarine is packed in many different shaped containers. Most margarine sold in the United States is in the form of ¼ lb (0.11 kg) sticks. These are commonly wrapped on a Benhil machine (Benz and Hilgers). It wraps 50–60 packages per minute and has separate molding and wrapping units. The margarine is dropped into a stainless steel hopper where it is compacted and pressed towards the extrusion orifice. Molding chambers are mounted in a revolving drum. The block of margarine is expelled and wrapped. Since there are several molding chambers on the machine, the cycle of operations is continuously repeated. The machine works well with parchment or aluminum foil laminates. It is equipped with coding devices. In most cases, automatic cartoning and boxing lines are integrated in-line. These boxes then go by chutes to a storage room for distribution.

Milk solids are still used in margarine in most countries to impart desirable flavor and aroma, to eliminate a fatty taste, to stabilize the emulsion and to retard oxidative flavor development. They help the formation of brown sediment during frying which is a characteristic of butter. Milk fractions are ripened through use of bacterial or chemical additives, and adjusted to the optimum pH (about 5.4–6.0). If pH drops as low as 4.3–4.5, a fishy taste can develop.

Where salt is employed it is added principally for taste and also to help preserve the product. Tastes vary among and within countries. Salt content may be as low as 0.1% and as high as 5%.

Some countries permit chemical preservatives to be added to prevent hydrolytic rancidity formation. Many different chemicals have been tried but most now use sorbic acid in concentrations of about 0.1% or less. Its use provides good shelf-life at 15.5°C and 60.5% rh for up to 8 weeks.

Coloring of margarine was once a controversial subject. Some countries insist on the use of the natural coloring materials such as betacarotene. Others have prohibited the use of colors used in butter. Some dairy states in the United States passed legislation prohibiting any coloring matter to be added. Now it is allowed almost everywhere but a few states levy high taxes on the colored product.

Legal regulations on margarine packaging have been passed to regulate not only the purity, safety and composition of the product, but also to guarantee means for its ready identification. Some countries require indicator additives such as sesame oil or starch to be added. The sesame oil develops a strong red color when heated with hydrochloric acid and mixed with a few drops of alcoholic solution of furfural. Starch gives a characteristic blue color when it reacts with iodine.

Since margarine is now processed in vacuum or in nitrogen atmospheres to minimize oxidation, it is very desirable that the package protect against oxygen reentry. Control of evaporation of moisture is very important as the guaranteed minimum net weight of the package must be clearly indicated. Since the margarine is a pasteurized product and is susceptible to bacterial or mold attack, packaging materials should be free of contamination and should protect against contamination. Light, particularly ultraviolet light, is not good for margarine. It can cause a change in color, destroys vitamins and catalyzes oxidative rancidification. Like butter, margarine is very susceptible to absorption of foreign odors and must be protected against them and loss of its own aroma.

Package Forms and Packaging Materials

Margarine has been packed in containers of many different shapes and sizes. Bulk packages have included barrels, kegs, tubs and circular or oval containers. Present bulk packaging is primarily in large cubical or rectangular boxes. These have been parchment-lined wooden boxes but are mostly corrugated paperboard containers today. Parchment liners are now being replaced by polyethylene film bags and the containers themselves may be hot-melt coated with polyvinylidene chloride.

Retail packaging also has varied in the past with round rolls, pats and prints, as well as the now more popular cubes and bricks.

Some margarine packages are made strictly for economy. They appear to be attempts to duplicate the cheapest butter packages. Thus one finds a ¼ (0.11) or ½ lb (0.226 kg) chunk wrapped in cellophane or a ½ (0.226) or 1 lb (0.45 kg) block wrapped in butter parchment.

Some export packages are tinplate cans. The newer soft spreading margarines are being marketed in paperboard, plastic, or formed aluminum tubs and trays.

Those margarines packed in tin containers may be filled hot (pourable semiliquid) or cold (extruded solid). Containers must be of good quality and free of pinholes in order to avoid rust or corrosion. Cans may be enameled and often also include parchment liners. Most canned packages are vacuum packed or nitrogen-packed.

When the intimate wrap for margarine prints is a parchment paper, it is important that the paper be free of pinholes and materials that might be extracted by the margarine and cause adverse changes. Metals such as copper are particularly undesirable as they catalyze degradation reactions. Good grades, 25 lb (11.34 per ream, higher vegetable parchments or imitation parchments with a suitable level of moisture (not too dry but not more than 9%) are used. They should be free of mold spores and placed in close contact with the surface of the margarine to eliminate air pockets.

Paper when manufactured comes off the calendar stacks and driers mold spore free. Most paper manufacturers will guarantee mold-free stock out of their door, but will not guarantee the paper will be mold-free on receipt as they have no further control over possible recontamination. Coverters of packaging materials who coat or laminate papers follow the same rule. Careful handling and storage by the margarine packager and good plant sanitation will minimize chances for mold spore contamination. Sorbic acid or sorbate dusts may be put on wrappers or incorporated into coatings or adhesives for more positive control.

The better grades of margarine have been packaged with foil laminates as intimate wraps. These can be of various combinations depending upon length of storage, desired degree of protection and cost. Foil parchment and foil-wet-strength-tissue have become the most common materials; however, much experimentation has been done with film-foil-parchment combinations. Films have been polyvinyl chloride, polyvinyl acetate, polyvinyl propionate, polyethylene, polyamide, polypropylene and polyester.

The most popular overall packages have consisted of foil-wet-strength-tissue inner wraps and foil-paperboard laminated outer cartons. These result in a very high quality appearance of the package. In recent years rigid thermoformed polyvinyl chloride, polypropylene, high density polyethylene and acrylonitrile butadiene styrene plastic containers pigmented to make them light-proof have been introduced. The margarine is filled in a very soft state, a snap-on cover (high density polyethylene) is pressed on or a peelable lid is heat-sealed and the product is chilled in storage rooms.

Special varieties of margarine are produced for special uses. For example, regular table margarines are not the same as margarines made for pastry or for cake. Different blends of fats are used and different processing procedures. Cake margarines usually have a high percentage of hydrogenated fats, while pastry margarines have some added oils. Other margarines include cream margarine used for ice cream manufacture and whipped margarines, meat-flavored margarines and cheese-flavored margarines. The margarines made from peanut oil, cottonseed oil, sunflower seed oil, soybean oil, corn oil or safflower seed oil have greater percentages of polyunsaturated fatty acids in their molecules than butter, lard or palm oil margarines. These are believed by some medical authorities to be less likely to cause atherosclerosis (hardening of the arteries). This medical theory received a great deal of publicity and the consumer demand for the soft polyunsaturated margarines skyrocketed. To distinguish them from regular margarines new air-formed foil containers were designed. The containers have attractive colors and shapes so that they can be placed directly on the table. Again, snap-on lids are employed, although some work is being done to develop hermetically sealed containers with peelable membrane heat-sealed lids.

In India a margarine called *Vanaspati* is produced which is similar in consistency to ghee. Vanaspati is made from hydrogenated vegetable oils with a minimum of 5% sesame oil, added vitamin A and special flavorings. The product is packed in small $^1/_5$ lb (0.09 kg) tinplate containers or in 40 lb (18.14 kg) bulk tin containers.

Courtesy of Aluminum Foils, Ltd.

FIG. 8.1. FOIL-PARCHMENT LAMINATE FOR ENGLISH MARGARINE
Illustration of a 1 lb (0.45 kg) brick of margarine containing 10% butter packaged in a wax lamination of embossed aluminium foil and a 17 lb (7.25 kg) vegetable parchment.

COOKING FATS AND OILS

Solid Fats

Butter, lard, chicken fat, beef tallow and to some extent the solid fats of other animals such as sheep, goats, buffalo and the like have long been used as fats in cooking. Butter and lard have been used for shallow and deep frying while the other fats have been used chiefly in basting roasts and in making gravies. Chicken fat, lard and butter have also been used in baking as solid shortenings.

When the hydrogenation process was commercialized in 1911, it became possible to change vegetable oils into solid fats. A number of such products came on the market in competition with lard. The products

TABLE 8.1. MARGARINE TYPES SOLD

Type	Package	Comments	Volume Sold
Regular stick margarine	Four 4 oz (113.4 g) sticks to the pound (0.45 kg).	Widely used	66%
Soft margarine	8 oz (226.8 g) and 1 lb (0.45 kg) bowls.	Used as a table spread	20%
Diet imitation margarine	Four 4 oz (113.4 g) sticks to the pound (0.45 kg).	Half the fats and calories of regular margarine	2.7%
Whipped margarine	Six sticks to the pound (0.45 kg).	"Economy spread"	—
Low-fat spread	—	60% vegetable oil	5%
Liquid margarine	PE bottle.	—	1%
Unsalted margarine	Four 4 oz (113.4 g) sticks and soft tubs.	Growing market	20 million lb

Courtesy of Reynolds Metals Co.

FIG. 8.2. MARGARINE PRINT WRAP AND CARTON
Aluminum foil is used both in the intimate wrap for a single print ¼ lb (0.1 kg) stick and in the registered embossed protective carton.

TABLE 8.2. BRAND SHARE—TOTAL UNITED STATES MARGARINE MARKET (IN KG)

Brand	1975	1976	1977
Blue Bonnet	4.7	5.3	5.39
Fleischmann's	3.8	3.85	3.9
Total standard brands	8.5	9.16	9.29
Imperial	2.9	2.76	2.67
Promise	0.54	0.36	0.3
Total Lever Bros.	3.49	3.12	2.99
Parkay	5.48	5.57	5.57
Miracle	0.58	0.45	0.45
Total Kraft	6.07	6.0	6.0
Mazola	0.86	0.99	1.0
Nucoa	0.45	0.45	0.4
Total Corn Products	1.3	1.45	0.09
Total Mrs. Filbert's	2.4	2.2	2.22

Courtesy of Reynolds Metals Co.

FIG. 8.3. ALUMINUM TUB CONTAINER FOR SOFT MARGARINE
This formed aluminum container is attractive enough for table use and can later be reused as a candy cup or dish.

looked like lard but had a blander flavor. Advertising stressed better digestibility of the new vegetable fats. Most of the vegetable shortenings were packaged in tins with a key wind-tear-strip opening device and a lid shaped so that it could be replaced. The 1 lb (0.45 kg) tin was a popular size; however, ½ lb (0.226 kg) and 2 to 3 lb (0.9 to 1.36 kg) larger capacity containers were used later.

One of the latest packages developed for solid shortenings is a spiral-wound composite can. The body is made from a two-ply, skived, spiral-wound kraft paperboard with an outer coating of white-pigmented polyethylene. There is also an inner liner made from foil and polyolefin laminated to the kraft body. The top is a coated aluminum lid with a ring tab which allows the whole top lid to be removed. The bottom is 80 lb (36.28 kg) tinplate. With this package, which can be hermetically sealed, shortening can be shelf-stored for as long as 2 years.

Liquid Oils

In the section devoted to margarine it was pointed out that there are many types of edible oils available from a wide variety of sources. Most of these oils are also suitable for use in cooking both as frying oils and as salad oils. They can be modified if desired by blending, partial hydrogenation, interesterification or transesterification, or polymerization.

Blending.—Blending of several oils may be done for economic reasons, to improve color, flavor, odor, stability, digestibility, viscosity, melt point, smoke point, chill point or any other property. The only requirement for such blends is that the package be clearly labeled as to the compositions of its contents.

Packages and Packaging Requirements.—Most oils have some degree of unsaturation and are, therefore, susceptible to oxidative rancidity which is catalyzed by light, particularly the blue and ultraviolet wavelengths. For this reason they have traditionally been packaged in tins and colored glass bottles (amber to brown). Both types of containers are still used. Olive oil and corn oil are not as sensitive as some of the others. With today's improvements in refining and use of antioxidants, some packers feel they can safely use clear glass. Recently, with the development of good heat stabilizers that meet U.S. Food and Drug Administration (FDA) approval and prevent thermal degradation during extrusion, clear and colored polyvinyl chloride resins have been used to make blow-molded bottles for cooking and salad oils. Other plastics have also been used such as ABS modified styrenes and polycarbonates. Injection-molded polycarbonate bottles have been used in Japan for salad oils.

In the marketplace one finds the old-style lacquered tin cans now equipped with molded plastic caps and pour spouts. A ½ liter can of walnut oil packaged in France contains a tinplate top with a fitted plastic collar opening. Inside the collar is a high density polyethylene spout with a ring tab. By pulling on the ring tab the end of the spout can be torn open. A rigid plastic screw cap protects the spout and serves as a reclosure feature.

In another corner on a gourmet shelf one finds a series of clear glass bottles with conventional screw-on caps with a pigmented shrinkable film overlay neck band. They come in 8 (226.8 g) and 16 oz (453.6 g) sizes and contain refined salad oils: almond oil, rice bran oil, peanut oil, cottonseed oil, avocado oil, safflower seed oil, soybean oil, sesame seed oil, walnut oil and corn oil are listed.

PVC Packages.—The first types of rigid PVC sheet and film were produced by solvent casting and by calendering. Solvent casting involved high solvent costs and calendering equipment was very expensive ($1 to 4 million). Extrusion methods involved much less costly equipment and processing but unfortunately at extrusion temperatures the polymer degraded. Heat stabilizers were developed. By 1964 rigid clear bottles of PVC were available for use in nonfood products. In Europe, PVC plastic containers of all shapes were adopted for food use for a wide variety of products from beer and wine to salad oils.

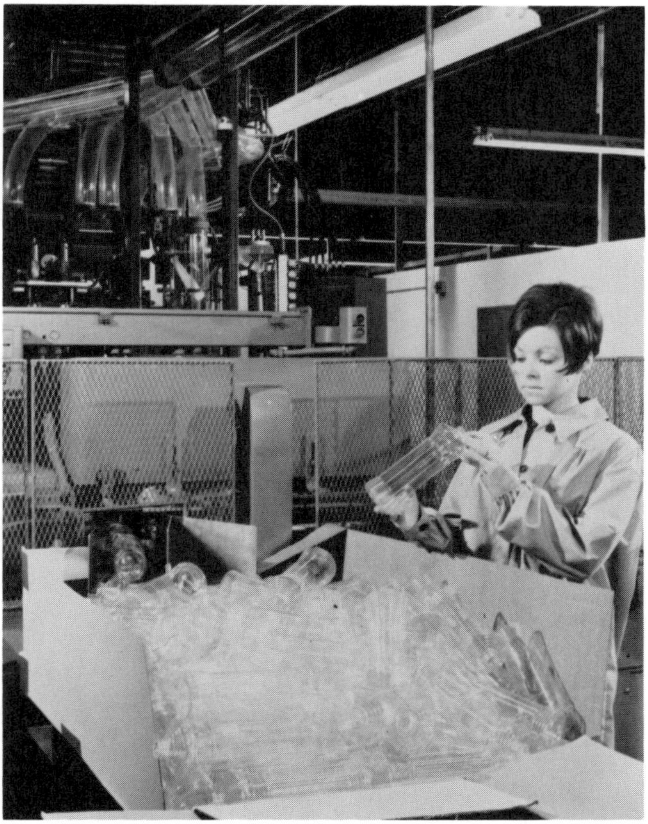

Courtesy of Air Reduction Co., Inc.

FIG. 8.4. PVC BOTTLE INSPECTION
Blow-molded PVC bottles made by the Air Reduction Co.

Blow-molded plastics can be formed in a variety of attractive and interesting as well as functional shapes. They have many advantages including their light weight and lack of fragility. The latter property assures no fragments chipping off into the product and there are less breakage and cleanup losses in chipping and in supermarkets. For light sensitive products the PVC can be colored, usually an amber to brown shade; other possible colors include opaque whites. Cap closures are usually screw thread and molded from a plastic such as polypropylene. Air is removed at time of bottling and polyvinylidene chloride liner is used in the cap to insure against oxygen reentry. Thickness of the plastic wall is controlled as thin as possible for economic reasons but strength and permeability requirements put a minimum limitation on how thin.

Courtesy of The Institute of Packaging (U.K.)

FIG. 8.5. SUPER DESIGN
The aesthetics of this smartly designed olive oil bottle makes it very salable.

Several United States companies are using polypropylene modified PVC packages. The Floria Packing Corporation of Boston is packing olive oil and wine vinegars in 4 (113.4), 8 (226.8), 12 (340.2) and 15 oz

Courtesy of Air Reduction Co., Inc.

FIG. 8.6. PVC BOTTLE FOR OLIVE OIL
Gloria Packing Corp., a division of C. Pappas, Inc., have adopted blow-molded PVC bottles for their olive oil and vinegar.

(425.25 g) containers; the Uddo and Taormina Corporation (Progresso Foods) of New York are packaging a grated cheese product.

The Society of the Plastics Industry estimate the market for PVC containers could be as high as 3 billion units by 1985 or over $500 million dollars. At present FDA limits usage to temperatures below 65.5°C.

Other Recent Trends.—Another approach has been the use of flexible cartons such as "Tetra-Paks," for edible oils. These are made from ethylene vinyl acetate wax-coated paper-polyethylene-paper-ionomer resin coating. The interior ionomer coating is heat-sealable and oil resistant. A strip of ionomer film is used to cover the raw edge and thus avoid oil wicking along the longitudinal seam of the package.

Flexible laminations of film-foil-films may be expected to be utilized as carton inner liners for edible oils. Interior plastic components will probably be ionomers, polypropylenes or polyamides. If made heavy-bodied

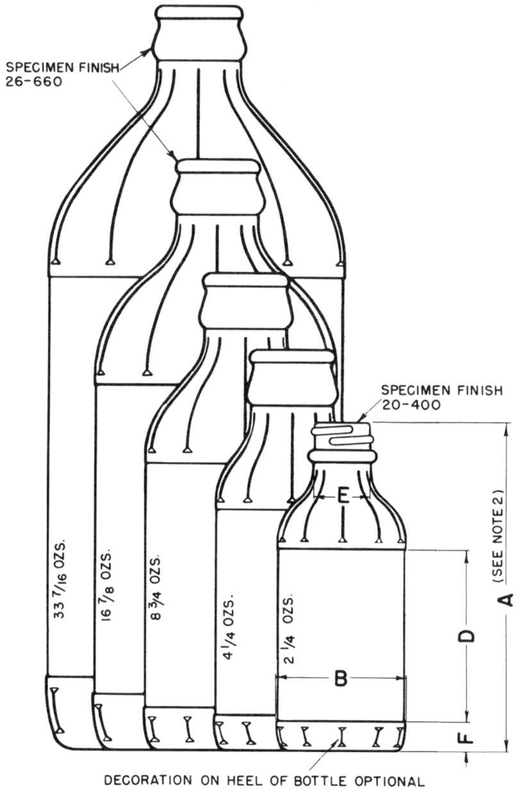

GCMI ITEM NO.	CAP. O'FLOW FL. OZS.	WT. MAX. OZS.	A	B MAX.	D	E	F	SPECIMEN FINISH
50-08	2 1/4	3	4 3/32	1 37/64	2 5/32	11/16	3/8	20-400
50-14	4 1/4	4 1/4	5	1 7/8	2 9/16	1	7/16	26-660
50-21	6 19/32	5 1/2	5 5/8	2 5/32	2 27/32	1	1/2	26-660
50-28	8 3/4	6 1/2	5 31/32	2 13/32	3 1/16	1	17/32	26-660
50-40	12 3/4	8 1/2	6 49/64	2 11/16	3 1/2	1	5/8	26-660
50-50	16 9/16	10 1/2	7 3/8	2 29/32	3 7/8	1	11/16	26-660
50-51	16 7/8	10 1/2	7 3/8	2 61/64	3 7/8	1	11/16	26-660
50-52	17 3/8	10 1/2	7 3/8	2 63/64	3 7/8	1	11/16	26-660
50-64	25 3/8	14	8 3/8	3 11/32	4 7/16	1	25/32	26-660
50-75	32 3/4	17	9 3/8	3 35/64	5	1	29/32	26-660
50-76	33 7/16	17	9 5/16	3 39/64	5	1	29/32	26-660
50-77	34	17	9 5/16	3 21/32	5	1	29/32	26-660
50-81	48 1/2	22	9 19/32	4 15/64	5 3/16	1	15/16	26-660
50-85	58 1/8	25 1/2	9 13/16	4 17/32	5 9/32	1 3/8	31/32	36-700
50-91	105 5/8	37	10 5/32	6 5/64	4 3/16	1 3/8	1 7/16	36-700

NOTES:—

1. WHEN OTHER FINISHES ARE USED, CAPACITY, WEIGHT AND HEIGHT SPECIFICATIONS ARE ADJUSTABLE WITHIN THE REQUIREMENTS OF THE FINISH USED. SEE NOTE 2.
2. HEIGHT (DIMENSION 'A') IS BASED UPON USE OF SPECIMEN FINISH SHOWN.
3. THE SPECIFICATIONS SHOWN MAY VARY MODERATELY ACCORDING TO COMMERCIAL TOLERANCES AND INDIVIDUAL MANUFACTURER'S PRACTICE.
4. 'B' DIMENSION IS VARIED TO MAINTAIN CAPACITY.

From Joslyn and Heid (1963)

FIG. 8.8. STANDARD DIMENSIONS FOR SHORT LINE, ROUND ROD LINE GLASS CONTAINERS

and properly folded, a fin sealed bag can stand by itself. In a suitably designed carton with tear-away upper portion it could be used as a pourable dispenser as is done in Europe for milk packages.

REFERENCES

ANDERSON, A.J.C. and WILLIAMS, P.N. 1965. Margarine, 2nd Edition. Pergamon Press, London.

ANON. 1967A. Plastics packs for margarine. Packaging *38* (451) 80–82.

ANON. 1967B. Ionomer resin coatings tailors Tetra Pak for edible oils—will motor oil, food follow? Packag. News *14* (1) 1.

ANON. 1967C. Margarine goes into PVC tubs. Packag. News *14* (12) 1–50.

ANON. 1967D. Fiber can for shortening. Packag. Innovator *2* (5) 5.

ANON. 1968A. Clear PVC bottles for dressings, oils. Food Eng. *40* (5) 51.

ANON. 1968B. Clean-pouring PVC bottle for edible oil. Eur. Packag. Dig. *58,* 3.

ANON. 1970A. Pre-printed polyvinyl chloride thermoformed margarine packs. Packaging *41* (481) 40–41.

ANON. 1970B. Easy-pour margarine from PVC. Packag. News *17* (6) 1.

ANON. 1978. No-plastics pack tackles fats. Packag. News *June* 1–2.

ANON. 1968C. PVC for food; it begins. Mod. Packag. *41* (8) 101.

CRUESS, W.V. 1958. Commercial Fruit and Vegetable Products, 4th Edition. McGraw-Hill, New York.

ECKEY, E.W. 1954. Vegetable Fats and Oils. Reinhold Publishing Co., New York.

EGGER, D.E. 1970. Cartoning System saves 25–50% time, labor. Food Process. *33* (4) 16–17.

GONZALEZ-QUIJANO, R.C. *et al.* 1969. Storage of Virgin Olive Oil in High Density Polyethylene Containers. J. Sci. Food Agric. *20* (10) 11–223.

JOSLYN, M.A. and HEID, J.L. 1963. Food Processing Operations, Vol. 2. AVI Publishing Co., Westport, Conn.

KIRSCHENAUR, H.G. 1966. Fats and Oils, 2nd Edition. Reinhold Publishing Co., New York.

9

Food Flavoring and Condiments

Nose, nose, jolly red nose,
And who gave thee this jolly red nose?
Nutmegs and ginger, cinnamon and cloves,
They gave me this jolly red nose.

The Knight of the Burning Pestle

In the Middle Ages salt was so highly regarded it was served at noble dinner tables in magnificent gold cellars. A man's social position was measured by whether he sat above or below the salt at banquet tables.

Wars have been fought over salt and battles lost for lack of it. Salt-making was undertaken at Syracuse, New York, because it became so scarce and expensive during the War of 1812. This production helped save the Union in the Civil War, whereas the South lost nearly all its salt production. By 1863 salt was literally worth its weight in gold in the South. Today, salt is so low-priced that cost of transportation is a major factor in its production.

Some authorities have suggested that the strong, widely held preference for salt is innate, a product of evolution during the days when our foraging ancestors lived off the land. As tropical animals, our ancestors lost much salt in their sweat and needed to replace it to prevent sodium deficiency in the blood.

Spices come mostly from the seeds or leaves of plants, but a few are exceptions. Ginger and turmeric come from roots; the clove is a flower bud from a tree. Cinnamon is the inner liner of a bark. The nutmeg is a fruit pit. Allspice and peppercorns are dried berries. The outer covering of the nutmeg gives us mace. The rarest of all spices, saffron, comes from three hair-like filaments (stigma) from each blossom of a crocus. It takes 225,000 stigma to make 1 lb of saffron. Other spices are also very tiny in size, such as the celery seed (750,000 to make 1 lb) or the poppy seed (900,000 to make 1 lb).

Spices were literally worth their weight in gold in early times. Rents, dowries, taxes and tributes could be paid with spices. When Rome was ransomed by Alaric the Goth, he demanded 30,000 lb of peppercorns as part of the treasure. Arab traders monopolized the trade in spices from the East and the price was exorbitant. King Edward I of England spent over $100,000 in 1 year for spices for his table. Europeans sought cheaper ways to obtain spices by establishing their own trade routes. Marco Polo in 1270 went overland to China and established an overland trade route, but when Constantinople fell to the Turks in 1453, Europe was completely cut off from the spice trade. The quest for new routes by sea led the Dutch, Portuguese, Spanish and English explorers to the rediscovery of the ancient sea lanes and led Columbus to the Americas.

SALT

Production

Salt mining is conducted at many different sites to reduce transportation costs. There are three basic methods for obtaining salt and all are used commercially.

Solar Evaporation of Sea Water.—Solar evaporation of the oceans and the Great Salt Lake is done by pumping water into ponds in which it is evaporated by solar heat. When the concentration has reached the saturation point by evaporation, the brine is moved by gravity to crystallizing ponds where the salt crystallizes out. The remaining liquid is returned to the sea (or lake). An entire year's supply is gathered during the summer and refined year-round. Final processing consists of drying, coarse screening, milling and fine screening. Final products are graded and then packaged. Additives are added to enhance free flowing or to improve nutritional values. For example, special salts with mineral additives are made for cattle feeding.

Mechanical Mining.—Underground mining of rock salt is the second oldest form of salt production. Underground deposits are the remnants of prehistoric inland seas which deposited their salts in vast beds when they were dried up by the sun and winds. Such salt beds formed during several geological periods and were covered with earth and rock to great depths, sometimes several thousand meters below the surface. Concrete-lined shafts are sunk to the salt bed and tunnels are cut into the bed—forming great rooms. Solid pillars of salt are left standing in these rooms to support the roof. By this procedure up to 65% of the salt can be excavated. The coarse lumps are sent to crushers and screeners below ground.

Some underground deposits are salt domes. Unlike the strata of the previously described deposits which may be from 2.4 to 5.48 m thick, salt domes may be a mile in diameter and up to 4876.8 m thick. Here the mining technique is similar but is called quarrying. Rooms are cut down vertically as well as horizontally and may reach depths from floor to ceiling as much as 27.4 m. This makes percentage removal higher. Final processing is the same.

Evaporated Brine.—In some mines it has been found more convenient to sink two holes to the salt bed and to pump water down one hole and brine up through the other. The brine goes to huge settling tanks where impurities are removed chemically. Then the brine is pumped to vacuum pans where three successive pans at successively lower pressures are steam-heated to boil the brine. The "pans" are three stories high, and as the agitated brine boils, cube-shaped crystals form and settle to the bottom where they are removed as a slurry. The slurry is pumped to a filter where it may be completely or partially dried. After drying it is milled, screened and graded.

Marketing Information

About 40 MMT (million metric tons) of dry salt or salt brines are produced in the United States annually. Of this amount 30 MMT are consumed by the chemical industry. About 5 MMT are used to control ice and snow during the winter. The food processing industry uses about 3 MMT, and 1 MMT reach home kitchens. Other quantities are consumed in animal feed mixtures. Salt is used by restaurants, hospitals and hotels, in swimming pools and in water softeners. There are an estimated 14,000 different uses; new uses are constantly being found.

Product Characteristics and Packaging Requirements

Salt is very soluble in water. In moist atmospheres, the surface of a dry salt crystal will absorb water from the air. The basic aim in salt packaging is to keep it dry, free flowing and clean. Ease of opening and dispensing together with reclosure features are very desirable in consumer packages.

Packaging

Bulk packaging of salt is usually in multiwalled bags in various sizes up to 100 lb (45.36 kg). The modern package contains a polyethylene liner.

The consumer package for fine grained table salts for many years has been a 26 oz (737 g) capacity spiral-wound paper canister with paper ends. A pivoting metal pour-spout was attached to the top lid. The opening way sealed against moisture and accidental spillage by a wafer of glassine paper. A printed paper label completed the package. A modernized version of this package has an injection-molded plastic opening in addition to the plastic pour-spout. The opening has perforations, making the can a large shaker dispenser.

A smaller 4 oz capacity spiral-wound canister with a decorative printed foil label (free of brand identification) is sold as a table-ready shaker dispenser. The top is molded plastic and is twisted to move a shield away from the perforated openings. Three of these salters are sold as a unit in a paperboard tray which carries brand identification. The whole unit is overwrapped with cellophane or various shrink films.

Other dispensers of blow-molded plastic have been designed as table-ready dispensers. A miniature version is marketed for picnic and patio use. Small individual servings have been available for several years in polyethylene-coated paper packets. One variety has fluted paper or fluted foilpaper laminate as one side of the packet. This adds stiffness and supposedly easier opening.

Future Trends

We have already seen a trend toward more individualized portion packaging and greater use of plastics. This undoubtedly will continue. More decorative packages for table use will be introduced. More varieties of salt can be expected, including finer or coarser grains and different flavors. Other constituents may be added, such as tenderizers and flavor intensifiers for use in patio cooking.

FRUIT AND CITRUS FLAVORINGS

The use of concentrated juices and extracts of fruit for flavoring of other foods is principally limited to beverages, candies and baked goods (see Chapters 10, 11 and 12). A few fruits and fruit flavorings are used in the preparation of condiments; these are treated later in this chapter. Some fruit extracts are used also for coloring principles. A few, such as lemon and orange extracts, are packaged for home use. Most, however, are packaged in bulk for industrial consumption. Over $60 million of extracts were sold in supermarkets in 1977.

Product Characteristics

The skin and peels of fruits carry most of the flavor, flesh and color. Some added flavoring may be extracted from seeds. Ideal flavor and

Courtesy of Morton Salt Co.

FIG. 9.1 UPDATED VERSION OF AN OLD PACKAGE
Morton Salt Company has moved with the times in adapting plastics to the pouring features of the package top and in modernizing their famous brand identification—the girl with the umbrella.

aroma is achieved when the fruit reaches complete ripeness and the sugar content reaches its maximum. The total flavor is a compound mixture and is derived from several parts of the whole fruit: some from extraction of the fruit flesh and some from aromatic constituents extracted from peel, seed or stems.

The flavoring essences are for the most part quite volatile and are chiefly alcohols, ketones, ethers, acids and esters. As such they are generally only slightly soluble in water but readily soluble in alcohol. For this reason alcohol is usually the extractive medium. Straight distillation or vacuum distillation of fruit juices and fruit extracts in order to concentrate the juice will volatilize nearly all of the flavor constituents. Some means must be used to capture them in order that they may be added back or used separately as flavorings. Different fruits have different sensitivities to heat. Only those temperatures that will not deteriorate juice quality can be used. On the other hand, it is difficult to capture highly volatile components in a vacuum system, especially when the

Courtesy of Morton Salt Co.

FIG. 9.2. TABLE DISPENSERS FOR SALT AND PEPPER
The use of imaginative package design has brought these points of sale packages to the table.

volatiles are mixtures of some 50–100 different chemical compounds, each of which may be present in as little as one part per billion and rarely more than 10 parts per million. Frequently the major flavor producing component is present in the smallest amount. For example, total esters in Concord grape juice are about 50 ppm concentration, but methylanthranilate, the major flavor contributor, is present only in about 1–1.5 ppm concentration. Present practice is to heat juices or pulp mixtures at atmospheric pressure at maximum permissible temperature for a brief period to volatilize essences and capture them, then the feed stock is quickly chilled and further concentration is done under vacuum at lower temperatures. When extracts are the goal, alcohol may be added to the original feed stock to create lower boiling azeotropic mixtures and to help capture the essences.

Essence recovery is achieved through condensation of fractionated vapors and by scrubbing of noncondensable gases. Alcohol is recovered and reused.

Alcoholic extracts containing 8% alcohol can be further concentrated by freezing. The ice crystals will be almost totally free of essence and alcohol. Added sugar, however, makes this procedure more difficult.

Packaging Requirements

Most fruit essences are fairly stable at room temperature and can be stored for several months. Color principles may be light-sensitive, requiring protection. The chief package requirement, however, is prevention of losses of flavor and aroma through evaporation or permeation. The package material also must be resistant to deterioration by water containing some fruit acids and some ethyl alcohol.

Packages

Bulk packaging of fruit essences is in large glass containers or in paperboard or fiberboard drums with polyethylene liners. For long-term storage the product is usually frozen, in which case either the latter mentioned package or tin cans are used.

Retail packaging of essences is almost exclusively in glass bottles. Most of these are now enclosed in printed paperboard cartons.

Future Trends

While there has been little if anything done in this direction for direct retail packaging of essences, it might be possible to absorb an essence onto an inert water soluble porous carrier. Freeze drying or some other means would be used to remove water, and the result might be a highly concentrated free flowing granular solid or powder.

VINEGAR

Vinegar is well documented in ancient writings, so again we must assume its origin is lost in the prehistoric era. In the Old Testament Ruth "dips a bit of bread into vinegar." Roman legions put wine vinegar into their drinking water to purify it.

Vinegar is a fermentation product. Its name comes from the French *vinaigre*, meaning "sour wine." This is probably the classical source of vinegar. A wine that is allowed to ferment too long will convert some or all of its alcohol content to acetic acid.

Any sugar- or starch-containing material can be fermented to an alcohol and then to a vinegar, provided the sugar content exceeds 9%.

Vinegar was a byproduct of winemakers and brewers until about the 17th century, when vinegar-making was established as a separate industry in France.

Product Definitions

Cider Vinegar.—Cider vinegar or apple vinegar is the product made from fermentation of apples. About 7.5% of the total United States apple crop is diverted to vinegar production. In the United States it is the only product that may be labeled "Vinegar" without qualification. In addition to acetic acid, it must contain (when undiluted) at least 1.6 g of apple solids per 100 cc and of the latter not more than 50% may be reducing sugars.

Grape or Wine Vinegar.—Grape or wine vinegar is the product made by fermentation of grapes. In the United States it must contain more than 1 g of grape solids and 0.3 g of grape ash per 100 cc.

Spirit, Distilled or Grain Vinegar.—Spirit vinegars are made from fermentation of distilled alcohols.

Malt Vinegar.—Starchy cereal grains, usually barley, may be converted to sugars using an enzyme, diastase, which is contained in malt.

Other vinegars are made from various fruit culls including oranges, prunes, peaches, pears, pineapples and bananas. Sweet potatoes make a pleasant vinegar, whereas white potatoes produce a disagreeably-flavored vinegar which must be distilled. Low grade honey will make a vinegar if yeast is added to convert it to alcohol. Some red-colored vinegars are made from grapes with the skins included during fermentation, or they may be made from bad batches of red wine which have turned sour.

There are also a number of flavored vinegars produced. These are specialty products, such as "spiced vinegar," tarragon vinegar, chili vinegar, rosemary vinegar and fine herb vinegar.

Manufacture

Vinegar manufacture from fruits consists of crushing, grating, or chopping the fruits, or gathering the already pressed pomace from juice manufacture and fermenting the mass. Cereals or starchy vegetables are crushed and heated with water to gelatinize the starch and make a mash. The mash is then cooled and mixed with malt. The malt converts starch to maltose (sugar). Yeasts are added which convert the sugars to alcohols. Care must be taken to use pure strains of yeasts, as wild strains can cause undesirable side effects. Sulfur dioxide is added to the fermentation process to control wild yeasts and increase yield. Since fermentation creates heat, some cooling may be required. When fermentation to alcohol is complete (not needed when wine is used), the liquor is separated, diluted and sent to different reactor tanks where acetic acid producing bacteria *(Mycoderma aceti)* are added. These convert the ethyl alcohol to acetic acid. Again, temperature control is desirable (29.49°C being optimum), and since heat is generated cooling is required.

After fermentation to vinegar and clarification, the product is aged in barrels to eliminate harsh flavors and to develop full bouquet. Final filtration leaves the product ready for packaging.

Package Requirements

Vinegar is corrosive to metals and is adversely affected by dissolved metallic salts such as iron, copper or tin, hence metal containers are to be avoided. It also is susceptible to contamination by further growth of vinegar bacteria, other wild strains of bacteria or certain types of insects. Pasteurization, micropore filtration or addition of sulfur dioxide helps prevent growth of microorganisms.

Packaging

Bulk packaging of vinegar has always been in wooden barrels or large glass bottles. Oak formerly was used but it has largely been replaced by spruce. Interiors of the barrels are heavily coated with paraffin. Polyethylene-lined drums should serve just as well, since polyethylene cap liners have been used in closures for bottles with no difficulty.

Retail packaging has been in clear glass bottles. Screw cap closures have largely replaced the crown cap. Metallic caps must have liners that will not allow penetration of volatile acid with subsequent corrosion. Some bottles have shrinkable plastic covers applied over their caps and necks.

The latest trends in vinegar packaging are the use of plastic bottles. Polystyrene has been used in England, and blow-molded PVC bottles are now being evaluated.

VEGETABLE FLAVORINGS

There are several hundred different flavoring materials which come from plants, trees and shrubs other than the fruits and berries previously described. Some are used in cooking as direct additives, as in the case of the spices and herbs. Others are extracted or distilled and used as flavorings or coloring materials for prepared foods and beverages. In the latter type the botanical product is extracted, with or without heat, with mixtures of alcohol and water. In commercial dealings the various seasonings are classified as spices, aromatic seeds and herbaceous plants (herbs). The most commonly used of each of these classifications are shown as follows. Black pepper is the most widely used spice in the United States. Mustard seed ranks as a close second in volume consumed and is, by far, the most consumed aromatic seed.

Spices	*Aromatic Seeds*	*Herbs*
Allspice	Anise	Basil
Cassia	Caraway	Garlic
Cinnamon	Cardamon	Laurel (bay)
Cloves	Celery	Marjoram
Ginger	Coriander	Mint
Mace	Cumin	Onion
Nutmeg[1]	Dill	Oregano
Pepper (black)	Fennel	Parsley
Pepper (white)	Fenugreek	Rosemary
Pepper (capsicums)[2]	Mustard	Sage
Turmeric	Poppy	Saffron
	Sesame	Savory
	Star anise	Tarragon
		Thyme

Product Characteristics

Two basic characteristics are common to all herbs and spices: they will spoil if not dried as they are subject to molds and enzyme actions, and they will lose flavor if not kept cool and in closed air tight containers, for their principal flavor constituents are volatile essential oils. On the other hand, freezing of spices is not recommended as they are said to lose flavor. Many spices are chopped fine or made into grains or powders. If allowed to absorb moisture they will cake and not flow freely from the package. This is a great disadvantage, as usually very small quantities are used. Insect or rodent infestation is also a problem that must be dealt with in shipping and storage of spices.

Product Preparation

Spices, seeds and herbs are dried at or near the point of harvest, else they might spoil in shipment. On arrival at a destined port they are inspected, graded, partially cleaned and sold to spicemakers. The spicemakers grind, blend and package the product for distribution and sale to consumer and industrial markets. Nearly 40% of the spices produced are consumed in manufactured products such as cured meats, baked goods,

[1] Nutmeg was introduced in England in the 12th century. From that time on, many people would carry around nutmeg graters to be sure of having their own supply. This perhaps was because of the habit of adding a little ground nutmeg into wine and ale. Today, nutmeg graters are a collectible item with some varieties worth up to $800.

[2] Cayenne, chili, red, paprika peppers.

pickles, canned foods, beverages and condiments. Oleoresinous extracts may also be made as described earlier.

Package Forms and Materials

In early days one could buy dried herbs in loose bundles tied with string or in loosely woven cloth bags. The finer imported spices were sold in glass or earthenware jars, in tinplate boxes or in metal-lined wooden chests. Small retail packages consisting of paperboard cartons with paper liners were used for many years in grocery stores. Small glass jars with screw-on plastic tops replaced the cartons and probably are responsible for the major upswing in spice consumption. Spices in glass do not lose flavor so rapidly and are easy to dispense through molded plastic, perforated inserts in the mouth of the jars. Some dried vegetable products (some of which are freeze-dried) are now marketed from the spice shelf in identical packaging. These include powdered, granulated, minced garlic or onion, chopped chives, chopped green sweet peppers, celery flakes, grated lemon or orange peel, and mixed vegetable flakes. Other mixtures sold include flavored salts, flavored sugars, apple pie spice, barbecue spice, chili powder, crab or shrimp spice, curry powder, herb seasoning, Italian seasoning, mixed pickling spice, poultry seasoning and pumpkin pie spice. One also finds cream of tartar, meat tenderizers, monosodium glutamate and charcoal flavoring.

Recent Trends

There are only a few novelty packages on the market but perhaps they are indicating new trends in spice packaging.

A polystyrene vial with a rigid plastic friction lid cover is available containing 1 g of dried whole saffron. This is packaged in Spain. A similar product from France contains 0.5 g of powdered saffron in a glass vial with a cork closure.

A meat tenderizer is available in solution in a polyethylene squeeze bottle. A lemon peel flavorant (for cocktails) is packaged in a squeeze bottle with atomizing top. Two ounces of the product containing 85% alcohol are packed in a four-ounce bottle. The air left in the bottle atomizes the product when the bottle is squeezed.

An injection molded polypropylene jar is being used to package a dry mustard powder. Four ounces of the mustard powder are now packaged in a jar which weighs only 20 g. The old glass jar that was replaced weighed 100 g. The unbreakable plastic jar doesn't require dividers in shipper cartons. Thus the user saves three ways: on shipping weight and volume, size of shipper and breakage. The jar has a recessed bottom to

FIG. 9.3. PAPRIKA IN GLASS RETAINS FLAVOR AND AROMA
Brand identification is maintained by McCormick and Company through standardized bottle shapes, distinctive plastic tops and foil label design.

Courtesy of McCormick and Co., Inc.

accommodate the top of another jar when stacking on the shelf and a protruding ring just below the lid to protect paper labels against damage.

Many spices are also being sold in single portion aluminum foil laminates for easy use in cooking. They are usually contained in four-sided sealed pouches. Dried onions are packed into such pouches for consumer use.

Vanilla Flavorings

Cortez' expedition to Mexico City in the early 1500s resulted not only in the discovery of chocolate (Chapter 11) but also in the discovery of the

FIG. 9.4. SPICE MIXES Preformulated spice mixes are packed in foil pouches for seasoning special foods such as Mexican style tacos.

Courtesy of McCormick and Co., Inc.

vanilla bean. Actually the chocolate drink enjoyed by Montezuma was heavily flavored with vanilla. Despite zooming vanilla bean prices due mainly to Madagascar using the power of its 80% monopoly, extract use has been steadily increasing.

The vanilla is a type of orchid and has a beautiful lime-yellow bloom that lasts less than a day. The raw bean is picked when ripe but has very little flavor. Beans are dried in the sun, then "sweated" to initiate enzymatic reactions and a slight fermentation. This process is repeated for several months until full cure has been attained. Beans are then totally dried, inspected, graded and packed in waxed paper-lined tin boxes containing about 7–8 kg.

Vanilla flavor is extracted from the dried beans by soaking in alcohol at room temperature. Three repeated soakings each for five days are made. The residue is washed with water and pressed to get the last traces of flavor. The diluted wash waters are blended with the concentrated extracts. The extract is aged in glass or stainless steel containers and then bottled. Instant powdered vanilla is made by mixing the extract with sugar.

SYNTHETIC FLAVORINGS

Chemists have been studying natural food flavorants for many years and have successfully purified and identified many of their component ingredients. In many instances these have proven to be relatively simple compounds, but even some of the most complicated ones have been synthesized successfully. The natural flavorants are blends; for example, over 124 different compounds have been identified in orange aroma and over 137 in strawberry flavor. Synthetic anise flavor is now used in place of natural anise in the United States by a factor of ten to one. Synthetic vanillin consumption is three times that of natural vanilla. Synthetic strawberry and grape flavors are used in several-fold ratio over their natural counterparts. Most synthetic flavorants are consumed by the food industry and are, therefore, packaged in bulk. A few are formulated into alcoholic extracts, bottled and sold for home-cooking.

PICKLED FOODS

Pickles are undoubtedly as old as vinegar and spices. We know that cucumbers have been cultivated for more than 3000 years, and sauerkraut dates back to ancient China. The Egyptians' use of vinegar in cookery and their knowledge of fermentation processes leads to the conclusion that pickling also originated before written history. Even in Roman days pickling factories existed in Spain. They preserved their meat in brine, soaking it later in milk to remove most of the salt.

A wide variety of fruits and vegetables can be preserved by pickling. They may be preserved as a single pack, that is, all of one kind, or they may be a mixed pack of several types. When coarsely chopped, they are used in relishes or chutneys; when finely comminuted they are added to sauces, spreads and dressings. Within each category there is a wide variety of possible combinations of sizes, cuts, pickling process and final composition. For example, cucumber pickles range in size from tiny cocktail gherkins to large whole dill or kosher-style pickles. They may be packed as long quarter strips, as cross-cut chips, in odd chunks or chipped as a relish. They may be finally processed as sweet, sour, dill, bread-and-butter, or mustard-style pickles. Similar variations can be applied to

TABLE 9.1. SPICES AND EXTRACT SALES IN SUPERMARKETS

	SALES 1977 Dollar Volume (millions)	1976 Dollar Volume (millions)	Percentage of Difference	*PROFIT* % of Dept. Gross Profit	Gross Profit Dollars (millions)	*ASSORTMENT* Items/ Brands/ Sizes at Whse.	*MARGIN* Avg Gross Margin (% of retail)
Extract and Food Coloring	$ 62.89	$ 59.22	6.20	12.85	$ 21.04	29	33.5
Lemon extract	2.47	2.32	6.47	0.58	0.95	2	38.3
Vanilla extract	39.46	38.32	2.97	7.70	12.59	6	31.9
All other extracts (incl. grenadine) and bitters)	16.03	13.93	15.08	3.53	5.79	17	36.1
Food coloring	4.93	4.65	6.02	1.04	1.72	4	34.8
Pepper	71.51	66.19	8.04	13.43	21.97	16	30.7
Black	65.35	60.38	8.23	12.03	19.67	13	30.1
White	1.23	1.16	6.03	0.28	0.46	1	37.7
Red/Cayenne	4.93	4.65	6.02	1.12	1.83	2	37.2
Whole and Ground Spices	76.45	78.95	− 3.16	17.09	27.95	34	36.6
Cloves	8.63	9.29	− 7.09	1.94	3.18	3	36.8
Cinnamon	22.19	23.22	− 4.43	4.71	7.70	4	34.7
Nutmeg	7.40	8.12	− 8.86	1.66	2.72	1	36.7
Paprika	11.10	11.61	− 4.38	2.49	4.07	2	36.7
All other whole and ground spices and seeds	27.13	26.71	1.57	6.29	10.28	24	37.9
Leaf Spices and Herbs	34.52	33.68	2.49	8.08	13.20	19	38.2
Basil	4.93	4.65	6.02	1.16	1.90	1	38.5
Oregano	9.86	10.45	− 5.64	2.18	3.56	3	36.1
Thyme	2.47	2.32	6.47	0.61	1.00	2	40.3
All other leaf spices and herbs	17.26	16.26	6.15	4.13	6.75	13	39.1
Vegetable Seasonings	83.85	82.46	1.69	17.94	29.33	29	35.0
Bell Pepper Flakes	1.23	1.16	6.03	0.29	0.48	1	39.1
Parsley flakes	12.33	12.77	− 3.44	2.68	4.38	3	35.5

Chopped chives	3.70	4.65	−20.42	0.78	1.27	2	34.4
Celery flakes and seeds, etc.	4.93	4.65	6.02	1.17	1.92	3	38.9
Onion flakes and juice, etc.	27.13	26.71	1.57	5.74	9.39	12	34.6
Garlic flakes and powder, etc.	30.83	39.03	6.20	6.47	10.58	6	34.3
All other vegetable seasoning	3.70	3.49	6.02	0.81	1.32	2	35.6
Table Salt	37.39	37.15	0.65	4.17	6.81	7	18.2
Regular	21.16	24.38	−13.20	2.41	3.94	4	18.6
Iodized	16.23	12.77	27.09	1.76	2.87	3	17.7
All Other Salt	4.93	4.65	6.02	0.83	1.35	5	27.4
Other Seasoning and Flavoring	140.56	140.49	0.05	25.61	41.88	56	29.8
Chili powder	11.10	10.45	6.22	2.21	3.62	4	32.6
Garlic salt	20.96	17.42	20.32	4.26	6.96	5	33.2
Monsodium glutamate (MSG)	13.56	15.10	−10.19	2.06	3.38	3	24.9
Seasoned pepper	6.17	8.12	−24.00	1.15	1.89	4	30.6
Meat tenderizers	13.56	18.58	−27.01	2.31	3.77	6	27.8
Charcoal flavoring Artificial smoke	2.47	2.32	6.47	0.45	0.73	1	29.5
Seasoned salt	28.36	29.03	− 2.30	4.84	7.91	10	27.9
Special-purpose Seasonings and seasoning mixes	13.56	13.93	− 2.65	2.70	4.42	8	32.6
Cinnamon-n-sugar	1.23	1.16	6.03	0.26	0.43	1	35.1
Horseradish (shelf stable)	1.23	1.16	6.03	0.18	0.29	1	23.9
All other seasoning and flavoring	28.36	23.22	22.14	5.19	8.48	13	29.9
Total	$512.10	$502.79	1.85	100.00	$163.52	195	31.9

Extracted from Chain Store Age, Supermarkets, July 1978, page 159.

each of the other fruits and vegetables listed, although fruits are usually pickled sweet and spicy.

Pickles in the form of relishes and condiments accounted for $400 million worth of business in grocery stores in the United States in 1978. This figure does not include sauerkraut, which is treated as a canned vegetable.

Cucumber Pickles and Relishes

Manufacture and Product Characteristics.—Pickling varieties should be firm in flesh and picked when underripe. Bruising is to be avoided. A short length of stem is left attached when harvesting. After sorting and preliminary grading for size, the pickles are cured in a 10% brine during which the cucumbers ferment and cure. This process takes from 4 to 6 weeks. Salt concentration must be maintained and replenished as needed during the brining operation. Sometimes about 1% dextrose sugar is added to control the character of the fermentation. Air is excluded as much as possible to allow the lactic acid bacteria to multiply. In the early stages carbon dioxide gas is generated and must be allowed to escape. When cured, pickles are translucent and have a yellow or olive green color. They are then stored in a higher concentrated brine until ready for final processing.

Final processing consists of a careful grading for size, a water soak to remove salt and then the addition of vinegar. Water soaking takes several days and several separate baths. Alum or calcium chloride additives in the final wash bath are used to harden the texture, and turmeric is used to improve the color. "Vinegaring" consists of soaking the pickles in one or more baths of vinegar of the desired strength to give a final equilibrium concentration of 2.5–3.5% acetic acid. If packed at this point, the pickles are "sour pickles."

Sweet pickles differ only in the final vinegar soak, where a spiced sweet vinegar is used.

Dill Pickles.—Dill pickles may be categorized as fresh pasteurized, genuine and imitation. Genuine dills consist of larger size pickles which are fermented in 40–50 gal. (151.4–189.27 liters) barrels with dill weed, dill vinegar, salt and dill spice. Sale of refrigerated pickles is showing excellent growth in the supermarket.

The dill weed is placed on the bottoms of clean barrels and these barrels are then filled half full with cucumbers. Mixed dill spice and salt is then added. After shaking, more dill spice, salt and dill vinegar is added on top. The barrels are then filled with water or a 29° salinometer brine. They are then placed on the side with bung up. Fermentation takes place best at 26.6°–29.4°C for 6–8 weeks.

In order to insure adequate shelf-life, the pickles are usually pasteurized. The pickles are placed in jars and covered with brine from the original fermentation process. Cloudiness in the brine is removed by filtration. The brine is heated to 71.1°–76.6°C and poured over the pickles. After capping and sealing, they are pasteurized at 73.5°C for 15 min followed by prompt cooling.

Most pickles are packed in glass jars. If cans are to be used they must be heavily lacquered.

Packaging.—Pickles can be canned in heavily lacquered cans if the cans are heated to 93.3°C for about 10 min prior to closing. An additional process at 93.3°C for about 10 min is an extra precaution.

Most pickles are packed in glass jars and sealed, preferably using a vacuum. Vacuum is attained by steam injection into the head space just prior to capping. Brines are usually filtered to remove cloudiness, or they may be replaced with fresh brines. Jars come in 6 (177.4), 10 (295.7), 16 (473.2), and 20 fl oz (591.5 ml) sizes. The larger half gallon (1.9 liter) and gallon (5.78 liters) sizes are mostly used by restaurants and institutions. Lacquered metal caps are used with cork or coated paperboard liners with a hot-melt sealing compound around the rim.

Plastic bottles as well as pouches have been used for pickle products. Since dill pickles are perishable unless pasteurized, this unit is intended for a short shelf-life. An additional problem in using plastic for pickles is the possibility of acetic acid diffusion through the walls of the package. Unless pickles are pasteurized, they rely on acetic acid for bacteriological stability. Loss of acetic acid through the package walls can reduce shelf-life and cause contamination with adjacent products. Other materials used for pickle pouches include polyamide-polyethylene pouches.

Other Pickles and Relishes.—There are only minor differences in the handling, processing and packaging of pickles and relishes made from other fruits and vegetables.

Recent Trends

An 8 fl oz (236.6 ml) jar of miniature ears of corn in a pickled brine is being replaced in the New York market with a polyethylene-coated polyester bag in a printed paperboard carton.

The development of suitable flexible materials for sterilizing will probably cause significant development in flexible pickle packages, either in transparent film pouches or packages fabricated from aluminum foil laminates.

A novel use in pickle packaging has been the adaptation of the "Pure-Pak" milk carton to package dill pickle slices and sweet pickle relish. An inner aluminum foil lining is used to insure flavor retention and adequate shelf-life.

Courtesy of E.I. du Pont de Nemours and Co.

FIG. 9.5. "PURE-PACK" MILK CARTON NOW A PICKLE PACKAGE
This version of a familiar carton is adapted to pickles through the addition of a foil lining.

OLIVES

Olives are not only one of the oldest fruits used by man but they are one of the first to have been processed to improve their usefulness. The fruit grows only in climates that are similar to the Mediterranean, consequently most of the world crop comes from Spain, Italy, Greece and Portugal. There is a substantial crop produced in California—over 2.5 million cases of canned ripe olives are processed there. Another 1.5 million cases of green olives are imported annually. About $65 million worth of olives were sold in 1977.

Product Characteristics

Fresh olives are extremely bitter and unpalatable. The bitterness can be removed by dilute alkali at room temperature. The classical method was to treat them with wood ashes (more likely a solution of potash obtained from leached wood ashes). Such fruit is palatable but will spoil.

Processing

Green or ripe olives can be pickled or canned. The chief differences are at what stage they are picked and how they are lye-treated. Both types are stored in brine until ready for processing. Sometimes a partial fermentation occurs. The fruits are lye-processed in repeated baths of dilute lye to set color and texture and to remove bitterness. Fermentation is done in large vats of brine (sometimes sugar is added). This is a lactic acid fermentation similar to that for cucumber pickles.

Final Packing.—After fermentation the fruits may take several alternate paths. They may be canned or, after being pitted and stuffed, may be pickled in fresh brine and packed, or they may be pickled and packed whole or chopped.

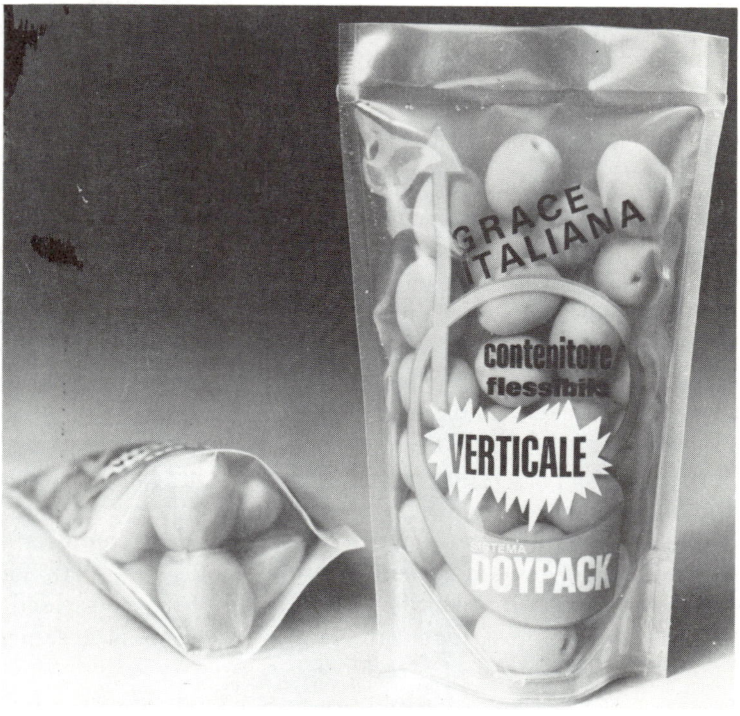

Courtesy of Istituto Italiano Imballaggio

FIG. 9.6. "COYPACK" STAND-UP OLIVE PACKAGE
An Oscar winner for 1968 at the Italian Packaging Institute was this flexible film gussetted pouch which, because of its construction, will stand without topping.

Canned olives are packed in brine, exhausted, sealed and retorted at 115.5°C for about 60 min. Pickled olives are packed in brine and lactic acid and bottled with care to exclude air.

Packages.—Enameled tinplate cans are used for canning. Glass bottles and jars are used for pickled olives. It is reasonable to expect that plastic bottles or jars may eventually be used for this purpose.

An interesting concept for olive packaging is the "Doypack" concept marketed extensively in Europe. Polyester-polyethylene laminates are used to form a stand-up pouch containing a gusset. A novel feature of this package is that the gusset can be made from a different laminate than the body of the pack. The entire package can now be formed on a form-fill machine.

SAUERKRAUT

Sauerkraut is the fifth largest volume vegetable canned in the United States. It is also very popular in northern and eastern Europe. It is one of the few pickled foods that is normally cooked before eating and served as a main course vegetable rather than as a condiment.

Product Characteristics and Manufacture

Sound firm heads of cabbage are selected; green, leafy, open-headed cabbage is not suitable. After a slight wilting the outer leaves are stripped and the core is reamed and shredded. Then the entire cabbage is cut into thin shreds. The shreds are mixed with salt (about 2½% by weight) and placed in a fermentation vat. Heavy pressure is applied which, together with the salt, extracts the cabbage juice, and soon the liquid covers all of the shredded cabbage. First stage fermentation converts the sugars to lactic and acetic acid, ethyl alcohol and mannitol and evolves carbon dioxide gas. Second stage fermentation sets in when acidity reaches one per cent with no evolution of gas. Then, when acidity gets still higher, a third stage fermentation begins with evolution of carbon dioxide and continues until fermentation is complete. Total acidity reaches more than 2% and contains lactic and acetic acids in about 24:1 ratio. After fermentaion the sauerkraut may be stored in the fermentation vats almost indefinitely, so long as air is excluded and temperature is maintained at a cool level.

Packing and Package Forms

Most sauerkraut is canned, some put-up in jars and an increasing segment of the market is put in plastic bags and sold refrigerated. Canned or jarred kraut is heated to 71.1°C to purge carbon dioxide and to pasteurize. The hot kraut is placed in the container, filled with brine, juice or plain hot water to adjust salinity or acidity, and then the container is exhausted and closed. After a brief further processing to bring the contents to 82.2°C, the containers are cooled and labeled.

Bagged sauerkraut is sold in printed low density polyethylene film bags containing 1 (0.45), 1½ (0.68) or 2 lb (0.9 kg). Since the kraut has not been pasteurized, chemical stabilizers and bleaches, such as sodium bisulfite and sodium benzoate, may be added.

Recent Trends

Considerable effort has been made in Germany and the United States to develop a polyester-aluminum foil-polyolefin laminate flexible package. The foil makes an attractive package and shuts out harmful light. Since the kraut can be pasteurized in the package, no preservatives are needed. This package has been sold commercially in Germany for several years and is finding wide acceptance.

PICKLED WALNUTS

Green (unripe) walnuts are picked before they have formed a hard shell. They are softened by steam, pricked with a needle and steeped in a pickling liquor containing vinegar, sugar, water, mushroom ketchup, caramel and spices. They are then drained, packed in jars and covered with malt vinegar. Shelf-life is up to 2 years or more.

CAPERS

Capers are the flower buds of *capparis spinosa* which have a unique flavor. They are grown in France and brined there. Pickling merely consists of draining the brine and adding malt vinegar. The product is usually sold in small screw-top glass jars; however, a new package has just been introduced. It consists of green-colored plastic bottles. These are lighter weight and use only about half as much shelf space. Furthermore, breakage is eliminated.

CONDIMENT SAUCES (VINEGAR-BASED)

There are several basic types of sauces used as condiments. They can be thick or thin, vinegar or oil-based, and sweet, sour, salty, spicy or peppery. Ketchup would be an example of a thick sauce, while Worcestershire sauce would be a thin one. The aforementioned sauces and various fruit chutneys are principally vinegar-based with little or no oil, whereas most salad dressings are oil-based.

Courtesy of Aluminum Foils, Ltd.

FIG. 9.7. A FOIL LABELED GLASS BOTTLE FOR A THIN SPICY CONDIMENT SAUCE
Thin condiment sauce is shown packaged in a traditional glass bottle. The foil label allows wiping the bottle clean and gives a quality look.

Tomato Sauces

Tomato sauces are made from tomato puree, sugar and vinegar and vary primarily in their secondary components. Because variations are almost infinite, many countries have established legal or industrial standards of definition.

Packaging.—Tomato puree is shipped in bulk to institutions or to manufacturers of other tomato products. At one time this was done in 50 gal. barrels with vinegar added as a preservative. In more recent times

Thick Fruit Sauces

Thick fruit sauces are made from finely minced fruits, vegetables and spices suspended in sugar, salt and vinegar. Vegetable gums or flour starches are added to give body.

After finely comminuting the ingredients, all are brought to a boil. Thickeners and final flavorings are added and the product is simmered for about 30 min. Some sauces are matured in oak casks for several months. Finally the sauces are bottled and capped, usually using a hot fill.

Thin Sauces

In order to keep them thin and watery in consistency, thin sauces are made from vinegar extracts of the various ingredients. The ingredients are ground, macerated or crushed, boiled in vinegar and then filtered or strained to remove solids. The different extracts are then blended, matured and bottled.

Packaging

As with thick sauces, the thin sauces are packed hot in bottles and capped so as to create a partial vacuum. In former years a cork-covered glass stopper was used. The modern bottle has a rolled metal or molded plastic screw cap with a hot melt-coated paperboard liner. Sometimes the bottle neck is made very narrow, as with hot pepper sauce, and sometimes it is restricted by means of a plastic insert to prevent measuring out too large a dose.

Recent Trends

Thin sauces have been packaged in blown polyethylene bottles and the newer PVC bottles. Some bottles that formerly were overwrapped with paper are now being overwrapped in clear plastic film.

Thick sauces have been packaged in aluminum collapsible tubes in Europe, but this has not caught on in the United States, probably because of the "toothpaste" image.

Polyethylene squeeze bottles have also been used to package thick sauces, sometimes in attractive or novelty shapes.

Mint Sauce

Mint sauce is a thin sauce but contains finely chopped mint leaves. These are steeped in vinegar, sugar and salt and then bottled. If the mint is mold-free, no trouble will occur in storage. Otherwise liquor should be pasteurized.

No. 10 (6 lb, 10 oz) (3.00 kg) and 5 gal. (18.9 liters) heavy tinplate cans have been used. Cans are filled with hot product (about 82.2°C or more) and then sealed. For retail markets 6 (177.4), 8 (236.6), 16 (473.2), and 20 fl oz (591.5 ml) cans are used with a short sterilization at 100°C after filling and closing. Tomato paste is treated and packaged in the same manner as the puree in cans.

Catsup or ketchup is made in the same way thick sauces are made and is bottled hot after deaerating. A postpasteurization at 82.2°C for 45 min is used if the product has been allowed to cool to 71.1°C.

Chili sauce is more like a chutney, as it contains pieces of tomato. A wider mouth bottle is needed. If filled above 85°C, no further processing is necessary, otherwise the bottled sauce is pasteurized at 82.2°–93.3°C as with catsup.

Recent Trends.—Tomato sauces are being packaged in collapsible metal tubes in Europe and in cellophane-polyethylene-aluminum foil-polyethylene or cellophane-polyethylene packets in the United States. Some polyethylene squeeze bottles with dispensing caps have also been tried. Polyethylene liners do a good job for fiberboard bulk shippers.

Rigid PVC bottles may well move into this market in the future if problems of hot fill and oxygen barrier can be resolved. The use of aluminum foil labels has added sales appeal to the traditional glass bottle.

MUSTARD SAUCES

There are numerous theories as to the origin of the name mustard.

Brewer's Dictionary of Phrase and Fable (Harper & Row, 1971) mentions the legend of Dijon, the French town generally thought to produce the finest mustard. They say there that "in 1382 Philip the Bold, the Duke of Burgundy, granted to the town of Dijon armorial bearings with the motto *Moult Me Tarde* (*Multum ardeo*, I ardently desire). The arms and motto, engraved on the principal gate, were adopted as a trademark by the mustard merchants and got shortened into *moulttarde* (to burn mush)."

Much more likely (and more widely accepted) is the theory, also mentioned in Brewer's Dictionary, that the word is derived from *mout*, the French word for must, a word much used by winemakers. Mout or must is the grape juice used as the basis for making wine. It is known as must from the time of its pressing until it ferments and can properly be called wine; the original liquid used in making a paste of mustard seeds to produce the moutarde of France was the freshly pressed juice.

Through the centuries, there have been a surprising number of literary references to mustard, beginning with the Bible: "If ye have faith as a grain of mustard seed, ye shall say unto this mountain. Remove hence to yonder place; and it shall remove"; and "The kingdom of heaven is like to a grain of mustard seed."

Shakespeare made frequent reference to mustard. In Henry V, Part II, Falstaff says, "He's a good wit! hand him, baboon! His wit's as thick as Tewksbury mustard." And, in "The Taming of the Shrew," "What say you to a piece of beef and mustard?"

The association of roast beef and mustard is recurrent. Anatole France, in "La Revolte des Anges," noted that "A tale without love is like beef without mustard; insipid dish."

Boswell, in his "Life of Johnson," 1791, recorded that "Johnson's conversation was by much too strong for a person accustomed to obsequiousness and flattery; it is mustard in a young child's mouth."

One of the most interesting and curious terms in the American language is "to cut the mustard." It means to accomplish something in an expert manner; to score, to gain respect. In early America, "the proper mustard" meant something genuine and not watered-down.

Mustard has been used since antiquity as a healing device for everything from headaches and the common cold to a bad case of the vapors and spider attacks. History has it that as far back as 500 years before Christ it was considered a proper antidote for the scorpion's bite.

Over the years it has been said to be good for ailments involving the "shoulders, chest, hips and loins," and mustard oil was recommended for the likes of "rigid sinews and numbness in the loins." As anyone who was a child not too many decades ago must know, mustard plasters applied to the chest would cure coughs and colds overnight (and, one might add, leave you hairless and skinless, once they were removed).

Such French mustard as that of Dijon or Meaux, bearing such names as Grey Poupon, Maille, Pikarome, Pommery and so on, go especially well with such foods a la française as roast leg of lamb, French charcuterie such as *saucisson a l'ail*, or garlic sausages, French style, *andouillettes* (chitterling sausages), *boudin noir* (blood sausages) and so on. And it is present in mayonnaise sauces of French origin (most are); half the world's mustards are produced in Dijon.

German mustards such as Dusseldorf are highly compatible with smoked German pork products. Creole mustard, notably Zatarain's brand, made in New Orleans, has no peer as an ingredient for remoulade sauce (it bears little resemblance to the classic French version) used on shrimp, and for which the crescent city is named.

The bright yellow mustard, whose color is due to the addition of turmeric, is made in America and most often found in ball parks. It has no challengers when it comes to something to smear on hot dogs.

Finally, the ideal condiment for roast beef, English style, is freshly made mustard paste made from English-made powdered mustard. This paste should be allowed to stand before using.

Mustard plays an important role in Indian cookery, principally in the form of black mustard seeds and in mustard oil. Mustard oil makes a fine dressing for freshly cooked, hot, drained green beans.

One of the great mustard specialties, incidentally, is the famed mustard fruit, or *mostarda di frutta*, of Cremona. It goes well as a side dish with boiled beef and the like. Let it be said that one of the finest brands of that frutta available in this country is made in Switzerland and bears the Ferrara label. It is made with natural flavors and no preservatives.

In England, one of the most famous personages in mustard lore is a Mrs. Clement of Durham (some say Lancashire). Although mustard paste had been used for many years, this 18th century woman produced a mustard made with a finer powder and a brighter appearance. In 1730, she started to peddle her wares, traveling on a pack horse from house to house. As she became more famous for her "Durham mustard," she accumulated a small fortune. George I is said to have been one of her patrons.

A hundred years later at the beginning of the 19th century, according to Mr. Stobart, "a young miller from Norwich, Jeremiah Colman, began to take an interest in mustard and by the middle of the century he had set up a factory." Today, the name Colman is virtually synonymous with English mustard.

The preparation of mustard and the various kinds of seeds used are confusing. There are black mustard seeds, yellow mustard seeds, and brown mustard seeds, and even the experts in the field have trouble sorting them for the public mind. They go by names such as *nigra*, *juncea* and *alba*.

Prepared mustard is made from ground brown and yellow mustard seeds, vinegar, salt, and spices (turmeric, clove, pimiento, and cayenne pepper). The ingredients are mixed to a fine homogeneous fluid paste and are packed in jars. In France and England, collapsible tubes are used as well as plastic squeeze bottles and glass jars. One of the newer packages comes with a rigid plastic pump in the lid by means of which the mustard can be dispensed.

Individual servings of mustard as well as ketchup are packaged in flexible envelopes. There are made of cellophane-polyethylene. A PVDC coating is almost always added for barrier protection. Foil-based laminates are superior to the others used in product protection. A vital point in this concept is the use of a proper primer or adhesive capable of withstanding the delaminating effect of mustard oil. The innerply is an important consideration; however, the primer must be capable of holding the laminate together. Other single service concepts include thermoformed PVC trays with aluminum foil peelable lids.

Courtesy of Reynolds Metals Co.

FIG. 9.8. INDIVIDUAL KETCHUP SERVING
Cellophane-polyethylene-foil-polyethylene pouches have been used for individual servings of ketchup and similar condiment sauces for a number of years.

TABLE 9.2. TIME REQUIRED FOR ODORS TO PENETRATE 1 MIL MYLAR POLYESTER AND 0.8 MIL CELLOPHANE FILMS[1]

	Time in Hours for First Olfactory Perception	
Source of Odor	Mylar	300 MSAD Cellophane
Vanillin	141	20.0
Pinene	140	20.0
Methyl salicylate	116	24.0
Ethyl butyrate	20	0.5
Propylene diamine	92	1.0

[1] Nagel and Wilkins (1957). Reprinted with permission of Food Technology *11*, 180–182.

Piccalilli

Piccalilli is made by combining a prepared mustard sauce with a chopped pickled relish made from various vegetables, such as cauliflower, small cucumbers and onions. Other spices are added, along with some cornstarch for thickening. The product is packed hot in jars.

CONDIMENT SAUCES (OIL-BASED)

Mayonnaise-type Salad Dressings

These dressings are thickened emulsions of oil in water using egg yolk as the emulsifier. Most any edible oil is suitable, but vegetable oils predominate. Sugar, salt, vinegar, spices and thickeners are the remaining ingredients. Spices include mustard and tarragon. Some countries allow dried egg to be used, others artificial coloring and preservatives. Mayonnaise is usually a little thicker-bodied and stronger-spiced than "salad dressing" and the latter contains more sugar, more starch and less oil.

Sandwich spreads are made by mixing a salad dressing with some fine chopped brined vegetables, usually onions, cherkins, cauliflower, olives and pimiento.

Courtesy of E.I. du Pont de Nemours and Co.

FIG. 9.9. INSTITUTIONAL MAYONNAISE PACKAGE
The "Pure-Pak" foil-lined carton is adaptable to 1 gal. (3.78 liters) capacity institutional mayonnaise packs.

Since they may contain rancidity-prone edible oils, salad dressings should be kept cool and away from light and oxygen. For this reason also they are regarded as having only a 3- to 6-month shelf-life. Batches are date-coded. Ingredients are premixed, pumped, emulsified in a colloid mill and then filled at about 120 packages per minute.

Packaging.—Mayonnaise products are packed in 5 (147.8), 7 fl oz (207.0 ml), ¼ (0.95), ½ (1.9), and 1 gal. (3.78 l) wide-mouth clean glass jars, with tightly fitted presterilized metal screw caps or with resilient-coated paperboard liners. PVC and wax coatings are used. A steam injection applied just before capping permits a vacuum pack. After closing, jars are labeled, then stored at 2°±3.5°C. Recent European trends include the use of thermoformed plastic tubs of mayonnaise with peelable aluminum foil lids.

The "Pure-Pak" foil-lined carton is being used in the institutional food market for mayonnaise. The new carton has no exposed paperboard edge along the longitudinal side seam inside the carton. This results in an airtight package and assures freshness and flavor retention.

Other Salad Dressings

There are many varieties of so-called salad dressings which are primarily oil based. These run from the simplest spiced oil and vinegar dressings to more complicated thickened French dressing, Thousand Islands dressing and Roquefort dressing. They are variations on thick sauces, thin sauces and salad dressings with one thing in common—high oil content. They may be thin, without emulsifiers, or they may be thickened and contain emulsifiers. They are almost always packaged in glass bottles like the thick sauces, but there are already indications that blown PVC and blown polyethylene bottles may be used extensively for these products.

Horseradish Sauce

Horseradish is a pungent root. Preparation consists of trimming, peeling and chopping to a fine particle size. This is blended with water, edible oil, powdered skimmed milk, sugar, flour, a small amount of thickener and mustard. The blend is boiled, homogenized and cooled. Using this as a cream base, a variety of sauces can be prepared. For example, a relish can be made using two parts of the cream to one part of minced horseradish root with added vinegar and salt, or the cream or minced product can be blended with a vinegar-diluted mayonnaise. The final product is packaged in screw-top jars similar to those of mayonnaise or pickled products.

A horseradish pickle relish is made by chopping the root and simmering it in vinegar, salt and sugar for almost 30 min., then filling in bottles or jars. Red horseradish is made by the addition of beets to the mixture.

REFERENCES

ANDRES, C. 1978. Portion packed flavours offer cost savings. Food Process. *39* (2) 92–3.

ANON. 1966. Twin-lid spice can closure and the machine that applies it. Mod. Packag. *40* (3) 60.

ANON. 1967A. Polypropylene spice packs. Packag. Rev. *87* (11) 36.

ANON. 1967B. Drop-at-a-time colours from drop-shaped vials. Packag. News *14*, 32.

ANON. 1968. Methodology of flavor evaluation. Spec. Rprt. Packag. Inst. *66–1*, (D 2146.2/25, 847).

ANON. 1971A. Aerosol foods: Holly Farms introduces aerosol barbecue sauce. Aerosol Age *16* (1) 16–17.

ANON. 1971B. New plastic package for salt designed for table use. Packag. Eng. *16* (7) 17.

ANON. 1971C. Packaging liquid flavours. Flavour Ind. *2* (11) 615–6.

BANNAR, R. 1978. Ketchup in a plastic bag? . . . Heinz says "yes". Food Eng. *50* (6) 142–3.

BINSTEAD, R., DEVEY, J.D. and DAKIN, J.C. 1962. Pickle and Sauce Making, 2nd Edition. Food Trade Press, London.

ELGIN, P.N. 1975. Packaging storage, and handling flavouring materials. S. Afg. Food Rev. *2* (4) 65–9.

FREEMAN, F. 1978. Soups and sauces: A packaging challenge. Food Mfg. *53* (6) 45–49.

GRIFFIN, R.C. 1967. Flexible packaging for pasteurized foods. Food Eng. *39* (7) 76–79.

HALL, R.L. 1968. Food flavors; benefits and problems. Food Technol. *22* (11) 55–58.

LOPEZ, A. 1969. A Complete Course in Canning, 9th Edition. Canning Trade, Baltimore.

MELILLO, D. 1968. Creating carbonated beverage flavors. Food Technol. *22* (11) 65–68.

POTTER, N.H. 1968. Food Science. AVI Publishing Co., Westport, Conn.

10

Beverages

> With the continuing changes in the beverage market, it seems fair to say that by the start of the next decade . . . beverages on supermarket shelves will look a lot different than they do today. . .
>
> *Modern Packaging, July, 1978*

WATER

Man has always had need for water to quench his thirst. When men roamed the earth in small numbers, pure water abounded in lakes, rivers and streams. As he multiplied in number and began to live in permanent settlements, he began to pollute this supply. Today, vast sums of money are spent in trying to purify used water, to find new pure sources and even to recover it from the salty sea.

Exclusive of packaged distilled or deionized water used for scientific or industrial reasons, potable water has been canned for military or emergency use. Military specifications have been drawn up covering the procedures and requirements. In Vietnam special packages of water in double-walled plastic bags were airdropped to front line positions.

In areas where civic water supplies are not particularly "tasty," pure mineral spring water can be bottled and sold. Following is a listing of presently bottled water by country and brand name:

Country	*Brand Name*
France[1]	Badoit
	Evian
	Perrier
	Vichy Celestins
	Vittel Grande Source
	Volvic
Belgium[2]	Spa Reine

Country	*Brand Name*
England	Schweppes Malvern Water
Finland	Oy Hartwell Ab
	Pyynikki
Germany	Apollinaris Water Bad Neuenahr
Hungary	Kristalyviz Natural Mineral Water
Italy	Acqua Minerale Naturale Sant'Elena
	Acqua Fiuggi
	San Pellegrino
Portugal	Aqua do Luso
Scotland	Jaggers Sparkling Scottish Water
Sweden	Ramlosa, Royal Swedish Mineral Water
Switzerland	Henniez-Lithinee
	Valser St. Petersquelle
	Silber-quell
	Walzhauser Mineral Water
Cyprus	Madari Mineral Water (Comet Farms Ltd.)
	Moutoullas Mineral Table Water
Soviet Union	Borjomi
	Narzan
	Essentuki, Nos. 4, 17 and 20
Thailand	Polaris
Canada	Sparkling Spring Water Ltd.
	St. Justin Mineral Water
Mexico	Manantiales Penafiel
Andorra	Aigua d'Andorra

Perhaps one of the fastest growing industries in the United States is bottled water. Extremely hard to package and very sensitive to flavor and/or odor pick-up, water is now a big business. In offices, bottled water from a cooler is contained in large size reverse-mounted bottles. Production of these has now stopped domestically and they are now being made in Mexico. The new Mexican bottle production was deemed necessary because of the increasing use of bag-in-box in the United States for office cooler applications.

[1] United States law prohibits the importation of plastic bottles of water. In France, PVC bottles are more popular than glass. Evian was the first bottler to market water in aluminum cans, an operation they have since ceased due to the high cost.
[2] Available in a uniquely designed PVC bottle.

Various firms bottle and sell "branch water" for mixing with whiskey. In Europe mineral water has been bottled and sold for many years.[3] Here the connotation is a little different from that used in the United States. European mineral water ranges from a normal clear spring water as we know it to a highly flavored mineral water from medicinal springs often containing some carbonation—for example, Vichy water. A noteworthy trend was the change in packaging of French mineral water from glass to PVC. This represents an enormous new volume for rigid PVC.

Mineral water is also being packed in 20 centiliter capacity miniature bottles of PVC. These are triangular in shape and sealed with polyethylene caps. They are replacing the one piece aluminum cans that have been used previously. Air France now uses these packages and claims that they are lightweight, economical and capable of being crushed flat for disposal purposes.

Mineral water consumption per person in the United Kingdom is 0.1 liters vs 52.0 liters in France. To find the cause of this difference one has to return to the time of the Industrial Revolution.[3] At that time the relative importance of agriculture in the United Kingdom waned and people moved to the cities. As the urban population increased, it became necessary to organize the previously random arrangements for water supply and sewage to prevent disease. Since then, the majority of the population in the United Kingdom has had the benefit of good water supplies and freedom from water-transmitted diseases.

France has a history of agricultural rather than industrial development, with the population more diffusely distributed across the country. The transmission of disease through sparse populations is less likely, and so the need for a controlled clean water supply is not so acute. With the purity of the supply uncertain, it was not wise to drink unboiled water, so the development of a market for pure bottled water was easy. Once the market was established, custom caused it to continue.

Bottled water (carbonated and still) is now a rapidly growing market in the United States. Although per capita consumption figures are not yet available (supermarket sales reached $36 million in 1977), a combination of desiring a pollutant-free water and wishing to obtain the product's reported health-giving properties has caused many Americans to drink imported brands of mineral water such as Perrier[4] (France), Apollinaris (West Germany) and Evian (France). Still-water consumption is also on the upswing. Mostly domestic, distilled and purified waters are now available on many supermarket shelves. Recent packaging innovations include the six-pack "perrier" container and 1 gal. (3.78 liters) high-

[3]In 1793, Jacob Schweppe arrived in England from Geneva, Switzerland, and began making artificial mineral water. At first it was sold through drugstores, but when the tax on patent medicines was lifted it began to be used as a mixer.

density PE containers for many domestic still waters. The biggest problem facing bottled waters has been out-of-stocks. Because of its size and weight, bottled water presents logistical problems in warehouses and stores.

TABLE 10.1. MINERAL WATER CONSUMPTION

Country	Liters (million)	Liters Per Head
United Kingdom	4	0.1
France	2704	52
Germany	1302	21
United States	not available	not available
	(rapidly growing market)	

For many Americans, imported mineral water has become an acceptable alternative to alcoholic beverages. A combination of successful advertising coupled with the surging bottled-water market has brought the product to new outlets—fast food stores and supermarkets. This is particularly true of the rather unique Perrier squat water bottle.[4] The unusually shaped bottle has been partially responsible for Perrier's success in the marketplace.

Ice

Ice is sold in the United States in many stores or from vending machines. Large solid blocks are rarely, if ever, packaged, but chipped ice or ice cubes are bagged to prevent soilage and for easier stacking and carrying. At one time double-walled paper sacks were used with wire ties. These had the disadvantage of weakening when wet if the ice were allowed to thaw somewhat. The paper bag has been largely replaced by a low density polyethylene bag with a draw string closure. This package is not affected by melting and gives an added advantage: the consumer can see the product and be sure there is no visible dirt.

AROMATIC INFUSIONS

An aromatic infusion is a beverage prepared by steeping a material containing coloring and flavoring matter in boiling water. Usually the solid residue is filtered or strained out, but sometimes it is allowed to settle instead, as is the case of Greek or Turkish style coffee. Some infusions are used in the preparation of other products such as soft

[4]Perrier is the leading water sold in supermarkets. In 1978, Perrier anticipated importing 80 million ($25 million sales) of its distinctive, lean-shaped green glass bottle.

drinks, liqueurs and food flavorings. Others are consumed as beverages immediately after preparation. This latter type is sold to the consumer in a dry form. It is pointless to ship water when not necessary, and in the case of infusions the proportion of water used is high. Furthermore, delicate flavors and aromas are best captured when the beverage is freshly prepared. Reheating of a previously prepared infusion results in an inferior concoction. Warmed-over coffee or a reused tea bag leave much to be desired.

Coffee

The coffee bean apparently originated in Ethiopia and spread throughout the Arab world. Legend has it that a goatherd named Kaldi discovered it in 850 A.D. Because the prophet Mohammed had instructed his followers not to consume wine, coffee became known as "the wine of Islam." The Arabs introduced it to the Turks in 1554, who brought it to Venice in 1615. There is a fine 18th century engraving by Lepicie of Boucher's picture of men, women and children drinking the then-new beverage coffee. "Charming coffee," it goes, "Your agreeable liquor/ calms the ecstasy of Bacchus/Your divine fire dissipates from the table/ Both loathing and overindulgence."

Today, coffee is grown in 57 different countries and imported in significant quantities by about 25 of the nongrowing countries. In March 1977 the U.S. Department of Agriculture (USDA) estimated world coffee production in 1977–78 to be 68.5 million bags. Because production of coffee beans can exceed demand in order to stabilize prices, 66 nations signed an International Coffee Agreement in 1962 which assigns export and import quotas.

Product Manufacture.—Coffee is picked as a berry which looks somewhat like a cranberry. The berries may be dried and then cleaned, or the seeds may be removed immediately (wet process). Fermentation solubilizes a mucilaginous layer which is then removed by washing; the seed is then dried. Milling removes the "parchment" shell leaving a gray-green colored bean. This is the form in which it is usually exported. The green beans are usually packed in jute bags of 60 kg weight. Coffees from different countries or different regions have different flavor and aroma characteristics. Green beans are imported and blended to arrive at a desired end result. Taste experts grade coffee and test blends. After blending, the coffee beans are roasted at temperatures of up to 213°C. Roasting is halted when the desired brown color is achieved. In the process the beans swell and the aromatic oils are driven to the surface. The coffee is then cooled either by water spray or forced air. Some coffee is sold in the bean and ground for the consumer in the retail market. Most coffee is ground and packaged for retail sale by the coffee blending house.

An increasing amount of coffee is now being processed by first brewing an extract from the roasted and ground coffee and then evaporating the extract to dryness. Soluble powdered coffees may be dried by drum or spray drying; however, the latter is best. Instant soluble coffees are now being marketed that have been foamed and freeze-dried.

Decaffeinated Coffee.—The method for decaffeinating coffee beans is more or less what it was when the process was first developed in Germany in the early 1900s by Dr. Ludwig Roselius. The son of a coffee merchant and taster who died at an early age, Dr. Roselius was convinced that it was the constant intake of caffeine that had caused his father's premature death. The method he developed used a combination of steam and benzine on green (unroasted) beans. A series of additional washings and steamings removed the benzine residue and, finally, the coffee was dried, roasted and ground.

Dr. Roselius called his new coffee "Sanka" from the French *sans caffeine*. After producing Sanka in Germany and France for several years, he brought it to this country in 1910. In 1928 his company was purchased by General Foods.

The chemical solvent used by almost every decaffeinating plant today is methylenechloride. This was substituted for the previously used trichloroethylene, on which the U.S. Food and Drug Administration (FDA) had proposed a ban because there was evidence that it was carcinogenic. Because it too is a chlorinated hydrocarbon, methylenechloride is also under suspicion as a possible cancer-inducing agent, but inhalation and ingestion tests on it have only just begun. Results of the ingestion tests will not be available until February 1980 and the inhalation tests will be completed in October 1981. Meanwhile, the FDA has placed the maximum level of permissible solvent residue in decaffeinated beans at 10 parts per million.

Among the companies that have decaffeinating plants in the United States are Tenco, a division of the Coca-Cola Company, which uses the residue caffeine in its world-famous beverage; General Foods, in four Maxwell House plants that turn out both Sanka and Brim; Standard Brands; and Nescafe, which makes only instant decaffeinated coffee.

In addition, there are several companies in Germany, Italy and Switzerland. Each company is reticent about discussing the precise method of decaffeination used. The basic differences have to do with whether the beans come into direct contact with the solvent or if the solvent is diluted in water or steam. The noncontact method is considered preferable not only because of possible taste impairment but also for reasons of health.

The best news in the decaffeinated field was announced by Coffex Ltd. in Switzerland, where a pure water process using no chemical solvents is said to have been perfected. For the time being Coffex will continue to use the old method in its plants in France, Germany and the Netherlands, but the new method is being used in its Schaffhausen, Switzerland, location. Because the new production is limited, Coffex has so far been able to supply coffee dealers only in Switzerland and Austria, but will start shipping to the United States shortly. There is a noticeable difference between green beans decaffeinated by the old process and the new. Those done by the water-only process are a darker, brighter green.

When the chemical solvents are used, there is a double problem in getting decaffeinated coffee that is as good as whole coffee. The caffeine contains much of the aroma and flavor of coffee. When it is removed, certain essential oils go with it. If not completely removed, the solvents will leave a trace of unpleasant flavor in the bean.

Coffee Additives.—With the price of coffee reaching very high levels in the world market, additives have been used in some blends to add flavor or otherwise permit the use of extenders. Chicory intensifies bitter flavor. Coffee blends containing chicory often claim "you can use less" to get the same strength of brew. Cereal extenders are also being used. Nestle's "Sunrise" blend of instant coffee contains both chicory and cereal. It is packed in wide-mouthed glass jars with screw caps that can be reused as kitchen containers.

Package Requirements.—There are four problems in the packaging of coffee: (1) the prevention of losses of flavor and aroma by evaporation or migration of essential oils, (2) the prevention of rancidification of the coffee acids by oxygen, (3) the evolution of carbon dioxide gas from the roasted bean, and (4) prevention of moisture absorption by the dried, roasted bean.

Packaging Materials and Package Forms.—The earliest types of coffee packages were papersacks in which the consumer carried home the whole roasted beans. Each householder had a coffee grinder which was used whenever a brew was desired. In-store grinding became common practice some time after the turn of the century. Later the paper sack was made double-walled; then it was lined with glassine. A major improvement came with the substitution of rubber hydrochloride film for the inner liner. The film provided an excellent grease barrier. Packaging of coffee in tins became prevalent in the 1920s and 1930s with perfection of vacuum-packing equipment. Nitrogen gas-flushed tins with domed lids were introduced in the late 1940s. For some reason, ground coffee in glass jars has not proved as popular as the can, except during wartime metal shortages. The introduction of instant soluble coffees brought the glass jar into prominence. While tinned coffee still accounts for about 85% of the total pack in the United States and in Europe, particularly Sweden, Norway, Finland, Denmark, England, Holland and Germany, the trend is away from the can and strongly in favor of flexible vacuum packages. The United States liter/head annual consumption of coffee is similar to German consumption (145 liters/head), and ahead of France (130 liters/head) and the United Kingdom (81 liters/head). Thus, many of the truly important packaging innovations for coffee have been developed in either the United States or western Europe.

Flexible Packaging.—A variety of "soft" and "hard" packs is now available, which consist of an inner flexible component that could be gas-flushed or vacuumized together with a printed paper outer component or a printed carton. Aluminum foil laminates have traditionally been used, but plain or metallized film laminates are now being increasingly used.

Additional coffee trends in flexible packaging include a foil laminate unit which uses a one-way valve system to allow the carbon dioxide to escape so that the coffee could be packed soon after grinding. In Britain,

Courtesy of British Cellophane Ltd.

FIG. 10.1. FLEXIBLE FILM PACKS FOR COFFEE
Instant coffee and ground coffee blends are packaged for caterers in laminates of printed "Cellophane" and polyethylene.

pouches formed on vertical form-fill machines and gas flushed are widely used, especially for the popular 250 g package.

Cans.—In the United States the standard 1 lb (0.45 kg) coffee tin has evolved from a rather wide and squat keywind strip opened can to a taller version that is opened by a standard can opener and carries a snap-on high density polyethylene reclosure lid. Vacuum pack has been the chief method.

Also in recent years Metal Box has carried out a number of market surveys on coffee in Britain, West Germany, Belgium, the Netherlands and France. This has revealed important differences among the five countries. In Britain there is a predominance of soluble coffee and a correspondingly greater use of glass bottles, but there is also a growing use of laminate bags in competition with vacuum-packed tins in the small R & G (Roast and Ground) sector.

West Germany presents an entirely different picture with an overwhelming preference for R & G coffee, associated with a strong tradition of *kaffee kultur*; there is a great use of 250 and 500 g easy-open cans as well as vacuum packs.

The Netherlands, with probably the highest per capita consumption of coffee in the EEC, has an equally strong tradition of coffee drinking, and the vacuum pack is the preferred method of retail presentation. R & G coffee accounts for over ¾ of the market—an enormous growth from only 16% in 1965.

In France, perhaps surprisingly, there is a slightly lower per capita consumption than in West Germany and the Benelux countries. A small proportion of sales are in cans but vacuum packs and laminate bags are also widely used; there is, too, a fast-growing instant coffee sector.

In Belgium the production of manufactured coffee is lower than in many other countries of mainland Europe, but imported brands are very important. Per capita consumption is almost as high as that of Holland at 6.14 kg. Again, vacuum packs are extremely important.

Despite the preponderance of cans in the market, the likelihood is that other packages will take over in the not too distant future. These will be the new flexipacks discussed before, composite can use and various pouch concepts. Cans will probably try to halt the trend by including easy-open devices and/or interior portion packaging. In the latter instance, lighter weight cans will function. A new and novel concept in coffee cans is the product marketed by the Nairobi Coffee & Tea Co., Ltd. They are selling their blends in attractive cans decorated with coffee themes. The plastic lids are held securely in place with wire fasteners in order to keep the contents as fresh as possible.

Glass.—Regular roasted and ground coffee is rarely sold in glass; the weight and breakage factors predominate over the initial low cost and reclosure features. Also large glass containers are not particularly suited to vacuum packaging. Gas flushing is much safer.

Instant (soluble) coffee manufacturers found glass jars quite satisfactory. Here the product unit size per dollar cost was much more favorable, and the reclosure feature became predominant as the container would be opened and closed many times. Glass still dominates this portion of the coffee market. Competition lies primarily in the shape of the bottle, the type of screw cap closure and the attractiveness of the label. Most glass jars have a Saran-coated, glassine membrane closure sealed over the mouth prior to capping. One novelty package is a glass percolator filled with instant coffee.

The dominance of the glass jar for soluble coffees is expected to persist in western Europe, possibly with more variation in shape and style being introduced. For cost saving reasons, particularly for private label packs and possibly more strongly associated with products containing additives, the increasing use of alternatives is predicted. Down trading in unit size is also likely to influence their situation.

With cartoning and flexible packaging machinery available, these alternatives could be adopted with modest capital outlay and the operation could be expanded as required on a modular basis. Initially regarded as refill packs for use with glass jars as storage containers in the home, consumer acceptance, particularly of lined cartons in Britain, is well established.

Plastic Jars or Containers.—If adequate barrier properties can be developed, plastic packaging may be used as a substitute for the can or the glass jar. Straight PVC plastics may prove satisfactory for short-term protection but migration of aroma may prove a problem. One major food company has reported development of a 2 lb (0.9 kg) capacity "plastic" can for roasted ground coffee with nitrogen flush and claims satisfactory shelf-life. It seems likely that the short-term plastics will be used primarily as coatings; however, they might be used as outer containers for inner portion packs where the latter would provide the needed barrier properties.

Composite Cans.—Several companies have advanced to test market with composite cans for roasted ground coffee.

These are foil-lined and will have either scored tops for easy opening with can openers or easy-open pull tabs. Both vacuum pack and nitrogen flush packages are being tested in both 1 (0.45) and 2 lb (0.9 kg) sizes. Consumer acceptance appears favorable. Savings of up to 25% are reported. One package maker has developed a complete system which includes multistage nitrogen flushing of the coffee right out of the grinder. So long as the residual oxygen in the coffee is brought below 2%, and preferably below 1%, the shelf-life can be expected to be satisfactory, i.e., 6 months or more. No problems with leakers have been reported. Conversion of can filling and closing machinery to composite cans is reported to be relatively inexpensive and involve only changes in the lid clincher.

Institutional Packs.—Packages used for institutional markets range from a 20 lb (9 kg) capacity vacuum-packed bag in a PVDC-coated corrugated shipper used for the military, and a 2 lb (0.9 kg) capacity nitrogen-flushed bag for restaurant brewers, to small unit portion packs of instant powder for making one cup. Nonmilitary institutional packs are used up more rapidly and therefore don't require as long a shelf-life. Two to three months is considered very adequate. A popular item for office coffee brewers is a complete kit including a coffee portion and filter, cream or noncream whitener (usually powdered), wooden stirrers and drinking cups, all in a paperboard carton. Many motels and hotels are adding single cup coffeemakers which preheat water. The guest opens a fin-sealed packet of instant coffee, a packet of noncream whitener and one of sugar and adds the boiling water.

Because of the significant rise in coffee prices, development is now being conducted in the area of changing package design. This is necessary to reflect a more expensive, different product. Coffee packers should:

Reduce Costs	Package to Reflect Changing Market
—reduce pack size	—segment brands more clearly
—use more economical materials	—use the pack to identify cheap vs luxury brands; go for an "up market" image which will more adequately reflect the price the consumer has to pay.
—use more economical closures	
—use simpler decoration	

Surface design, either label or print, is vital in communicating a brand to the buying public. Yet many of the label/print designs now on retail shelves had been conceived when coffee was a relatively inexpensive, everyday commodity. Some are downright dull.

But now that coffee is becoming more of a "special occasion" drink, the label should reflect this. With instant coffees, brand names could be a help: associations of "Royal" to suggest luxury, the use of gold labels which had more in common with wine labels, carton packs which use graphics to subtly point up an image of quality, and cans and/or composites with an up market appeal involving such props as a silver coffee pot.

Courtesy of Fresco Systems (U.S.)

FIG. 10.2. VACUUM-PACKED COFFEE
A display of coffee in vacuum-packed flexible containers.

Tea

According to legend, tea drinking began in China about 2737 B.C. "Somewhere, deep in the misty mountains . . . Ta Mo, the Bodhidharma, fifth reincarnation of Buddha, faced a stone wall and meditated. Spring mists became summer rains and the flower-covered hillsides turned green. Bamboo forests swayed like oceans in the soft warm breezes. And autumn came with crisp air and the leaves colored and fell, soundless to the earth. Winter came, its light snows dusting the wall and Ta Mo's coarse cloak. Still he meditated. Unblinking. Unsleeping. Unsearching. And the cycle of seasons began again. And again. The years passed by; the Bodhidharma remained. Finally his attention wavered and his chin dropped, his eyes closed to gentle sleep.

Courtesy of Istituto Italiano Imballagio

FIG. 10.3. VALVED COFFEE PACKAGE
An Italian design for a coffee package features a valve which vents gases evolved by the coffee after packaging. Without a valve the vacuum pack (on left) would become excessively distended (package on right). Laminate is polypropylene-aluminum foil-polyethylene.

"When Ta Mo woke, perhaps a day or a month, perhaps a year later, he was so angry with himself for failing in his task of meditation that he sliced off both his eyelids with a sharp knife and cast them to the ground. Immediately upon hitting the rich soil of Szechwan the saintly eyelids sprouted roots and grew into what we know as the tea bush, symbol of wakefulness."

The first record is in 350 A.D. when it was described in a Chinese dictionary. Its cultivation and use spread all over China, Japan, India and Malaysia under the encouragement of Buddhist priests who urged its use rather than alcoholic beverages. Cultivation of tea was brought to Java in 1684 and later in the 19th century to Formosa. The use of tea as a beverage was first brought to Europe by the Dutch in 1610. It reached Russia in 1618, France in 1648, England and American in 1650.

Tea was a popular commodity in 17th–19th century Europe. Tea caddies bore various designs and even now are an extremely collectible antique for many enthusiasts. They were made in different sizes, lined with tinfoil and were used to hold up to 5 lb (2.26 kg) and more of tea leaves.

Worldwide consumption of tea in 1977 amounted to about 272 g per capita. In 1976, over $95 million worth of tea was imported into the United States.

Product Manufacture.—The tea plant is an evergreen shrub which, if not pruned, would grown 9 m high. It looks somewhat like a myrtle with blossoms resembling a wild rose. It requires a moist, warm to temperate climate. Leaves are plucked at various times of the year depending upon the country, the climate and the desired end-product. Black and green teas are made from the same leaves. After picking they are dried in the sun until withered. This takes about 24 hr. Next they are rolled about 3 hr to break the leaf cells. After a short prefermentation the product is put through final processing. Green tea is steamed without further fermentation. Black tea is fermented thoroughly before "firing" (drying). Oolong tea is a compromise between the two others. The tea is then cut, sifted, sorted and packaged.

Product Characteristics.—Tea contains caffeine, tannin and essential oil. The caffeine acts as a stimulant, the tannin gives "body" and the essential oils the aroma and flavor. Some teas are brewed and then the infusion is evaporated to dryness to make an instant tea.

Packaging Requirements.—The main problem in packaging tea is to protect against moisture which can permit mold growth or fermentation, and to protect against loss of aroma by evaporation.

Package Materials and Package Forms.—Earliest packages were metal-lined wooden boxes. When tea first came to Europe it was so expensive it was weighed on apothecary scales and sold in papers (paper wrappers) (see Chapter 1). When the price came down to reasonable levels, tea was sold in tin boxes or in glass jars. Flexible packages were adopted as soon as metal foils were available for protection.

The packaging of tea in tea bags is one of the first examples of unit prepackaging. The classical method for making tea involved measuring loose tea into a teapot of boiling water and then allowing it to brew. The problem of keeping leaves out of the cup was at first ignored, then metal and china strainers were introduced. The first use of a porous cloth to hold the leaves in the pot is not known. The use of small perforated metal "tea balls" served the same purpose and an attached chain allowed the user to remove it when desired. Small tea balls were made for brewing single cups of tea. The first tea bag was undoubtedly a handmade packet of cloth tied with a thread or string. This method proved so popular it was commercialized. Eventually, porous wet-strength papers replaced the fine cheese cloth bags. Aluminum wire staples now attach the string to the bag and to the tag. Some bags are enclosed in individual paper packets and the tags are torn from the packet itself. One manufacturer puts the tea into a narrow bag that is folded back on itself and claims better brewing because the water can flow through. Modern transportation and distribution are more efficient, making the need for highly protective packaging less important. Simple printed paperboard boxes are used with paper liners for loose tea and paper packets for bagged tea. Many cartons containing tea bags are overwrapped with cellophane or polypropylene for added protection.

New Trends.—The latest trend in tea packaging is in instant tea powders. These are sold in glass jars, laminated foil pouches and in foil-lined composite cans. In 1978, mixes out-sold loose tea in supermarket sales. Mixes are instant teas combined with other flavorants such as lemon, pineapple, banana and the like. They are extremely popular for summer iced tea drinks. Another rapidly expanding market is presweetened iced tea in aluminum cans. Designed to compete with carbonated soft drinks, canned tea is a big summer seller.

Other Tea Drinks.—Teas can be brewed from a wide variety of aromatic plants using leaves, roots or blossoms. Sassafras root shavings are sold in polyethylene bags for this purpose. Other types include Verbena, Hibiscus, Peppermint, Fennel, Linden, Camomile and Rosehip. Tea bags containing the latter varieties are often sold in gourmet stores as five bags in a cellophane overwrap.

Beef or chicken bouillon tea is made by dissolving bouillon cubes in boiling water. The cubes are made from concentrated meat broths dried to a powder, then compressed. Instant bouillon powder or granules are also available. The cubes are individually wrapped in foil paper laminations to protect against moisture and rancidity. The wrapped cubes are packaged in glass or plastic jars with metal or plastic screw caps. Spiral-wound composite cans have also been used for this purpose. The instant product comes in a glass jar with a sealed, coated paper membrane and metal screw cap for a top closure.

FRUIT AND VEGETABLE JUICES

Just as he experimented with the preparation of foods, primitive man also experimented with beverages. There is little doubt that the first beverages other than water were the juices of fruits or berries which were consumed together with the fruit. Just when man learned to press the juice out of the fruit to drink as a separate beverage is not known. It undoubtedly was followed by the discovery of wines made from the fermented juices.

There's a strong case for claiming cider as the world's oldest alcoholic drink. Apples are, after all, mentioned as growing in the garden of Eden.

Courtesy of the Tea Council of the U.S.A., Inc.

FIG. 10.4. THE OLD AND THE NEW IN TEA
A metal tea caddy holds loose tea on the left, paper tea bags are in the metal container on the right, while instant tea is in the glass jar.

In Britain certainly it is one of the venerable beverages. The southwest of England has always been ideal apple country, and indeed many scholars now believe that the United Kingdom was the land of golden apples known to classical mythology as the Hesperides. They suggest that stories of journeys to these edge-of-the world islands are in fact folk memories of early voyages to Britain.

By the time the Celts were settled as the master race in Britain, the apple cult was an established part of the life and folklore of the southwest. In the Romano-British era it is worth remembering that the Isle of Avalon, to which King Arthur was ultimately transported, means literally Isle of Apples. Its supposed situation at the centre of the Somerset level is, even today, the heart of cider apple country.

Cider was quaffed regularly throughout the middle ages, and by Elizabethan times was an important export. It was included among the provisions of all West Country merchant ventures' ships leaving for the new world. Sir Walter Raleigh is reputed to have been particularly fond of the brew of his native country, preferring it to the finest vintages offered at court. That such a sophisticated courtier should have admitted cider to be his favourite tipple would have caused no raised eyebrows in his own time.

The myth that cider must be either a murkily crude liquor fit only for peasant palates, or else so innocuously sweet and fizzy as to be suitable only for nursery service, is of very recent origin. It can be traced to the rise of brewing as a national industry in the mid to late 19th century, and the vast improvements in communications in the same period which brought both choice and cheap wines from abroad to a wider market than ever before. Up to the 1850s owners of great estates in cider country saw no shame in setting the produce of their orchards before even the most honoured guests. During the Napoleonic Wars many of the greatest families saw it as their patriotic duty to eschew produce from the lands under Bonaparte control.

Most fruit juices ferment so readily that there has been no commercial packaging until comparatively recent times. By the mid 19th century, the knowledge gained by the canning industry was transferred to the bottling of fruit juices. Welch began bottling Concord grape juice in the United States in 1869. The Swiss bottled unfermented pasteurized apple juices about the same time or not long after. Large-scale packing of juices did not become commercial until about 1925. Domestic United States consumption of pasteurized single strength juices (principally tomato, citrus and pineapple) reached a peak of 1400.5 million liters in 1945, with per capita consumption the following year of 9.5 kg. In 1945, the introduction of frozen concentrated orange juice began to change the juice drinking habits of the nation. Canned juices continued to increase but at a lower rate. Most of this increase has been through the introduction of other juices and juice mixes. Over the years per capita consumption increased more than 16.3 kg. Of this amount, about 4% were chilled single strength juices, about 50% were frozen citrus concentrates, and about 46% were pasteurized single strength juices. This does not include the fruit juices used as flavorings in noncarbonated fruit or milk drinks.

In recent years lower-priced imitation fruit drinks have been introduced. This competition is being challenged by fruit juice processors who are bringing out formulated (modified) fruit juice and fruit juice concentrates. Vegetable juices have not had as much popularity as fruit juices. If it weren't for tomato juice which is classified as a vegetable, although botanically it is a fruit, the consumption of vegetable juices would be negligible. Some carrot juice and sauerkraut juice is packed. The remainder are either specialty diet-fad foods or used blended with tomato juice. Vegetable juices are also used as flavorants in baby foods and in soups.

Large-scale production of fruit juices in Europe started after World War II. Previous efforts were small-scale and sponsored mostly by teetotalers who sought to eliminate the drinking of brandy. Apple and grape juice were the first products. Citrus juices were packed mostly in Israel and Spain. Berry juices began in Germany and spread into surrounding countries.

Product Characteristics

Although many plants yield delicious juices, there are also many that do not. Different varieties of a single fruit may be desirable or undesirable for juice either through flavor differences, stability or difficulty in exacting or preparing the juice. For example, bitterness in orange juice is related to the root stock of the original tree and to the method of extraction. The flavor, aroma and color of juices are affected by the variety, degree of ripeness, soil and climate conditions, and mechanical conditions when the fruit is harvested.

Product Manufacture

Apple and some other fruit juices are expressed from healthy ripe fruit by hydraulic pressing. Solids are removed by straining, clarification or filtration. Continuous centrifuge-type processors have been developed and are being used in place of the older batch-type presses. After juices have been extracted and purified, there are several alternative further processes. If used as is, they are known as single-strength juices. These can also be diluted and processed into fruit juice beverages. Otherwise, the juices can be concentrated and frozen, or concentrated and dried. Concentrating can be done by vacuum evaporation or by freezing. When evaporation is used, volatile flavoring oils are stripped out and collected for later readdition.

Tomato Juice

For tomato juice preparation, firm and ripe tomatoes should be used. After arrival in the plant, the tomatoes are sorted and washed by soaking. This removes any residual dirt and foreign matter. In soaking, a continuous supply of fresh water is essential, and steam or compressed air may be added to agitate the fruit.

After the soak tank, the tomatoes are washed with high pressure sprays and then sorted to remove any undesirable fruit. They are then trimmed to remove rot or mold.

Hot or cold extraction methods may be used; however, hot extraction produces a superior quality juice. In hot extraction, the fruit may be chopped or crushed prior to heating, or the whole tomatoes may be heated. The tomatoes are inserted into large tanks equipped with an agitator. Steam coils are used to provide rapid heating. For cold extraction, the tomatoes are scalded prior to extracting, chopped and then extracted without heat.

Extraction involves either screw type or paddle type extractors. A shaker screen may also be used to remove any foreign material. Any dissolved or occluded air is removed by a deaerator. In some cases, the juice is homogenized at 1000–1500 psi at 65.5°C. This results in a thick bodied product with minimal setting of the solids.

Presterilization precedes final canning. This is necessary to kill *Bacillus thermoacidurans* which causes flat sour spoilage. The process involves heating the juice to 121.1°–135°C. The juice is then canned.

Standard fillers are used for filling tomato juice in cans. The cans employed should be plain hot dipped or electrolytic tinplate bodies and enameled electrolytic tinplate ends. Prior to filling, the cans are spray washed to remove any dust.

Salt may be added to the canned juice by tablets. The cans are then closed under vacuum at 87.7°C on a Panama or CR type closing machine. After closure, the cans are washed with a hot water spray.

Processing conditions depend on whether the juice is presterilized or heat-treated below 100°C. For presterilized juice, the minimum holding times are:

Closing Temperature	Minimum Holding Time
87.7°C	3 min
93.3°C	1 min

The cans are then water cooled until the temperature of the contents reaches 37.7°C, labeled and packed.

Packaging of Juices, Beverages and Concentrates

Single strength juices and fruit beverages are sold refrigerated on a local basis and pasteurized in cans or bottles for wider distribution. Refrigerated juices may be packed in glass jars, glass jugs or PE-coated paperboard cartons. HTST pasteurized juices are filled hot into clean enameled or lacquered cans or clean glass bottles or jugs. Closures can be screw caps, vacuum caps or crown caps. The heat of the juice pasteurizes the container. Bottle and can sizes vary from 4 (113.4) to 54 oz (1530.9 g) and jugs up to gallon (liter) sizes.

Heat processed orange juice has also been packaged in PVDC and foil-PE flexible pouches. These pouches offer superior protection over polyester-PE or cellophane-PE pouches. Juice in pouches made of the former maintain a quality comparable to canned juice during storage for at least 5 months at 2.2°C.

Concentrated beverage syrups are packaged in a wide variety of containers, from cans, plastic jugs and bag-in-box concepts, to huge drums and tanks.

Concentrated frozen juices are packaged in tinplate, aluminum, or composite cans. The latter are largely replacing the others due to lower costs. Easy-open tear strip devices are used extensively.

FIG. 10.5. BOILED FRUIT JUICE
This bottle featuring elegant foil top and body labeling protects the pure applejuice from light by using dark green glass.

Courtesy of the Council of Industrial Design (London)

New Trends.—The various containers used in Europe for milk packaging have also been tried for single strength juices. These include the "Zu-pak," the "Tetra-Brik," "Doypack" and "Roypack." Another concept, the "Expresso" carton by Akerlund and Rausing (Sweden), appears to be of definite interest to United States fruit juice packers. In 1978, a machine firm introduced equipment that sets up paper/foil laminate cartons, then fills and seals them for concentrated-fruit packers. The concept appeared on the United States supermarket shelf in late 1979.

Nonalcoholic cocktail mixes are now being sold in bottles, tinplate cans and paper-PE-foil-PE single service pouches. The consumer adds water and/or the alcoholic beverage. Several packages of cocktail mixes have also appeared in the thermoformed plastic single service tubs with peelable seal membranes.

A new glass beverage container combines a tear-drop shaped glass container set in a preprinted polyethylene base.

Small bottles and cans of juice are being cluster packaged in PVC shrink film overwrapped bundles of six units. Paperboard stiffeners are used to divide the two rows of three and give backbone support to the package. Prune juice is being marketed in a four 4 oz (113.8 g) bottle unit paperboard carrier. Bottles are closed with twist-off caps.

A novel package for frozen juice concentrates consists of a paperboard-PE-foil carton with plastic-coated sealing strips. The shield on the consumers end is constructed from polyester-PE and the other shield is pouch paper-PE. The carton is opened by zipping the strip from the end and lifting a perforated major flap which, when torn off, peels with it the plastic safety-shield.

INSTANT POWDERED BEVERAGES

Children have had flavored syrups to make beverages from for many years. The use of flavored powders or granules has not attained great commercial importance until relatively recent times. Adult instant beverages began with coffee substitutes like "Postum" in the 1920s and 1930s. The popularity of instant coffee, tea and fruit beverages has coincided with the development of improved methods of drying and dehydration perfected in the 1950s. Dried soups and sauce mixes created industrial demand for dried tomato juice and dehydrated fruit and vegetable purees. Such children's drinks as "Fizzies," "Koolaid" and "Tang" popularized dried fruit flavored beverages. In the latter instance it was soon discovered the drink appealed to adults as well. Similar products have been introduced not just as breakfast drinks or refreshers but also as exotically named cocktail mixes. Printed paper-polyethylene-aluminum foil-PVDC heat seal coating laminates are formed into fin sealed pouches with extremely high barrier protection. Multiple packets are sold in folded paperboard cartons.

Some of the instant drinks are sold in glass jars with screw cap closures. PVDC-coated glassine membrane closures are sealed across the mouth of the jar for added protection. Instant orange juice substitute is packaged in 18 (510.3) and 27 oz (765.45 g) jars as well as smaller sizes. It also is sold in 4⅔ oz (132 g) easy-open pop top cans. Tablet type drinks are packaged in formed foil packets with paper or paperboard laminated tops.

Methods of drying and dehydration are discussed in Chapter 2.

MILK DRINKS

Chocolate and vanilla flavored milk drinks and egg nogs have been available for many years. The major problem in developing fruit-flavored milk drinks has been the action of fruit acids on milk casein. Destabilization of casein during pasteurization renders the beverage unsaleable. Recent studies have indicated that addition of dibasic potassium phosphate will stabilize fruit flavored milk beverages during pasteurization

Courtesy of Glassine and Greaseproof Manufacturers Assoc.

FIG. 10.6. MILK SHAKE MIXES
Chocolate-flavored milk shake mixes are packaged in individual serving pouches of wax-laminated white opaque glassine paper. Ten pouches are packedin printed cartons (left). A special pack (center and right) includes four pouches and a plastic shaker.

(30 min at 62.7°C). Shelf-life of at least 2 weeks is possible when properly refrigerated.

For localized delivery, milk drinks are packaged in the same type of containers used for whole fresh milk (see chapter 5).

Some chocolate drinks are bottled and canned. If suitable stabilization techniques can be perfected, a wider line of milk drinks will become available.

CARBONATED SOFT DRINKS

The first carbonated beverages were effervescent mineral spring waters. When Joseph Priestley discovered how to make carbon dioxide gas in 1772, he began experimentation trying to duplicate natural fizzy waters. Carbonated water was bottled commercially in Switzerland and England about 1790. Similar experiments were conducted in the United States by a doctor who thought the product might have curative powers. The chemist that he hired did not like the flavor of plain soda water and added some fruit juice. This was the first carbonated soft drink. All others since produced are fundamentally the same. Yet this basic formula is now a $6 billion industry in the United States with over 5 billion cases of bottled or canned soft drinks sold annually. Not bad for an industry which until 1953 had only had bottled drinks. In that year, the first soft drinks were sold in cans.

Recent trends appear to indicate that steel cans will become more widely used for soft drinks. Industry reports show three-piece soft drink can shipments moving up only 3% from 1976 to 1977 (13.4 billion compared to 13.8 billion), while the two-piece can soared 55% in the same period (6.1 billion to 9.5 billion).

Courtesy of The Metal Box Co., Ltd.

FIG. 10.7. CARBONATED SOFT DRINKS
This popular United States package is now being used in the United Kingdom. It features an all-aluminum easy-open top.

Product Characteristics

A carbonated soft drink is principally water; therefore, much care is expended in selecting good quality water. Carbonation consists of dissolving carbon dioxide gas in the water. The gas may be from any pure source. Modern processors utilize pressurized cylinders of purified gas. The amount used varies depending upon the type of beverage. Ginger ales are highly carbonated with up to 4½ volumes of gas, whereas fruit flavored drinks may only contain 1 volume of dissolved gas. (A 5 volume solution requires a pressure of about 5 atmospheres and produces about 1% solution of CO_2 in water.) The lower the temperature, the more carbon dioxide can be dissolved at a given pressure. The dissolved carbon dioxide not only contributes flavor and "fizz," but also inhibits growth of microorganisms.

Sugar used to be a major ingredient of soft drinks because it was a major component of the fruit syrups used for flavoring. Fruit syrups normally contain about 50–60% sucrose. About 1½ fl oz (42.5 g) flavored syrup is used in a 7 oz (198.45 g) bottle of soft drink beverage. In recent years many diet or low-calorie soft drinks have been introduced in which artificial sweeteners have been substitued for the high calorie sugar carbohydrate.

Natural fruit flavorings were the original additives. Today, numerous synthetic flavorants have been developed by using the technique of gas chromatography to identify components.

Soft drinks can be made clear or cloudy, colored or colorless, with or without natural fruit juices added. Where juices are included, homogenization is usually employed to stabilize particle size and distribution of pulp. Clear beverages are filtered or otherwise clarified. Coloring can be caramel, fruit colorings or certified food colorants. Although noncarbonated beverages may be pasteurized to prevent spoilage, carbonated beverages depend upon the cleanliness of the container, the acidity of the formulation and added sodium benzoate preservative to prevent spoilage. Long-term storage can result in flavor changes due to the action of light, heat, oxygen, enzymes or impurities introduced by the water or sugar used. Off-tastes and precipitation of solids are the usual symptoms.

Packaging Requirements and Packages

Carbonated beverages require a container that will hold pressure, not deteriorate and not contribute off-flavors. Traditionally this has meant the use of glass bottles, but in recent years special can coatings have been developed which permit the use of tinplate and aluminum cans. Bottles were more economical if used repeatedly, consequently the industry developed a system of collecting used bottles by refunding a deposit paid at point of purchase. Used bottles were washed to remove labels and sanitize, rinsed, sterilized with chlorine solutions, and then refilled, capped and relabeled. In very recent times lighter weight one-trip glass bottles have been developed which are discarded after one use. It was expected that with the continuing increase in labor costs the one trip bottle would ultimately displace the multitrip bottle; however, anti-litter laws are now beginning to affect this trend. Flint, amber and green glass have all been used for different types of soft drinks. An English ginger beer is bottled in a ceramic squat bottle with a crown closure. Product differentiation has been attained through distinctive shapes, distinctive labels, distinctive product names and intensive brand advertising.

In early days there was a variety of closures used for the bottled carbonated beverages. One consisted of a glass marble which was forced against a rubber ring in the neck by the gas pressure. A protruding rigid wire loop was used to pull the marble into position. To open, the loop was struck a sharp blow. This released the pressure with a loud "pop." The bottles became known as pop bottles and the contents became soda pop. The crown cork closure has already been described in Chapter 1. Screw stoppers relied on a rubber ring for tight closure. Another popular closure was the swing lever stopper. Here a porcelain stopper was held down against a rubber seat by a strong wire lever. Such bottles are now collectors' items.

The soft drink industry was a very strong influence in the development of labels and labeling machines for both bottles and cans. Bottle labels not only had to be attractive and resist coming loose in ice water, but also they had to disintegrate readily in the hot alkaline label-removing baths. Some soft drink bottlers used fired-on glass pigments for permanent labels and others used raised letters formed in the bottle blowing operation.

Soft drinks are even entering the political scene. Pepsi-Cola is exclusive to the U.S.S.R. while Coca-Cola has entered the Chinese market. The two largest selling cold drinks are thus opposing each other in the communist market. Will other soft drink bottlers enter the political races? Only time will tell whether the struggle to carbonate man's stomach will enter a new stage.

New Trends

There is a trend now toward returnable bottles. Twist-off crown bottle caps are being used. Aluminum and tinplate cans incorporate easy-open pop top lids. Labels are dispensed with, in favor of direct printing on the can metal.

Several interesting trends in soft drink packaging have appeared in the last 10 years. These include:

(1) Use of plastic bottles (PET).

(2) Use of plastic-coated glass bottles.

(3) New shaped cans.
(4) New closures.
(5) New bottom/lighter cans.

Courtesy of Reynolds Metals Co.

FIG. 10.8. PALLET SHIPMENTS OF SOFT DRINK BOTTLES
The "Reynolon Pallet Overwrapping System," developed and marketed by Reynolds Metals Company, enables one person to over-wrap pallets of glass bottles and other commercial and industrial palletized loads in just 3 min.

It is now possible to buy the three top brands of soft drinks in 2 liter plastic PET (polyethylene terephthalate) bottles. Consumer acceptance of these new plastic bottles is well established. Within one short year,

PET bottles have moved to a point where 1978 consumption has been estimated to be 34 million kg.

Glass bottles are being coated with "Surlyn" as a means of decreasing breakage in returnables. DuPont has developed a super-strong Surlyn for returnable bottles and it has stood up well to all distribution challenges. By using the coating, glass-bottle plants have cut down on breakage.

Can sizes and shapes introduced include the short, stubby ½ pint (.236 liter) and the slim 7 oz (207 ml) Pepsi. New closures include Alcoa's pop-top "Saturn," W.H. Hutchinson and Son's "Smoothie" aluminum roll-on cap and Owens-Illinois' first plastic cap for carbonated beverages, "Obrist." Cans are also becoming lighter with new bottom design.

ALCOHOLIC BEVERAGES

Fermented Wines and Beers

The invention of beer is lost in antiquity. Mead was produced from fermented honey before the middle Stone Age period.

It has been estimated that 40% of all grains raised in primitive agriculture were for brewing into beers or wines. Wine from the grape is known to be older than 4000 B.C. and probably predates 6000 B.C. Fermented date juice was popular in the Mediterranean and African areas. Fermented milk was of Asian origin, as was also rice wine. The South American Indians manufactured a wine from the cassava called *paiwari* and the Aztecs made a drink called *pulque* from the agave plant.

The excellent table wines of today date back to the labors of the Cistercian monks in the medieval monasteries, who perfected many arts in the growing and processing of foods and beverages. They established vineyards all over Europe and particularly in the German and French river valleys where climate and soil were most suitable. Some French vineyards date back to 630 A.D.

The method for making carbonated wine was known for a long time but no one could bottle it. Finally in 1670 a Benedictine monk, Dom Perignon, invented a cork that would hold the pressure.

Manufacture of Beers and Wines

Generally speaking, beers are made from grain and have a low alcoholic content, around 5% by volume, whereas wines are made from fruit and range as high as 14% alcohol. There is some overlapping of this definition for there are rice wines, barley wines and ginger beers. Fermented honey is called *mead*; fermented milk, *koumiss*; fermented apples, *cider*; and fermented pears, *perry*. Although other fruits have been used to make wine, the grape is by far the most popular and the most often used.

United States consumption of wine is about 1514 million liters a year. By 1980, more wine will be sold in the United States than distilled spirits. The per person drinking of wine (man, woman and child) is around 6.8 liters a year. It is expected to top 11.7 liters by 1985. Already a retail market of nearly $3 billion, United States drinking of wines is projected to grow 6% a year into the mid 1980s. The top domestic supplier is California (70%) with New York a distant second. The leader in imports is Italy.

United States beer consumption is over 150 million barrels. The top U.S. brewery (Anheuser-Busch, Inc.) sold over $2 billion of beer in 1977. In terms of volume, Californians topped the list with 16.8 million barrels; New York, 12 million barrels and Texas, 11.7 million barrels. The total consumption rate for malt beverages for the United States was 859.3 liters per person (over 21) or 14th in international beer consumption.

Imported beers are now becoming a factor, amounting to 2.6% of the U.S. beer market in 1979. As the popularity of these stronger-flavored beers grows, more and more brewers will emulate the Miller Brewing Co., which has started brewing Löwenbrau domestically.

Beer

Barley is the chief grain used in beer. Rice and corn are used where a pale color is desired. Barley malt is made by soaking the grain in water and allowing it to sprout. This develops the enzymes which convert starch into sugars. The malt proteins contribute flavor. The green malt is dried to halt further sprouting. After the dried malt has been milled to crush the husks, it is combined with water and cooked. Corn and rice mash are also added. The cooking action converts the starches to sugars. The converted mash is called wort. The sugar-laden wort is strained and boiled with hops and dextrose to coagulate hydrolyzed protein and albumin and to destroy enzymes and sterilize it. The hot wort is strained, cooled and mixed with a pure yeast. The yeast fermentation reaction converts the sugars to alcohol and carbon dioxide. The beer is cooled to 3.8°C and filtered. Next it is carbonated with carbon dioxide gas. With some beers a secondary fermentation is allowed. Carbonation may consist of several purgings and take 2 weeks in order to insure oxygen displacement and complete absorption. After a chill-proofing and a final filtration the beer is ready for bottling, canning or casking. Lager beer is aged after fermentation. Bock beer is darker and sweeter because the malt is roasted and caramelized. Most regular beers are made with bottom-fermenting yeasts. The yeast sinks to the bottom of the tank.

Ale, porter and stout are made from top-fermenting yeasts.

Wine

The type, characteristics and quality of wine are determined by many factors: the type of soil, the variety of grape, the location of the vineyard, the climate, the ripeness of the grapes, their sugar content, their acid content, how they are crushed, how they are fermented, and how the wine is aged. Grapes are harvested, inspected, crushed, destemmed and pumped to fermentation vats. The crushed grapes called "must" may be treated with sulfur dioxide to kill wild yeasts and then fermented by adding a carefully prepared pure culture of yeast. For white wines the must is pressed to remove "pomace" (seeds and skins) before fermentation; for red wines pressing is done after fermentation is nearly complete. Red wines are fermented in open vats, white wines in closed vats and at lower temperatures. If sugar is to be added it is done before or during fermentation. In both cases a preliminary period of settling or ageing is required to remove suspended material, off-tastes, off-colors and off-odors. Further fermentation may occur. Final ageing is usually done in wooden casks or barrels to develop flavor and bouquet.

Fortified wines owe their higher alcohol content to the addition of brandy or other spirits. Sweetened wines need to be pasteurized or specially filtered to avoid spoilage.

Spanish dessert wines are world renowned. The Spanish Malaga and Sherry wines are superb. Portugal is famous for Muscatel, Port and Madeira wines.

The characteristic flavor of Sauternes is achieved by allowing the grapes to be attacked by a special mold, *Botrytis cinerea*. This loosens the skin and dries the grape, increasing its sugar content. The moldy grapes must be harvested promptly and selectively. Usually only two pickings are employed for the highest quality. The mold contributes flavor and aroma, and helps keep acidity down and alcohol content up. Sauternes are aged 2–3 years in barrels and continue to age in the bottle.

Sparkling wines are first completely fermented and then drawn off and blended. Sugar and a new yeast culture are added for final fermentation in the bottle. Bottled wine is held in warm rooms during the second fermentation which may last 1 year. When fermentation is complete, the yeast dies. It is then necessary to remove the yeast sediment. Bottles are held upside down to allow the sediment to settle on the cork, the bottle is chilled and the neck area frozen. The cork and plug of frozen wine are removed and discarded. Losses are replaced and the bottle recorked. Sugar syrup and brandy are added just prior to the final corking to adjust flavor and alcohol content.

Some wines are carbonated artificially rather than by natural fermentation, such as sparkling Rosé and Burgundy.

Vermouths are base wines that are fortified and then flavored with spices. They derive their name from "Wermut," the German word for wormwood, one of the first used flavorants. Almost any kind of herb, spice or flavorant may be used, from allspice and aloe to vanilla, wormwood, yarrow and zedoary root.

Still and sparkling wines are also made from apples, cherries, plums, currants, blackberries, raspberries and other berries to produce sweet dessert wines. Other wines have been made from Concord grapes, pineapples, pears, pomegranates, oranges, grapefruit, figs, dates and honey. In all instances the basic manufacturing processes are essentially the same.

United States wine production is centered in California, New York and Ohio. Most United States wines are attempts to duplicate the finest European types. However, a few unique wines have been developed including the Californian "Angelica" and the eastern "Scuppernong," "Concord" and "Catawba" wines.

Packaging of Wines and Beers

Beer.—The classic package for beer was a keg or barrel which was tapped and drawn-off into mugs for immediate consumption. With the development of modern cooperage, metal barrels replaced the wooden kegs and draught beer became more and more common as central breweries distributed the product. Draught beer was used up promptly and rarely had time to deteriorate or spoil. When distribution began to cover larger distances some of the beer was pasteurized. This was accomplished by heating the beer to 60°C for a few minutes to destroy yeasts or other bacteria. When a bottle was developed which could hold the pressure of carbonation, this became a popular package. The crown cap had a very strong influence on the success of bottled beer, as the closure had to be able to withstand pasteurization also.

No one knows who first thought of putting beer into tinplate cans. Even before the sale of the first canned beer by the Krueger Company of Newark in 1935, both Pabst and Anheuser-Busch had tried it experimentally in the late 1920s. But Prohibition and the search for a suitable can liner caused a delay in the introduction of canned beer. When the Gottfried Krueger Brewing Company of Newark, N.J., introduced their 12 oz canned beer, they used an American Can Co. container. Within 5 months of the package's introduction, the company was running 550% of its pre-can production.

While all this was happening, other brewers were watching Krueger closely. Before long the Northampton Brewing Company of Northamp-

Courtesy of Reynolds Metals Co.

FIG. 10.9. BEER PACKAGES IN GLASS AND ALUMINUM
The 2 piece aluminum can is a popular package for beer. This brand of beer also features a brown glass bottle with a unique shape. A simple 6 pack carrier for cans is also shown.

ton, Pa., followed with their "Tru Blue Ale" and "Tru Blue White Seal Beer." The containers were made by National Can Co. Pabst was the first national company to use cans, not for their "Blue Ribbon" brand but for an "Export" brand. In case the development failed, their leading name would not be injured.

The first cans introduced had a flat top. Opened only by a "church key," they had an unbroken top as opposed to the modern ring tab device.

In late 1935, Schlitz developed an entirely different style of can. Resembling a metal bottle, it had a round body and a conical top sealed with a crown cork. Called "cone top" or "spout top," the can looked like a bottle and filled like a bottle. Consumer resistance to drinking beer in a metal container was reduced and the conversion costs to can filling equipment was partially eliminated. The first cans used outside the United States were all "cone tops."

To the designer, the "cone top" is a poorly designed container for holding beer. Although it did open easily, the package looked unattractive and resembled a tin of metal polish. And the can's unusual design has lingered on in people's memory. Even today, most people believe that the cone top was the first design introduced for beer. Yet the flat top preceded it by about 1 year.

In Europe, breweries were closely following developments in the United States. In January, 1936, the Felinfoel Brewery in the United Kingdom introduced their "Pale Ale" in cone tops. This was the first European beer to become available in metal cans. Although canned beer was slow to catch on with the United Kingdom consumer and the development was promoted heavily by the tinplate manufacturers, other British breweries soon followed the Felinfoel experiment. In the pre-World War II years, Simonds, Watney's and McEwan's introduced canned beer. But success was more rapid in the United States than in any other part of the world. This new form of packaging was popular as early as 1936 and more brewers expressed willingness to install the necessary can equipment.

In 1937, a quart cone top designed and made by Crown, Cork and Seal was introduced. Shortly afterwards, American Can produced a quart flat top. But the death of the "cone tops" had already started. In 1938, the firm that originally introduced the cone, Con Can, was making "flat tops." The Old Dutch Brewery in Brooklyn, N.Y., was the first to use them.

World War II halted can production in both Europe and the United States. No more cans were made in Europe in the early 1950s. This was unfortunate for the can manufacturers since much of their investment in marketing and publicity went to waste.

In 1940, Crown, Cork and Seal introduced their "Crowntainer" can. It was a new style of cone top. Schmidt's was the first to use this new two-piece can. The original cone tops had been three-piece constructions with a base, a main body and a cap. The Crowntainer had only one seam rather than two, so it reduced by ½ the chance of beer leaking away. The can was made of tinless black plate covered with powdered aluminum and had a permanent glazed appearance. Used during the war, the Crowntainer helped convert ordinary men and women to drinking beer from cans.

However, as the 1950s crept by, the cone top gradually became obsolete. The last cones were filled in the late 1950s and the last were sold in 1960. The era of the flat top was finally here.

Now manufacturers sought ways to improve the quality of their product. Steel had always been used for beer cans, but in 1959, Coors and Gunther experimented with aluminum for the first time. Both breweries eventually produced a 7 oz container. Their experiments were a success. During the 1960s, many brewers were converted away from steel.

Sometime in the late 1950s, the party can was introduced. Two sizes of flat top party cans were marketed, one holding 4 pints and one holding 7 pints. Many of the original cans had a hole in the top, sealed by a rubber bung which was removed to pour out the beer. Some had two holes next to each other. Although the party can was a success in the United Kingdom and Europe, it failed in the United States. Americans like to drink beer cold. In order to chill a gallon container, it had to go into a refrigerator. The large can was simply too clumsy to store properly. The price of the beer was also expensive and there were problems with the pump device developed to go with the cans. In 1973, Koch's was the last gallon can taken out of production.

In 1960, an aluminum top was developed for use with a steel can. Schlitz was the first to use this can. The first can with a paper label appeared in 1959 in Hawaii. Production problems were evident and these "Shiny Steinies" were soon withdrawn from the market. Development between 1960 and 1970 was basically concerned with opening methods. Simply carrying hole punchers around to pierce the tops was not satisfactory. In 1962, a "lift tab" was developed. When the tab was pulled off, it left a hole through which the beer could easily be poured. The "rip top" opening system involves ripping the tab off the can. This has caused problems with environmentalists and Oregon has even banned the use of the rip top completely. The most popular alternatives devised do not involve pulling anything entirely away from the body of the can.

The "push top" has two buttons on top of the can. Both are pressed inside the can but remain attached to the top. The beer is then poured out through the larger of the two holes. Continental's "Envir-o-Can" has a laminated foil strip attached to the top which is pulled up to reveal holes that have been prepunched in the top.

Recently, Reynolds Metals "Stay-On-Tab" has been introduced. It is a patented beverage can-opening device that improves pouring and stays on the can to help reduce litter.

The latest method for producing a two-piece can has been the "drawn and ironed" system. A cup is formed from a flat sheet of aluminum. This cup is ironed into a seamless body with very thin walls but a thicker bottom. The body becomes more rigid when filled, due to the pressure of the beer inside. In the mid 1970s, the "crimped" or "knicked in" can was introduced. This involves a reduction in the diameter of the top and bottom of the can, making space saving possible, which in turn makes multiple packaging much easier.

Beer can development is still continuing. Plastic cans and pouches have not yet been a success in the marketplace. ICI's "Merolite" pouch (inflated with CO_2) has not met with widespread acceptance. Neither has

Courtesy of Reynolds Metals Co.

FIG. 10.10. FOIL LAMINATED "COLD PACKS"
Foil laminated carriers for 12 cans of beer come in different designs. Carrying handles are an added convenience. The carrier is closed on all sides to take advantage of the fact that foil keeps the contents cold longer.

the plastic bottle marketed by the Pripp brewery, a Swedish firm. Even the plastic beer can called "Dumpies" and sold in the United Kingdom by Bass Charrington failed. The container, balanced on three squat points, was quite unattractive.

Wine.—As with beer, wines were once packaged in wooden kegs or casks. The corked glass bottle became the traditional package when glass bottles became cheap enough to be used for the purpose. Recent trends indicate a growing acceptance of plastic stoppers instead of corks for all wine bottle closures. In the United States nearly all wine is sold in bottles whereas in France, Italy and Spain sizable quantities are sold in bulk directly to the consumer.

The Harvey Wine Museum (Bristol, UK.).—The history of wine packaging is graphically illustrated in a permanent exhibit located in the United Kingdom and sponsored by Harvey's of Bristol.

There is a comprehensive collection of silver decanter labels (or bottle tickets) in the Main Gallery. It is based on the famous E.J. Pratt collection of more than 500 items which Harveys acquired in 1960 and has since added to, with 200 outstanding items on permanent display.

Decanter labels have been used for over 200 years to indicate the contents of bottles and later decanters. They were often made of parchment, bone, wood or ivory. With growing affluence and the introduction of elegant cut glass decanters in the second half of the 18th century, silver labels became the vogue.

The collection is displayed to show the development of decanter label design. Pride of place goes to rare specimens from the period 1734 to 1769, when the use of labels was confined to a few members of the aristocracy and very wealthy families.

Adjacent to the label display is a collection of tastevins, splendid tasting cups still used in France and Germany to examine wine drawn from the cask. The earliest tastevins on show date from 1677 (a two-handled style made in Paris) and 1690 (also two-handled and a fine example of the English William and Mary period).

An ancient stone amphora marks the entrance to the bottle gallery, which traces the history of the bottle through the centuries. The bottle developed not merely as a convenient container for wine; technical advances in glassmaking also led to the development of sparkling wines.

Glass was introduced to Britain, along with the wine itself, by the Romans. But stone amphoras were used for storage, transport and maturing of wines, so the real development of the bottle was delayed until the 17th century. With the bottle emerged the corkscrew, developed in the 17th century. Until then, if corks were used they had to be wedge-shaped and tied down with string. A really airtight seal was not possible until the corkscrew made it feasible for the cork to be pushed right down the bottle neck and later withdrawn cleanly. From that time, wine could be kept for long periods in the bottle and the art of maturing vintages—unknown since Roman times—was rediscovered all over Europe. Exhibits show the evolution of the bottle since its beginnings around 1630, with excellent examples of squat and dark mid 17th century bottles and some enamelled earthenware sack jugs of the same period. Machine-moulded bottles appeared first in the 1830s but the bottle-making did not become fully mechanised until 1900.

In France, government regulations exist which define the requirements that must be met in order that a wine be permitted to use a label identifying the particular region of its origin (*Appellation d'origine*). These regulations, known as appellations controlees, are administered by the Institut National des Appellations d'origine de Vins et des Eaux-de-Vie. They specify the geographic area to which a label may refer, the varieties of grape which may be used, the minimum alcohol content, and the maximum yield per acre. Unofficially the vintners of Bordeaux have had a voluntary type of appellation controlee for centuries. The wines of the Medoc were classified in 1855 into five "growths" or vineyards. The larger vineyards were called chateaux. Wines bottled at the vineyard carried its name, e.g., Chateaux d'Yquem.

Wine is difficult to package in metal, as this prevents proper aging and adversely affects flavor, color and aroma. Wine bottles come in a myriad of shapes, sizes and colors. The sizes range from individual serving "splits" to 5 gal. (18.9 liters) capacity "demijohns." They may be short, squat and rounded or they may be tall, narrow-necked and slender. They may be over wrapped with straw or woven "basketry." They may be made from clear white glass or colored glass. Carbonated wines are made of very heavy glass to withstand the pressures.

Riesling wines and other Rhine wines are produced in an area near Bonn which includes the village of Hochheim. The word "hock" has been used to describe German white wines and probably is derived from this village name.

Moselle wines are bottled in similar shaped bottles made of blue or greenish blue glass. Wines from the Franconia area of Germany are put in squat, broad-bottomed green glass bottles.

French red Burgundies and Chablis wines are bottled in green or brownish green bottles similar in shape to but lighter weight than champagne bottles.

Clarets from the Bordeaux region are in dark-green glass bottles with straight cylindrical sides and necks. This is also used for California red wines, Spanish sherries and Portuguese ports, although the last two may use brown glass as well. French and California Sauternes and French Graves wines are bottled in clear white or light green glass of the same shape as the claret bottles.

Champagnes and other sparkling wines are bottled in heavy green glass with indented bottoms.

Vermouths are usually put in green bottles of the claret shape. Italian red wines such as Chianti wines are nearly always put in round-bottomed green glass flasks with woven straw covers. Italian white wines are usually more conventionally bottled.

The glass bottle is traditional for wine packaging. The trend toward economy is reflected in the preference for larger sizes (quarts, liters and half gallons). Plastics have been tried in Europe with some success. The "Zu-pack" container has been used for this purpose. White wines employ a paper-polyethylene laminate, while red wines need the light barrier afforded by aluminum foil. In France, plastics were first tried in vending machines in factories. Here the worker could buy *vin ordinaire* and not have to carry a bottle to and from home. Within 3 years of the first trial, major wine producers converted to blow-molded PVC plastic bottles. The first plastic bottles used permitted a maximum shelf-life of 6 months, certainly not a fraction of the life in glass but enough for volume consumption of ordinary quality beverages.

Small individual servings of wine in plastic containers such as the tetrahedral-shaped package and the fin-sealed pouch are being used for airline serving. Made in France, the concept is not new and was first used on United States airlines in 1978.

DISTILLED SPIRITS

Although distillation was known in Aristotle's day, the distillation of spiritous liquors is believed to have been perfected in Europe sometime in the Middle Ages. (It is known that Arrack was distilled in India as far back as 800 B.C.) In northern European countries, where grapes do not flourish, fermented grains were distilled to produce whiskies, gins, vodka, schnapps, aquavit and the like. The Welsh were distilling mead in the 6th century A.D. Irish and Scotch distillers were active as early as 1170 A.D. In more southern climates the grape predominated, and wines were distilled into brandies in Italy as early as the 9th century. Other fruits, grains, or sugar- or starch-bearing plants have been utilized including sugar cane (rum), potatoes (vodka), plums (slivovitz), cassava (paiwari) and cactus (pulque).

In the United States and Canada over 1514 million liters of distilled spirits are consumed annually. White goods (vodka, gin and white rum) have shared in the trend away from brown whiskeys.

Brandy

Brandy, *weinbrand*, *branntwein*, *aqua vini* or *eaux-de-vie*, is the product of distillation of wine. Several successive distillate fractions may be drawn-off, blended and redistilled. Finally, the distillate is aged in wooden casks for up to 20 years or more. Brandies may derive their color and flavor from the original grapes (and wine), or flavorants and colorants may be added. They may be consumed as a beverage directly or used to fortify wines.

Eaux de Vie.—The "waters of life" are clear, white, fruit brandies made not from chemicals, colorings and sugar, as are the cordials, but from tons of pure, ripe fruit. Twenty-three kilograms of fresh strawberries are needed to make a ½ liter of strawberry *eau de vie*, which is usually known by its French name, *fraise*.

The taste and aroma of a true *eau de vie* are unforgettable. Uncorking a good *poire*—pear brandy—will fill a room with the bouquet of an orchard of pear trees. The taste is intense and fiery. It will appall someone who likes and is expecting a sweet cordial. For the sophisticated palate, though, fruit brandy is a great flavor experience.

A true fruit brandy must be made from the fruit whose name it bears, except for those fruits, such as wild berries, which contain so little natural sugar that they must be macerated in alcohol first. The alcohol is usually clear, unaged cognac. *Eau de vie* distillers, unlike their colleagues the winemakers, are forbidden to chaptalize (add sugar) to the fermenting fruit to increase alcohol.

All distillates are clear when they come from the still. That includes scotch and bourbon as well as cognac and *eau de vie.* The color comes from the aging barrels or, in the case of cheap products, caramel. The eau de vie producers, unlike the whisky and cognac makers, want to keep the original clear absence of color. So they age the young fruit brandy in glass demijohns, fiberglass or stainless-steel containers. A few start out the aging process in wood to get a special taste, then shift over to glass to avoid getting any color from the barrels.

The first distilled drinks probably were made from wine. Then came grain and finally fruit. Actually, the alchemists were said to have used fruit distillates in their potions and elixirs far back in the Middle Ages. Undoubtedly, fruit brandies were at least in part an outgrowth of the need to get rid of large quantities of fruit in an age when markets were difficult to reach, just as the first settlers in the American middle west discovered that it was easier to send whisky than grain back through the Cumberland Gap.

Most of the good *eaux de vie* come from northeastern France, Switzerland and the Black Forest region of Germany.

It is probably only fitting that such elegant drinks are produced in some of the most picturesque parts of the Old World. There are commercial distilleries but there are also small, family distilleries that sell their product at the door or to the larger shippers.

The most famous of all the *eaux de vie* is *kirsch*, which is made from cherries and which almost no one, not even in France, calls *cerise*. Cherries are a "stone" fruit (as opposed to "seed" fruits). They are fermented with the stones or pits, and the trick is to crush the fruit without damaging the pit, which is full of bitter flavors. Twenty-three kilograms or more of cherries are needed to make a single bottle of kirsch.

The next most important, or at least best known, *eaux de vie* are made from bartlett pears *(poire Williams)*, raspberries *(framboise)*, yellow plums *(mirabelle)* and purple plums *(quetsche)*. The pears provide the most spectacularly packaged fruit brandy. Some producers literally tie the bottles around tiny pear buds in the trees in the orchards each spring. The pears grow to maturity in the bottles, which are then filled with pear brandy.

Framboise, made from raspberries, has an intensely sweet "nose," redolent of ripe fruit, which makes its sharp taste come as a shock. However, *framboise* is not as powerful either in taste or smell as *poire* and *kirsch*.

The *eau de vie* producers pride themselves on being able to make a brandy out of almost anything that grows and has some sugar in it.

Fruit brandies are made from peaches, apricots, apples, blackberries, red and white currants, rose hips, black currants (cassis), mountain cranberries, gentian root, acacia flowers and quince. There is also an *eau de vie* made from eglantine, a shrub related to the rose, which also bears the slightly racy name of *gratte-cul*.

Whiskey.—Whiskeys are made from fermented grain mashes. The type of grain used, how it is malted and the later treatment determines the type of whiskey. Irish whiskey is made from malted and unmalted barley and from oats, wheat and rye. The malted barley is dried over peat fires before fermentation. The whiskey is given three successive distillations before ageing.

Scotch whiskey is made mostly from malted barley. It too is cured over peat fires. Scotch is distilled only once, then blended. Both Irish and Scotch are distilled at between 160 and 190 proof and aged in used charred oaken casks before bottling.

In the United States, whiskies are made from any grain or mixture of grains, but one grain must make up at least 51% of the mash. Rye whiskey comes from a mash containing 51% rye grain, bourbon whiskey from a mash containing 51% corn. Corn whiskey requires a mash to be 80% corn. When distillation is taken off at 190 proof or more, the product is called grain neutral spirits and contains 95% alcohol by volume. United States whiskeys are by definition less than 160 proof. When exactly 100 proof and aged 4 years, the whiskey may be labeled "Bottled-in-Bond." It must not have been blended or diluted with water, grain neutral spirits or other whiskies. Whiskies other than 100 proof may be labeled straight whiskey, for example straight rye whiskey. The straight whiskies must be 2 years old and may be diluted with water to desired proof but may not be blended with grain neutral spirits or other whiskies. Blends of straight whiskies are mixtures of straight whiskies with no other additives. Spirit blends or blended whiskies must be 25–40% straight whiskey with the balance grain neutral spirits. The label must include the word blended and cannot use the word whiskey without the qualifying adjective. Most United States straight whiskies are diluted to under 100 proof. All must be aged in new charred oak barrels.

"Proof" is a legal definition of alcohol content. The definition differs with each country. In the United States, a proof spirit is 50% ethyl alcohol by volume with the alcohol having a specific gravity of 0.7939 at 15.5°C. In Holland, a proof spirit is 50% by volume absolute alcohol at

15°C. In England a proof spirit at 15°C has 12/13 the density of distilled water.

Gin.—Gin is a distilled spirit made from corn, barley, malt and rye. Its name is foreshortened from the French word for juniper, *genievre.* In Holland and Germany it is called *schnapps.* In early days, it was called *Hollands, geneva* and gin.

English gin is distilled from a mash of 75% corn, 15% malt and 10% rye and is redistilled with added juniper berries and coriander seed. American gin is 85, 12 and 3% mash, and the flavorings are not added until the last distillation. Dutch gin contains 2 parts of rye to 1 part of barley malt. Juniper is added after the first distillation. There are three distillations in all. Dutch gin has a more "grainy" flavor.

Gins can be sweetened with flavored syrups to produce cordials, whereas liqueurs are more complex blends with flavored wines and spirits. Compound gins may contain added grain neutral spirits. Vodka is very similar to gin except that it does not contain juniper flavoring and may include potatoes as well as corn and rye malt in the mash.

Rum is made by distilling sugar cane residues and molasses. It can also be made from sweet potatoes and from sugar beets.

Packaging of Spirits

As with wines, distilled spirits first were packaged in wooden casks, later in glass bottles. The most important parts of the package were the label to identify the contents and the closure to prevent pilferage. It is prohibited by law to reuse a bottle that contained spirits and the glass is so marked. Most bottles used for spirits are colorless or pale green. The effect of light on the stability of various liqueurs was studied by Connor (1966) by storing them in sunlight for 3 months in flint, amber and green glass bottles. Color was measured as a change in absorbance at 580 and 650 μ wavelengths. Sloe Gin and Cherry Liqueur lost some of their red coloring and in general the green glass bottle offered the best protection. Triple-Sec in flint glass developed an off-flavor which may have been related to an increase in aldehyde content from 56 to 80 mg per liter.

The most recent improvements in the packaging of spirits has been in the closures used. These have evolved from cork of earlier days to molded plastic screw caps and roll-on metal caps. Clear or colored plastic shrink films are used over the screw caps as a means of preventing pilferage. The metal caps have to be destroyed, in part at least, to remove them the first time.

Christmas gifts of spirits have been promoted through imaginative packaging. This has been through the use of decorative decanter bottles or through use of printed cartons or of printed carton overwraps. The latter have become extremely attractive through use of embossed foil laminates.

Courtesy of Natl. Distillers Prod. Co., A Div. of Natl. Distillers and Chemical Corp.

FIG. 10.11. CHRISTMAS GIFT PACKAGING OF FINE WHISKEY
A straight Kentucky bourbon whiskey is shown in its holiday packaging, a deluxe, cut crystal-type decanter in an eye-catching oval shape. The gift wrap presents a flocked foil cap over a moire embossed foil shell.

The American Can Company and other firms have developed a vinyl interior coating for their cemented side seam aluminum can, which enables them to package alcoholic beverages in cans with a shelf-life of more than 1 year. The cans contain mixed cocktails and each 8 fl oz (237 ml) can holds about three normal drinks. The cans have an easy open top. Heublein, Inc., is marketing the cocktail mixes. Other firms have introduced similar packages. In 1978, a Japanese firm announced the development of powdered alcohol mix. By reconstituting with water, an alcoholic beverage was obtained from a dry mix.

Alcohol Beverage Regulations

The Bureau of Alcohol, Tobacco and Firearms was established by Treasury Department Order No. 221, effective July 1, 1972. The order transferred the functions, powers and duties arising under laws relating to alcohol, tobacco, firearms and explosives from the Internal Revenue Service to the Bureau completely. On February 21, 1976, Treasury Department Order No. 221-3 (Revision 1) delineated to the Bureau those functions, powers and duties relating only to criminal violations and forfeitures of the federal wagering law provisions of the Internal Revenue Code of 1954.

Headquarters of the bureau are located in Washington but since the bureau is decentralized, most of its personnel are stationed throughout the country where many of its operational functions are performed. The director, who is located in Washington, D.C. and who supervises the entire bureau, is appointed by the Secretary of the Treasury and is under the supervision of the Assistant Secretary (Enforcement, Operations and Tariff Affairs).

The bureau is responsible for enforcing and administering firearms, explosives and criminal violation and forfeiture aspects of the federal wagering laws, as well as those covering alcohol beverages and tobacco products.

The objective of the bureau's programs is to maximize voluntary compliance with these laws, and to minimize willful or involuntary violations of the laws. To achieve these goals, the bureau is divided into two basic functions: criminal enforcement and regulatory enforcement. The objectives of the criminal enforcement activity are to eliminate illegal possession and use of firearms, destructive devices and explosives, to suppress the traffic in illicit distilled spirits, to enforce the criminal violation and forfeiture aspects of the federal wagering laws, and to cooperate with state and local law enforcement agencies to reduce crime and violence. The regulatory enforcement activity determines and assures full collection of revenue due from legal alcohol and tobacco industries, fulfills the bureau's responsibility in the prevention of commercial bribery, consumer deception and other improper trade practices in the distilled spirits industry, and assists other federal, state and local governmental agencies in the resolution of problems relating to industrial development, ecology and revenue protection.

Wine.—Under the Federal Alcohol Administration Act (FAAA), wine standards of identity were established for wines. They are:

I Grape wines: table and dessert wines
II Sparkling grape wines, champagne and crackling wine
III Carbonated grape wine
IV Citrus and berry wine
V Fruit wines
VI Other agricultural wines
VII Aperitif wines
VIII Imitations
IX Retsina wine

The labeling requirements of wine are covered in Subpart D. Certain mandatory information is required and must appear on the wine label. This includes:

(1) Brand name.
(2) Class and type.
(3) Name and address of bottler and packer.
(4) Alcoholic content.
(5) Net contents in English and/or metric units.
(6) Foreign wines must have name and address of importer.

In Paragraph 4.39, an entire section appears detailing prohibited practices relative to the labeling of wine. No misleading brand names are allowed. No information can appear on the label that is in type smaller than 8-point Gothic caps, except on less than half pint bottles. All required data must be in English. All statements that are "disparaging of a competitor's products" are specifically prohibited. Misleading containers are prohibited from use and fill tolerances are given for containers from 4.9 gal. (18.5 liters) to $^4/_5$ quart (0.72 liters).

Liquor.—The labeling and advertising of distilled spirits is covered in Part 5, Subparts A-H of Title 27, Part 4. Subpart C details ten standards of identity for distilled spirits. They are:

I Neutral spirits or alcohol
II Bourbon and corn, blended and Canadian Whiskeys
III Gin
IV Brandy
V Applejack
VI Rum
VII Tequila
VIII Cordials
IX Flavored liquors
X Imitations

Courtesy of Natl. Distillers Prod. Co., A Div. of Natl. Distillers and Chemical Corp.

FIG. 10.12. BOTTLE FILLING LINE FOR GIN
A view of a bottling room in a distilled spirits plant showing inspection in the foreground and casing operations in the background.

Subpart D deals with the labeling requirements of distilled spirits. Labeling is essentially similiar to wine, except that the percent of neutral spirits and the state of distillation must be added. The label must remain intact except for vigorous washing with water and solvents. Gift cartons must also bear mandatory information as must any opaque outer container other than a shipper.

REFERENCES

ALDEISON, M.G. 1970. Packaging problems of carbonated soft drinks and fruit juices. Food Mfg. *45* (8) 67–72.

ALMARKER, C.A. and WALLENBERG, K.E.S. 1968. New pouch method for ground coffee. Mod. Packag. *41* (6) 146–148.

ANON. 1967A. Wine cartoning gets underway—two systems now ready to go using rectangular packs. Packag. News *14* (1) 2.

ANON. 1967B. Nitrogen packaging of coffee by patented process preserves freshness for over six months. Packag. News *14* (1) 11–12.

ANON. 1967C. New automatic machine for the production of teabags. Food Packaging Design Technol. April *40.*

ANON. 1967D. Bag-in-box for sherry. Packag. News *14* (4) 6–60.

ANON. 1967E. PE bumps glass for fountain jug favor. Packag. Innovator *2* (3) 6.

ANON. 1967F. Close up: Beer. Packag. Eng. *12* (8) 75–79.

ANON. 1967G. Beer drinkers want the one-trip bottle. Packag. Rev. *87* (10) 12–13.

ANON. 1968A. Shake-up package in cocktail-mix market. Mod. Packag. *41* (3) 97.

ANON. 1968B. Portion-pack for liquor. Mod. Packag. *41* (3) 93.

ANON. 1968C. Sophistication in liquor half-gallons. Mod. Packag. *41* (5) 136–137.

ANON. 1968D. Gas flushing increases beer can shelf-life. Packag. Rev. *88* (6) 22.

ANON. 1968E. Problems of plastics bottles for beer. Int. Brewers J. *104* (1233) 41–43.

ANON. 1968F. A review of pasteurization and sterilization of beer. Bottler Packer *42* (8) 53–56.

ANON. 1969. 'Multivation' in coffee packaging. World Coffee & Tea *10* (8) 13–37 (December).

ANON. 1970A. Bottling. Food Process. Ind. *39* (461) 35–51 (February).

ANON. 1970B. A scientific report on testing plastic beer bottles. Bottler Pakr. *44* (2) 64–66.

ANON. 1970C. Pop Bottles: Explosives in the market basket? Consumer Rprt. *35* (3) 158–160.

ANON. 1970D. Composite film for packaging coffee. Neue Verpack *23* (4) 458 (German).

ANON. 1970E. Coffee in a tea bag. Mod. Packag. *43* (5) 87.

ANON. 1970F. Juice in a box. Food Eng. *42* (7) 56–7.

ANON. 1970G. Distillers CO_2 used in coffee packaging. Flavour Ind. *1* (9) 612.

ANON. 1970H. Packaging beer and wines. Packag. Rev. *90* (11) 25–27.

ANON. 1971A. Packaged water assumes new shape. Mod. Packag. *44* (6) 42–3.

ANON. 1971B. Packs (citrus) juice in PVC. Food Eng. *43* (6) 48–9.

ANON. 1972. Plastic cans for coffee. Eur. Packag. Dig. *99*, 4 (Jan.).

ANON. 1976. Bury company claims first 'plastic' tea bags. Paper *185* (6) 270 (March).

ANON. 1977. Records drink powder growth, expanded market for canisters. Food Drug Packag. *37* (7) 39 (October).

ANON. 1978A. Quebec can tax. Tin Int. *51* (6) 236.

ANON. 1978B. Aerosal beverage concentrates; a novel alternative to powders. Food Prod. Dev. *12* (8) 86.

BISHOP, T. 1968. The no-deposit bottle success story. Bottle Packer *42* (2) 90–96.

BULLEN, P.L.R. 1969. Factors affecting air in the neck of bottled beer. Bottler Packr. *43* (11) 94–98.

JOSLYN, M.A. and HEID, J.. 1963. Food Processing Operations, Vol. 2. AVI Publishing Co., Westport, Conn.

JURCZANG, E. 1970. Pouch-in-carton cuts package cost without cutting shelf life. Packag. Eng. *15* (6) 68–71.

PINTO, A. 1978. Containers vie for beverage market. Mod. Packag. *51* (9) 17–24.

ROBBINS, R.H. 1970. Trends in technology of fruit juice processing. Food Mfg. *45* (8) 65–66.

SACHAROW, S. 1967. Market potential perks up for coffee flex packs. Paper, Film, Foil Converter 41 (6) 41–43.

11

Sugar, Chocolate and Confections

> Rabbit said, "Honey or condensed milk with your bread?" (Pooh) was so excited that he said, "Both," and then, so as not to seem greedy, he added, "but don't bother about the bread, please."
>
> A.A. Milne, "Winnie the Pooh"

Primitive man had knowledge of sweet fruits, berries and honey. Most ancient confections were mixtures of honey, fruits and flour pastes, often molded into shapes having religious significance. Ancient Egyptian temples show hieroglyphics of early candymaking. Centuries of experimentation by cooks, sugar bakers and chocolate millers have resulted in literally thousands of delectable combinations of ingredients.

Packaging of candies has grown from nothing, when candy was kept and displayed in open trays, to a definite and important role in protecting the product against damage, spoilage and vermin infestation. The nature of the candy industry in modern times has limited the use of the best packaging materials. The majority of candy manufacturers have been small and regional in operation. Many have specialized in only a few types of candy. This has limited volume consumption of packaging materials. Also, although candy has grown to have a large overall sales volume the unit price has been low. This has limited the cost of packaging materials. In more recent years, a major trend has developed where large diversified food processors are assimilating candy companies through mergers. This is resulting in a careful scrutiny of traditional packaging practices and the application of modern packaging know-how and materials to be well-defined product requirements. Per capita consumption of candy has declined over the last few years. Still, over $1.5 billion worth of candy and gum were sold in 1977. About $50 million worth of sugar-free mints are sold annually.

SUGAR

Cultivation of sugarcane probably originated in Asia. The troops of Alexander the Great encountered it in their invasion of India in about 300 B.C. In the 7th century A.D. Chinese and east Indians were extracting syrup from the cane, but the Arabs and Egyptians are credited with the first recrystallization processes. Arab caravans and Venetian merchants built a trade in eastern sugar, which by the 13th century had become a virtual monopoly. Returning Crusaders, who acquired a taste for sugar in their travels, helped create and build a demand for the product. The Venetians perfected methods for refining sugar and became famous for their sugar loaves and their artistic sugar sculptures. From the 15th century other nations began to compete in the cultivation of sugar cane. It was spread around the world by the Portuguese and Spanish explorers and brought to the Americas by Columbus.

Beet sugar was discovered by a chemist in 1747, although Californian Indian tribes were reported in 1775 to have been obtaining sugar in this way. Napoleon Bonaparte established the first sugar beet industry in Europe. Other sources of sugar include sorghum, cane and sugar maples.

In 1919, a 90 ft storage tank used to store molasses from the West Indies before it was distilled into rum failed freeing 2.3 million gal. of molasses weighing 13,500 MT. Twenty-one people died in Boston's North End from the wave which crested at 50 ft. This incident is called, "The Boston Molasses Disaster."

From its early beginnings, sugar has grown to the status of a commodity and world consumption is measured in billions of kilograms. In the United States alone over 11 MMT of sugar are consumed annually. In 1978, 4.23 MMT were imported (see Table 11.1).

Product Properties and Packaging Requirements

Sugar is refined to several purities running from white to dark brown. Molasses is a byproduct and is used in candy as a flavoring. The color and flavor of the brown sugars are due to the presence of dissolved nonsugars in the thin layer of syrup that coats each crystal. Because of this syrup constituent, brown sugars must be kept in a moist environment, preferably 65–70% rh, to prevent caking into a solid lump. White sugar on the other hand is a dry but slightly hygroscopic product. To prevent formation of a moist syrup on the crystal surfaces, which could cause caking, environmental moisture should be kept low, preferably 30%, and the environment should not fluctuate excessively in either temperature or relative humidity.

TABLE 11.1. UNITED STATES SUGAR SUPPLIERS (1978 Estimate of Impacts in MM MT)

Dominican Republic	0.94
Philippines	0.73
Brazil	0.36
Peru	0.23
Argentina	0.18
Guatemala	0.18
West Indies	0.18
El Salvador	0.14
Panama	0.14
Nicaragua	0.13
Colombia	0.13
Australia	0.11
Costa Rica	0.10
Belize	0.09
Canada	0.09
Mauritius	0.08
Taiwan	0.07
Swaziland	0.06
Malagasy	0.02
Malawi	0.02
Mozambique	0.02
South Africa	0.02
Thailand	0.02
Others	0.19
Estimate Total	4.23

Packages and Package Materials

Granulated white sugar is sold at retail in doubled-walled glued-end-flap paper bags containing 5 and 10 lb (2.3 to 4.5 kg) and in waxed-paper lined paperboard folded cartons containing 2 lb (0.9 kg). Cubed granulated sugar lumps are packaged in 1½ lb (0.67 kg) cartons, sometimes with a transparent window. A few specialty items, such as colored or flavored sugar crystals, are sold in small glass or clear plastic containers made from polystyrene or cellulose acetate. Small paper canisters and small opaque molded polyethylene canisters containing from 1 to 4 oz (28.3 to 113.2 g) have been marketed for picnic or patio use. For institutional or restaurant feeding, polyethylene-coated paper packets holding about 1 tsp. (4.93 ml) have been used for a number of years, and they are now available in supermarkets, 50 packets to the folded carton. Powdered white sugars and granulated brown sugars are packaged in 1 lb (0.45 kg) folded cartons with waxed paper liners. The greater the need for protection, the heavier the liner. Maple sugar is largely sold as a confection because of its strong flavor, and is packaged as such.

Recent Trends

Among new items moving onto the shelves are 1 lb (0.45 kg) boxes of "Brownulated" sugar in foil overwraps and individual servings of white granulated sugar in paper tubes. The paper tubes are each about 0.95 cm in diameter and 8.89 cm long made from polyethylene-coated paper. They come 50 to a carton, and are said to be superior to the flat packets in ease of opening and dispensing.

TABLE 11.2. WHERE THE SUGAR GOES (United States deliveries of refined sugar, 1977, in thousands of MT)

Industrial Users:	
Other	542
Dairy products	567
Processed foods	746
Confectionery products	951
Bakery and cereal products	1386
Beverages	2454
Total	6646
Nonindustrial Users:	
Restaurants	69
Others	148
Grocers and supermarkets	1288
Wholesale dealers	2095
Total	3600

Nature food stores are carrying raw sugar or Demarera sugar in unprinted polyethylene bags. These are semirefined light-brown pourable coarse-grain cane sugars.

Artificial sweeteners are found on the sugar shelf, packaged in glass and polyethylene bottles (liquids), paper packets (powders) and in glass vials (pills).

SWEET SYRUPS

Historically, sweet syrups predate crystalline sugar products. They can be pressed from a variety of plant juices or saps. Today, the most common are cane sugar syrup made from redissolved refined cane sugar and corn (sugar) syrup. Additives include various flavorings and some artificial colorings. Pure maple syrup is so powerful in flavor that it usually is diluted with cane or corn syrup. Honey is marketed as is, a viscous pourable syrup, or as a whipped "butter." Recent years have seen a collection of waffle and pancake syrups with added butter.

Molasses.—is the residual syrup left after the crystallization of sugar from the juices of cane, sweet sorghum, or beets. Cane molasses is sweeter than beet because of a higher content of invert sugars. However complete refining can strip nearly all the sugar, then the product is called black strap molasses. In Britain, a sweeter, partially decolorized product is called treacle. Most molasses production is used as a cattle food supplement or is fermented to industrial alcohol or rum.

FIG. 11.1. DELUXE SUGAR REQUIRES SPECIAL PACKAGING
This standard 1 lb (0.45 kg) carton has been given a new golden image and a special liner of waxed wax-laminated white opaque glassine to provide an extra moisture barrier for the superfine sugar. The tinier sugar crystals have greater surface area and need the added protection to prevent caking and lumping.

Courtesy of Glassine and Greaseproof Div., Amer. Paper Inst.

Packages and Packaging Materials

Most pourable syrups sold at retail are packaged in glass bottles. Bottle shapes have been the means for brand identification in addition to labeling. Honey is sold in jars. Two excellent varieties of honey sold in England are heather, which is gathered in Scotland and Dartmoor, and wild raspberry from Scotland. They are sold in glass bottles with screw-on closures. Plastic bottles of blown polyethylene or clear polyvinyl chloride are now coming on the scene as they result in less breakage. Individual servings, so popular in restaurants and institutions, are now being made available for home use. These include clear polyethylene and polyethylene-coated paperboard tetrahedral containers and shallow cups molded from polystyrene, polypropylene, polyethylene, polyvinyl chloride or ABS copolymer. These are closed with peelable membrane lids, some of which are vacuum metallized plastic films and others foil laminates. In either case, the plastic selected must be compatible with the cup to effect a seal. Fin-sealed pouches are also used. These can be made from cellophane-polyethylene-foil-polyethylene. In large quantities syrups are being marketed in foil-paperboard-polyethylene laminates formed into containers called "Zupack." Here the laminate is formed into a fin-sealed pouch-like structure, which then is folded along prescored creases into a rectangular box shape. Gallon (3.78 liters) quantity blow-molded polyethylene bottles are available, and still larger quantities can be packaged in a "Cubitainer" made from a cube-shaped corrugated shipper with a polyethylene liner that has a molded polyethylene bottle neck sealed to the liner. Due to the flexibility of the latter the bottle neck is retractable.

TABLE 11.3. STAGES IN SUGAR PRODUCTION

Growers	Processors	Refiners	Products
Corn →	Milled and heated with water. Pulp separated from husk, then enzymes added to produce sugar syrup. →		Syrup
Sugarcane →	Crushed, heated and purified to produce raw sugar crystals. →	Heated, washed and filtered to remove impurities. →	Refined sugar
Sugar beet →	Washed, sliced and leached yielding a sugary juice. →	Heated and concentrated. →	

CHOCOLATE AND CHOCOLATE PRODUCTS

Chocolate

The Aztecs, Mayans and Incas of Central and South America raised cacao trees and produced chocolate before the first Spanish conquistadores set foot on the soil of the New World. It was Montezuma's favorite drink. He never entered his harem without partaking from a golden cup which led to the belief, still prevalent as late as 1712, that chocolate was an aphrodisiac. Hernando Cortes brought cacao beans to Spain in 1528 and taught the royal court how to prepare the Aztec beverage. With the addition of sugar the drink became a flavor sensation and soon spread to other royal courts. Before long lesser members of court circles were congregating in "chocolate houses" where the drink could be bought, despite its high cost. Through the 1600s and 1700s little more was accomplished. In 1748, an Italian book by Concina discussed the question as to whether chocolate is a drink or a food and whether the church can allow its use during Lent and times of fasting. The medical uses of chocolate was also discussed. At the time of its publication, the book created a considerable amount of interest. In 1828, cocoa was invented. The defatted product could be mixed with sugar, chocolate and cocoa butter to form a moldable and more palatable food. Another important event was the discovery of milk chocolate in 1876. These two accomplishments made chocolate a much more popular confection. Its consumption has grown from a few handfuls of cacao beans to more than 270 million kg of beans annually in the United States alone. One of the most famous chocolate firms in the world is Cadbury. It was founded in 1824 when young John Cadbury opened a tea and coffee shop next door to his father's soft-goods store in Birmingham. He sold cocoa as a sideline beverage. In 1879 John's sons built their revolutionary "factory in a garden" in nearby Bournville, an estate that provided vast recreational facilities and later housing and other welfare benefits for employees.

The company started its famous milk chocolate in 1905, a product that soon broke what the company says was the "near monopoly" of the business by the Swiss.

The standard of chocolate in the United Kingdom, Australia, Canada, much of Africa and India is determined by what the Cadbury Company invented and developed. Hershey chocolate is blander and sweeter than Cadbury's and this taste is almost peculiar to the United States.

Production of Cocoa and Chocolate Bars.—Ghana's humid climate is right for cocoa trees, which thrive only in a narrow band 12.4 km on either side of the equator. The cocoa producer owns his land or shares in a community farm belonging to his clan or village. The farms may be as large as 20 ha, but most are far smaller. The soil must be well-drained and fertilized to curb fungal diseases. The seedlings, which will mature into producing trees in 3 to 4 years, must be shaded from the sun by larger trees until their own foliage develops. As many as 1000 cocoa trees may be planted on 1 ha. Mature trees produce up to 6000 flowers, of which maybe a dozen evolve into a yellowish or orange pod about 3.24 to 4.24 cm long and 2.16 to 3.24 cm wide—a process that takes 6 to 8 months.

The farmer, using a long pole with a sharp knife attached, cuts the ripe pods from the trunk. Each pod is broken open with a machete or mallet and the 20 to 50 cocoa beans contained in the gummy pulp are removed. The pearly white or purple beans soon turn brown once exposed to the air.

Then, the mass is dried in the sun to shrivel the pulp and free the beans. The Ghana farmer prefers then to ferment the beans for up to 1 week by placing them in many layered sandwiches between banana or palm leaves.

The farmer, Ghana's chief foreign-exchange earner may sell his beans to the state-controlled Ghana Cocoa Marketing Board for roughly 1/3 of the going world market price, about $3.00 a kilogram.

State-controlled marketing systems exist in most tropical commodity areas, including Brazil, Nigeria and the Ivory Coast. The marketing boards arbitrarily set the prices paid to the farmers; the boards themselves reap higher profits if world prices rise rather than passing them on to the growers. The system eliminates any profit incentive for the farmer. In Ghana, the farmer's natural frustration with the marketing system is compounded by the fact that he will be paid in *cedis*, the nation's currency that is practically worthless both in and out of Ghana.

Farmers who live near the border with the Ivory Coast, a thriving and well-managed nation, have the option of smuggling their beans across the boarder to willing buyers with hard and ready cash. This may explain why Ivory Coast cocoa exports have shot forward recently.

During the 1977–78 statistical crop year, Ghana's cocoa output reportedly fell to 268,000 MT from 320,000 the previous season, while the Ivory Coast, saw its "harvest" shoot up from 230,000 to 297,000 MT. Brazil's cocoa total last season rose to 283,000 MT from 234,000 in 1976–77.

Hershey remains the biggest buyer of cocoa beans, taking an estimated 35% of the 202,446 MT imported into this country last year, up from 161,177 in 1977, but the bean is not the only source of chocolate products. Americans consumed 536,000 MT of cocoa and chocolate products last year, some made from other cocoa byproducts imported from producing countries.

It takes a variety of beans to produce a consistent grade of product. Hershey, for example, may be buying and blending a score of different beans on any given day, the product arriving by rail from the Port of Philadelphia or perhaps from some warehouses in Boston, New Orleans or even San Francisco.

These beans fall into two basic groups, the ordinary "base" beans and the choice "flavor" cocoas. Unfortunately, most of the rather tiny supply of flavor beans come from the Caribbean Basin, Trinidad, Panama, Central America and lesser sources. Trinidad and Venezuela have struck it rich in oil and their cocoa workers won't grow the crops anymore. Ecuador, Brazil and other exporters are now increasingly selling cocoa products to manufacturers rather than the bean because it is more profitable and creates a domestic processing industry for local labor.

Hershey buys beans as well as byproducts, however, and turns them into cocoa and chocolate at its plant.

Under the building, sacks of beans weighing from 31.7 to 72.6 kg or more are unloaded from rail cars, cut open on the spot and poured into chutes and conveyor belts that will deliver them to silos segregated by place of origin. The silos have a capacity of 40.8 million kg, but an inspection of several silos on any day may find many of them empty.

The beans are cleaned, then roasted to remove moisture. They are then cracked by stainless steel rollers, adjusted to the size and grade of bean, which is why the beans are always kept segregated by size. Once hot air blows away the husks of the beans, the meat of the bean or nib is what is left. The nib is ground to produce chocolate liquor (it has nothing to do with alcohol) which is then cooled, solidified and sold without further processing as baker's chocolate. This dark and somewhat bitter powder is chocolate in its purest form. Sweet and bittersweet chocolates are made by adding sugar, cocoa butter and flavorings to chocolate liquor. Milk chocolate also contains whole milk solids. Lecithin is an optional additive to help emulsify fats.

The pressing of the chocolate liquor, produces two substances. The first is cocoa butter (55% of the nib), which is drained into troughs under the presses. The fat is used as a candy additive, as an ingredient in the making of other types of chocolate, and in the manufacture of pharmaceuticals and cosmetics. The butter-free mass is known in the trade as press cake. To make cocoa beverage powders, the press cake is combined with sugar and milk solids.

Bits or chips of chocolate are added to cakes, cookies and ice creams, or the chocolate may be sugar-coated so it will "melt in the mouth instead of your hand." A wide variety of ingredients are dipped in chocolate from ants and grasshoppers to fruits and nuts. Many candies and complex confectioneries are dipped or enrobed in chocolate.

To make chocolate, the chocolate liquors from a blend of base and flavor nibs are poured into a truck-trailer-sized vat with large amounts of cocoa butter, condensed milk and sugar, lots of sugar, and heated to about 54.4°C. For 72 hr the mixture is churned by huge granite finger-like rakes over granite sides and bottoms until it is as smooth as possible.

Once removed from the vats, the chocolate is quickly cooled to about 26.6°C and poured into molds on a conveyor belt that jiggles the warm mass to eliminate air bubbles.

Chocolate kisses are made by dropping the warm mixture from giant inverted toothpaste-tube like containers onto stainless steel conveyor belts that lead to wrapping machines.

A standard 1.2 oz (34 g) Hershey chocolate bar consists of 45% sugar, 25% condensed milk (ordinary milk would shorten its shelve life considerably), 16% cocoa butter and 14% cocoa liquor. Each bar also contains a trace of lecithin, an emulsifier that prevents spoilage and enhances smoothness, and vanillin, an artificial flavoring agent. Real vanilla beans are too costly. The standard Hershey chocolate almond bar,

1.05 oz (29.76 g), has the same ingredient ratios. Both vanillin bars contain about 180 calories.

Product Properties.—Although chocolate fat (cocoa butter) is not susceptible to rancidification, it does readily absorb odors and flavors. Thus, chocolate candies need protection against odor or flavor transfer. When chocolate is exposed to excess moisture or to sudden changes in temperature which cause condensation, sugar bloom results. Sugar bloom is a light gray coating of sugar crystals. Storage at humidities above 78% for milk chocolate or above 85% for plain chocolate is not recommended. Fat bloom is the formation of crystals of fat on the surface of the chocolate. Packaging is therefore concerned with protecting chocolate from the effects of heat and humidity, and from odor and flavor contamination.

The amount of protection which is built into packages for chocolate and chocolate confections is dependent upon how much cost the product can carry, the anticipated delay between manufacture and consumption, and the anticipated severity of conditions during shipping and storage. The most effective materials for providing maximum protection are aluminum foil and foil laminates, and the best chocolates are given this protection.

Packages and Package Materials.—Solid chocolate chunks or bars or chocolate covered bar candies, especially those containing nuts, are commonly wrapped in 11.34 kg per ream printed glassine paper. Larger bars may use a heavier weight 13.6 to 15.8 kg glassine. Where staining may be a problem, two 11.34 kg glassine sheets may be used with the inner an unprinted sleeve and the outer a complete wrap. Most flat bars use a glassine wrapper and a printed 22.68 kg paper outer sleeve. Aluminum foil is used for form-fitting wrappers on novelty items such as molded chocolate animals and eggs, and we all are familiar with the foil wrapped chocolate "kiss." Some chocolate bars are being wrapped with foil-wax-paper-wax laminations. Others that require longer shelf-life are being wrapped in a heat-sealed, foil-paper-polyethylene lamination.

Specialty holiday items are not only foil-wrapped, but often packed in attractive cartons with various types of closures, windows or overwraps.

Recent Trends.—A novel concept in Europe is to cast chocolate directly into foil containers. United States candymakers are taking a good look at this and are experimenting with direct casting into molded polyvinyl chloride or polystyrene trays. Barrier properties are improved by adding PVDC coatings to the trays and sealing a barrier film or foil laminate across the top. Pressure sensitive adhesives are also being tested because heat sealing always risks melting of chocolate components.

Candies with chocolate exteriors are not wrapped in transparent films for two reasons. The less expensive films do not give full protection

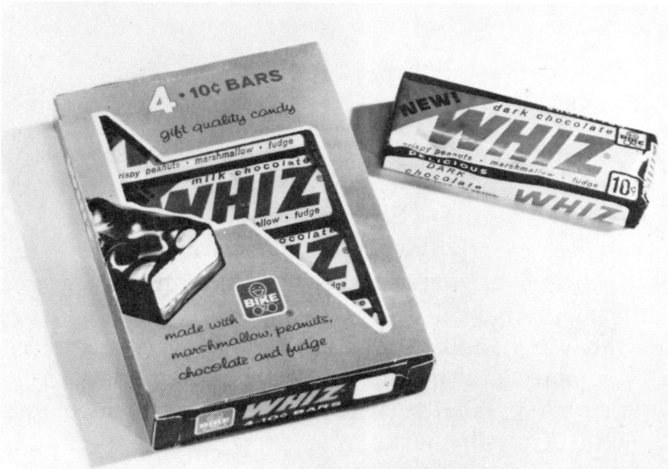

Courtesy of Glassine and Greaseproof Div., Amer. Paper Inst.

FIG. 11.2. DIE-CUT CARTON SHOWS WRAPPED BARS
A typical example of a complex candybar made of marshmallow, peanuts, fudge and chocolate covering. Covering can be either milk or dark chocolate. Bar wrapper is printed opaque glassine to hide unsightly chocolate stains and help protect aroma.

against foreign odors and some contain fat soluble plasticizers. More important, however, is the visibility factor. Finger pressures from handling causes chocolate to scuff or smear. A transparent film would allow this unsightly condition to be seen.

COCOA

Powdered cocoa is a relatively stable product that needs protection from moisture to prevent caking. It is sold in metal-end paperboard canisters. Individual servings in polyethylene-coated paper packets are now also available.

CHOCOLATE SYRUPS

Chocolate syrups contain so much water that they are susceptible to mold formation and bacterial contamination. For this reason they usually are canned. Refrigeration is recommended after the package is opened. Unit portion packages have been used to supply chocolate syrup toppings for ice cream or other desserts. Printed polyester-polyethylene film containing 1½ oz (42.5 g) in form-and-fill pouches can be heated in boiling water to soften the contents prior to use.

CANDY CONFECTIONS

Candies or confections are combinations of ingredients with one basic common property—a primarily sweet flavor. Cane sugar, other sugars and honey have for centuries been the primary sweeteners. Artificial low calorie sweeteners are relatively modern; low calorie candies have moved from the medicinal or dietetic shelf to close proximity with regular candies only in the past decade. Other primary ingredients of candies include butter, starch, milk, cocoa fat, chocolate, gelatin and eggs. A wide variety of secondary additives are used to contribute flavor, color, aroma, texture and stability. Finally, there are components which are added for contrast such as fruits or nuts.

Candies can be grouped into a number of basic types and often are sold in a simple form of one candy type. However, there are hundreds of possible modifications of the basic types. Chocolate, for example, is a basic candy, yet it is frequently used in combination with other candies.

TABLE 11.4. STORAGE TIME REQUIRED FOR PERCEPTIBLE TASTE TO APPEAR IN WRAPPED SWEET MILK CHOCOLATE BARS DUE TO ABSORPTION OF ODOR FROM OUTSIDE ATMOSPHERE[1]

Description of Wrapper	Time to Taste Turpentine	Peppermint
Aluminum foil (0.017 cm) coated with heat-sealing lacquer and laminated one side to moistureproof cellophane; coated surface heat-sealed along side seam and ends.	20 days	19 days
Aluminum foil (0.0017 cm) coated with heat-sealing lacquer; wrapper heat-sealed along side.	6 days	7 days
Plastic bleached greaseproof paper 10.2 kg; side seam and folds not sealed.	5 hours	3½ hours
Opaque greaseproof paper 13.6 kg, side seam and folds sealed.	5 hours	3½ hours
Control—no wrapper	½ hour	1 hour

[1] Alcoa aluminum foil, its properties and uses (1953). Chocolate bars stored in glass desiccators, one containing cotton moistened with turpentine and the other containing peppermint candy.

Product Properties Simple Candy Types

Hard Candies (Boiled).—Hard candies are made almost entirely of sugar. Sugar is dissoved in water and boiled with or without vacuum until the desired concentration is reached, usually judged by boiling point. Additives, including colorants, flavorants and medicinals, when used, are put in during cooling to minimize evaporation. If poured immediately into molds, clear hard candies result. If pulled or kneaded, the product will be not quite so clear and slightly higher in moisture content.

Medium Hard (Chewy) Candies.—Toffees and caramels are mixtures of sugar and glucose boiled together with added color and flavor and some fat to lubricate, emulsify and prevent the candy from sticking to the teeth. Toffees are hard because they are cooked to a higher temperature (lower moisture). Caramels contain flour. Both candies can be pulled, cast or cut. Modified caramels may include fruits or nuts.

Medium Soft Candies.—*Fondants and Creams.*—Sugar creams or fondants are like hard candies, but more water is left in and they are cooled rapidly to induce formation of a fine suspension of crystals of sugar in a saturated sugar syrup. Milk can be substituted for the water. If cream or butter is added it is called a modified fondant. Colors, flavors, fruits, nuts or vegetable products may also be added. Fondants are used as center components of coated or dipped candies such as chocolates, bonbons or candy bars.

Fudge.—Fudge is a "cross" between a caramel and a cream. It is a basic grainy fondant cream with further additions of cream, milk or butter. Nuts, fruits and flavorings are added.

Nougats.—Nougats contain honey, butter or cocoa butter, sugar, corn syrup, gelatin and egg white, and are cooked until like caramel.

Gums and Pastilles.—These candies are made from sugar, gum, water and flavoring and put in many shapes. The pastilles are usually softer than the gums due to a higher water content, but they are not as soft as jellies and are made rather rubbery by the addition of gelatin. The familiar gumdrop represents the other category.

Jellies and Pastes.—Jelly candies are made from water, sugars, fats, flavorings, color and thickening agents. Their consistency depends upon the percentage of ingredients, cooking time and holding time in starch molds. They include such products as jelly beans, Turkish paste and various gum drops or strings.

Soft Candies.—*Marshmallow.*—Marshmallow contains sugar, corn syrup, gelatin and egg white. Originally it also contained part of the root of the mallow plant. Now it contains a substitute vegetable colloid. It may be cooked or not, but it is always whipped to introduce air and develop a light consistency. Texture may be varied from short and crumbly to chewy or even semiliquid.

Marzipan.—Marzipan is a marshmallow formula plus almond paste. It is used extensively to mold and sculpture highly individualized decorative items.

Product Properties of Compound Candy Types

Filled hard candies are prepared by forming clear, pulled or grained hard candy into a hollow jacket, inserting any one of a variety of fillings, pinching the jacket shut and encasing the whole in another layer of hard candy. Fillings can be jam, preserves, fruits, berries, chocolate, peanut butter or any other desirable concoction. This category could also include chocolate-nut bars, peanut and other nut "brittles," candy coated popcorn, sugar glazed fruits and nuts, and enrobed centers.

Chocolate-coated Centers

Chocolate-coated centers consist of various creams, fruit jellies and nuts. In coating roasted nuts, the nuts must not be over-roasted since a bad taste will result. All the creams used should be fresh and of high quality. The fruits employed should be top grade glazed or preserved varieties.

In chocolate coating (enrobing), both machines and hand-dipping methods are used. Machines are used for chewy, hard centers and cream centers cast in starch. Fruits, nuts and hand-rolled centers are best hand-coated.

The modern enrobing machines require an experienced operator. Constant adjustment is necessary to insure a quality product. The chocolate is melted at 48.8°C in a steam bath or in a regular chocolate melter. It is then cooled to 30°C and taken over to the enrober. The machine operator then cools the chocolate further prior to dipping the centers. In order to prevent streaky or gray chocolates, the chocolate-coating room should be maintained between 15.5° and 18.8°C.

Hard chewy centers may be machine-coated by a 60.9 cm enrober with stringing and packing attachments. Stringing is used for the centers before they go into the machine. A cooling tunnel is also part of the in-line operation. After the center is dipped, the coated centers move to the cooling tunnel. They are then packed in stock boxes as they move along the belt after leaving the cooling tunnel.

Courtesy of Reynolds Metals Co.

FIG. 11.3. FOIL PROTECTS PEANUT BRITTLE
Aluminum foil heat sealable laminated overwrap not only provides an attractive package but also provides the vitally needed moisture barrier protection.

Delicate cream centers, nuts, fruits and cordials are hand-coated. Dipping tables are used to accommodate the workforce. Hand-dipping assures a firm finish as well as a sufficient coating laydown.

The packaging of chocolate centers is a delicate job. All chocolates must be handled carefully to avoid scratching. They should also be cupped. Setup boxes in a two layer box are traditionally used for quality chocolates. The top layer is fitted and supported with soft padding to avoid a crushed product. Also, the top layer is more tightly packed than the bottom one since the box tends to stretch more at the top.

Either thermoformed plastic trays or zig-zag paperboard inserts are used to partition the chocolate. The box is then checked for correct weight and a soft cushion mat is placed on top. If a waxpaper liner is used, it is clear moisture-proof cellophane or polypropylene film. Some lower priced confections are sold in small (less than ½ lb (0.23 kg)) units in small setup boxes or in plastic bags.

Product Properties: Complex Candy Types

This category includes complex confections: for example, a baked cookie with a layer of nougat topped with a layer of soft caramel and the whole enrobed in chocolate; or, an Easter egg with a marshmallow center containing chopped candied fruits and nuts, covered in chocolate, and decorated with sugar icing, marzipan or jellied ornaments.

Packaging Requirements

The packaging requirements of simple, compound and complex confections are dependent upon several factors: (1) the moisture content of the candy and hence its susceptibility to atmospheric moisture, (2) the fragility of the candy, (3) the susceptibility to rancidification (those containing nuts or butter), (4) the susceptibility of chocolate coatings to fat or sugar bloom, odor or flavor pickup, (5) susceptibility to loss of aroma or color, and (6) the need for protection against soilage, pilferage or infestation by insects or vermin.

Simple candies can be classified with respect to moisture sensitivity; other candies have to be judged individually under end-use conditions. A convenient classification for simple candies is according to their relative vapor pressures, which are in turn proportional to the residual moisture content and the hygroscopicity of the ingredients.

Confectioneries with Low Relative Vapor Pressure.—Hard candies have a relative vapor pressure (RVP) of less than 30%, toffees run about 48% and caramels a little higher. Low RVP candies absorb moisture from the air. This causes graining at the surface and can make them sticky; consequently, the package should provide a moisture barrier.

Confectioneries with High Relative Vapor Pressure.—The fondants, fudges and marzipan have relative vapor pressures ranging from 68 to 84%. This is definitely higher than atmospheric moisture, and the products tend to dry out. Packaging needs to prevent or slow down this action, however. In the event of sudden temperature changes, moisture sweating can produce stickiness or promote mold growth. Therefore, the package must have some ability to breathe.

Confectioneries with Medium Relative Vapor Pressures.—Relative vapor pressures of the gum candies range from about 51 to 66%, whereas the jellies range from about 59 to 75%. Since, except for the highest extreme, these are close to average atmospheric humidity, the candies do not tend to dry out and hence require little moisture protection. In fact, they are better off in a package that can breathe. Marshmallow and nougat candies are also in the range of medium relative vapor pressure. They require a little more protection than the gums and jellies in order to prevent drying, but the package also should be able to breathe a little as insurance against "sweating" or condensation of moisture.

Packaging Materials

In the application of packaging materials as direct wraps, bags, box liners, or box carton or tray overwraps, the packaging engineer has a wide selection of materials from which to choose. Table 11.5 lists some typical examples along with a general classification of their performance characteristics.

To show further the complexity of the problem of selection, Table 11.6 shows typical recommendations to the confectionery industry by manufacturers of cellophane films.

TABLE 11.5. PACKAGING MATERIALS FOR CANDIES

Material	Rating as a barrier to: Moisture	Odor and flavor	Oils and fats	Heat sealing
38.1 micron aluminum foil	Excellent	Excellent	Excellent	No
12.7 micron aluminum foil-25.4 micron polyethylene	Excellent	Excellent	Excellent	Yes
25.4 micron aluminum foil-25.4 micron polyethylene	Excellent	Excellent	Excellent	Yes
Coated 25.4 micron aluminum foil	Excellent	Excellent	Excellent	Yes
19.0 micron PVDC (Saran) film	Excellent	Excellent	Excellent	Yes
25.4 micron rubber hydrochloride film	Very good	Very good	Excellent	Yes
38.1 micron coated cellophane	Very good	Good	Excellent	Yes
50.8 micron polyethelene film	Good	Good	Very good	No
8.9 micron aluminum foil-tissue paper	Good	Good	Very good	No
19 kg/ream bleached Kraft paper, 3.6 kg/ream polyethylene	Good	Fair	Good	Yes
Glassine paper	Poor	Poor	Good	No

Polyethylene and polypropylene films are challenging cellophane in various end-use areas. While PE generally does not have the clarity or gas barrier characteristics of cellophane, it has excellent impact and tear strength, excellent heat-sealability, and is an excellent moisture barrier. It is softer than cellophane and, therefore, not quite as machineable. However, older equipment types can be modified to run polyethylenes and newer types are being designed to handle the softer films.

Another major advantage of polyethylene is its low temperature flexibility which allows storage and handling at temperatures below −45.5°C. On both a weight and area basis, polyethylene enjoys a price advantage over cellophanes, which accounts in no small part for the degree of its acceptance despite the shortcomings mentioned.

Polypropylene films are now receiving serious attention from the candy industry, as they have the optical clarity desired and can be made with more stiffness and better machineability than polyethylene. They are slightly less costly than cellophane on a weight basis and enjoy a substantial advantage on an area basis. Their major difficulties have been in their lack of good sealability. PVDC and vinyl coatings have been used to avoid this problem; however, composite films of polyethylene and polypropylene seem to be the answer. Polypropylenes have good ageing and low temperature properties, about equal to polyethylenes. They are superior in abrasion resistance, which is important where sharp sugar particles are present. Some polypropylenes have been specifically designed for twist-wrap, as they have the ability to lock in position after twisting due to the stretching of the film in the twisted area.

Shrink films are being used to an increasing extent for overwrapping candy boxes or for bundling multiple units. These include heat shrinkable polyethylene, polypropylene and polyvinyl chloride films. Aluminum foil laminates are being used more and more frequently as bar wraps, particularly for the most expensive products, or those requiring a high degree of protection. Foil-wax-paper or foil-polyethylene-paper-polyethylene combinations are used.

Package Types

Individual Wrappers.—Hard sugar candies and toffees are wrapped individually to protect them against graining or stickiness due to moisture pick-up. Many softer candies such as caramels or chocolate-covered confections are individually wrapped to avoid their sticking together or to prevent losses or transfers of flavor and aroma. There are several kinds of wrappers, and most of them today can be applied by automatic equipment. One of the most commonly used individual wraps is the twist wrap.

Both ends of the wrapper sleeve are twisted to effect a closure. It is adaptable to various shapes and sizes, and inner sleeves of contrasting or protective material can be inserted easily prior to closure. Over 300 million kg of candy are twist-wrapped annually in the United States. Heat is sometimes used to help insert the twist. From 1½ to 2½ full turns is the usual range of twists, and candies can be so wrapped at more than 1000 pieces in 1 min. Original twist wraps were plain tissues. These were supplanted by waxed tissues and then by cellophane films. Polyethylene and polypropylene films are now challenging cellophane.

Round or oval-shaped candies are individually wrapped in a bunch wrap where the ends are gathered underneath the piece and heat-sealed. The extra layers of wrapper insulate the candy from heat damage; however, on larger pieces a board-stiffener or tray may be inserted.

Rectangular pieces of candy can be wrapped with the caramel wrap. Here the ends are parcel-folded and the ears tucked back and heat-sealed underneath where the body overlap insulates the candy. The satchet wrap is a combination of the caramel wrap on one end and the twist wrap at the other. Larger bars of candy are often placed on a stiffener card or a U-board made from 18-point board with about 3 kg/rm wax coating. The wrapper can be similar to the caramel wrap or it can be made like a bag with a long body heat-seal and two closely trimmed end seals.

Courtesy of Extrudo Film Corp.

FIG. 11.4. POLYPROPYLENE TWIST WRAPS GIVE TIGHTER TWIST
A comparison of cast polypropylene film twist wraps on the left with cellophane twist wraps on the right shows that the former holds the twist better.

Round, hard candy lozenges are assembled in a cylindrical roll that is overwrapped and the ends folded or twisted. Ends are sealed shut with a hot melt or an end label. A printed paper sleeve is used for added strength and protection.

Strip pouches are multiple units of single pouches that have not been completely severed. The user tears off individual units as desired along perforated tear lines. This is popular for medicinals and for children's lollipops; however, the latter are also very often twist-wrapped.

Hard candies which have been individually wrapped usually require additional moisture protection. In earlier times this was provided by glass jars or metal tins with friction lids. Such packs are still available but usually as specialties with higher prices. The modern solution is to place the wrapped candies in an outer package which is most often a sealed bag made of moisture-proof cellophane, although here again polyethylene and polypropylene films are now being employed. Polyethylene bags in fiberboard boxes are also being used as wholesale shippers for hard candies and toffees.

Polyethylene bags 0.0025 to 0.0038 cm thick are being used as multi-packs for low price candy bars. They are closed with clips, heatsealed paper headers or plastic tape twist ties.

Cartons of various types are used for candies. They range from 15-point to 22-point paperboard. The board may be wax- or polyethylene-coated, or the carton may have a plain or waxed glassine liner. Window cartons may be used to give the buyer a view of the candy. Windows are made from polystyrene film, cellulose acetate or polypropylene. Thicknesses run from 0.0025 to 0.0038 cm. Some cartoned candies have a sealed cellophane overwrap with a tear tape opening device. This is very popular for cough drops.

Overwrapped cartons are used for six-pack and 10-pack units of candy bars. These are limited to higher priced candies.

Piece candy assortments where chocolate is a major component are aristocrats of the candy world and can run to whatever price the consumer will pay. Consequently, packages can run a gamut from rather simple setup boxes to extremely ornate constructions. Traditionally, each piece was placed in a 25 lb (11.34 kg) glassine-fluted paper cup. Candies would be arranged in layers with sectional dividers, trays and padded liners to prevent damage. The modern approach is to use molded polyvinyl chloride trays with individual cavities for each piece. These are less expensive, permit automatic loading and give better protection against damage.

Very aromatic or highly flavored pieces such as mints are often individually wrapped with plain foil to protect against flavor or odor transfer. Other candies such as caramels or fudge also are individually wrapped to protect against rancidity or desiccation. Quality chocolates are further protected by a plastic overwrap, which when sealed provides additional barrier protection as well as preventing soilage, scuffing and pilferage. Cellophane, again, has been the standard (140 to 195 gauge), however, again, 0.0025 cm polyethylene and 0.0029 cm polypropylene films are being used.

There are some chocolate assortments marketed in tinplate or aluminum cans and some in glass, but this is not very widespread. A few are

FIG. 11.5. RECLOSABLE FILM BAG WITH PIGGY BACK LABEL
This new form-and-fill bag for candies features a pull apart top seam plus a pressure-sensitive label which can be used to reseal the bag after opening.

Courtesy of E.I. du Pont de Nemours and Co.

being marketed in rigid 0.19 cm cellulose acetate trays, and a few are sold in thermoformed polyvinyl chloride or polystyrene trays. Trays are usually placed in cartons or boxes with film overwraps. Some specialty seasonal items are placed in expanded polystyrene foam trays with shrink film overwraps.

Convenience and Novelty Packages of Candy.—Because candy is a cheerful product, it is associated with children, holidays and special occasions. A wide variety of novelty packages has been developed to attract special impulse purchases or to introduce new packages via premium offers. Cartons can be printed to represent railroad cars or village buildings. Little plastic Santa Clauses can be filled with tiny candies. Glass animals filled with candies can become penny savings banks. Polystyrene hearts are brought out on St. Valentine's Day, and baskets of

Courtesy of King Candy Co.

FIG. 11.6. SET-UP BOX PIECE CANDY
Holiday motif is reflected in this heart-shaped box for piece candies. Note that some candies are foil-wrapped for extra protection. This package also features molded polyvinyl chloride trays with individual cavities of varied depth to separate the candies and to present a uniform level appearance.

Easter eggs come in a wide variety of shapes and sizes. Thanksgiving sees turkeys and cornucopias. Halloween sees witches and cats and pumpkins. Sugar candy or chocolate "cigarettes" and foil-wrapped "gold" chocolate coins are great favorites at all times. One popular Dutch-chocolate treat is a molded apple which breaks apart into segments like a navel orange. Plastic (PG or PVC) shrinkable film wraps can be used for multiunit bundling or for piggyback packaging, e.g., a candy cane attached to a bottle of dishwasher detergent. Colored foil containers with snap-on polyethylene lids make attractive candy dishes for parties. Even spiral-wound paper tubes can be overwrapped in bright foils and filled with candies. The limit is only the imagination of the package designer.

Frozen Confectionery

Because candies tend to be seasonal in demand, it would be advantageous if the manufacturers could store their product for long periods. This would enable them to stabilize production rates yet meet peak seasonal deliveries at Christmas, Valentine's Day, Easter, Mother's Day and Thanksgiving. Long-term storage would, of course, require that the candies remain fresh, free of staleness, rancidity, mold, or unsightly sugar or fat bloom. The only type of storage that meets these requirements is one where the product is brought to below 0°C and preferably as cold as −23.3°C. The freezing of confections does, however, introduce some additional problems. Where many candies are totally undamaged, some crack during freezing and some show an increased incidence of sugar blooming.

A considerable amount of work was done by Dr. J.G. Woodroof and others at the Georgia Experiment Station on these problems. They reported that spun candy chips, some chocolate-coated nuts, nut brittles, toffees and some creams may crack, but only the spun candies are seriously affected (shattering). The others return to normal condition when thawed. Whatever cracks are still present are extremely difficult to detect. Selectivity is the only way this can be avoided. Woodroof reported over 60 candy types that were not damaged by freezing, over 30 that were improved and about 20 that were impaired by freezing. Sugar bloom could be avoided by using moisture-proof wraps on the candies and thawing them slowly. Moisture-proof wraps were also needed to prevent drying out.

Packaging of candy that is to be frozen should therefore be capable of surviving freezing temperatures without cracking and must provide good moisture barriers. This rules out the conventional waxed papers and glassines. Although moisture barrier properties could be improved by using PVDC coated papers or cellophanes, the cold temperature conditions dictate a material that will not fracture in handling. This would have to be polyethylenes, polypropylenes or aluminum foils, or suitable combinations. Heat sealability is very desirable for the overwrap. Some heat shrinkable PVC, polyethylene and polypropylenes have been tried. Where economics permit and the individual candies can withstand the temperature, the individual wrappers may also be heat-sealed. The degree of protection required will be related to the types of candies, how much handling they will receive and how they are to be thawed.

Freezing of candies may be done on an increasing scale as manufacturers become more and more familiar with its advantages. A few supermarkets have tried selling chocolate assortments directly from the frozen food cabinet. The main deterrent will be added energy costs.

Other Confections

Candied Fruits and Nuts.—Fruits and nuts are frequently used as components in composite confectioneries. It has already been indicated how they can be dipped in chocolate. A few other techniques are also of interest.

Nuts can be "panned," that is, they are coated with a glaze made of sugar. Several layers usually have to be added. For thick coatings, as found in Jordan almonds, an occasional tumbling is inserted between coatings to knock-off rough spots. Final layers may be colored, and then the product may be wax-polished to a high luster.

Because of their high fat content, nuts are sensitive to oxidative rancidity and to extremes of moisture. Too much moisture will encourage mold, too little will embrittle the nut meat and encourage flavor loss.

Normal moisture is considered to be 3.5% for pecans and 2.0% or less for peanuts. If the nuts are allowed to absorb as much as 5½ to 6% moisture, they become soggy, tough and have a poor shelf-life.

If the *Aspergillus* type of mold is permitted to develop due to improper moisture level and storage conditions, a poison called aflatoxin can develop. There are several types of alfatoxins. While there is no direct evidence that aflatoxins are dangerous to humans, they do produce liver cancer in other species such as turkeys, ducks, trout, guinea pigs, rats, dogs, hamsters and mice. For this reason the amount of aflatoxin permitted in nut or cereal crops intended for human consumption is strictly regulated. There have been cases where candies containing aflatoxin-poisoned peanuts were condemned by the U.S. Food and Drug Administration (FDA). The Brazilian government has set up quality control laboratories to test for aflatoxins in exported Brazilian nuts.

Techniques of nut processing and roasting can help in establishing ideal moisture levels. Dipping the nuts in acetylated monoglyceride coatings helps retard moisture buildup. Refrigeration storage helps also to slow down the development of rancidity in nuts. For ideal shelf-life, packaging materials must be selected for essentially zero oxygen permeability, and the product should be vacuum-packed or nitrogen-flushed and held under refrigeration as much as possible. At low moisture levels, nut meats are brittle. The package must protect against physical damage during handling.

Ultraviolet light accelerates the oxidative rancidification of unsaturated fats and oils; consequently, a package for nuts will give better protection if opaque to these wavelengths of the spectrum.

Courtesy of E.I. du Pont de Nemours and Co.

FIG. 11.7. PEANUTS GAIN ADDED PROTECTION FROM COEXTRUDED FILM BAGS
These preformed bags are made from a coextruded film containing plies of polyethylene and "Surlyn" A ionomer.

Packages for bulk shipment and storage of nut meats are commonly polyethylene-lined drums or shipping cases. Vacuum-packed tins are used for deluxe nut-meat assortments and brittles. Nitrogen-flushed glass jars are also employed. Flexible packaging normally requires aluminum foil to attain the light and oxygen barrier needed; however, for peak volume holiday "quick turnover" sales, nut meats are often sold in small sealed cellophane or polymer-coated cellophane bags. This is one area where polyethylene will not readily compete due to its high permeability to oxygen. However, composite films containing PVDC coatings will prove satisfactory for short-term storage.

When the nut is incorporated within a compound or complex confection, the degree of protection required will depend upon whether the nut is exposed. If it is encased by the other candy components, particularly chocolate, less protection will be needed. If the nut meat is partially or totally exposed, the entire confection should be packaged to prevent deterioration of the nut meat.

Whole fruits such as cherries or kumquats, and fruit parts such as citrus slices or citrus peels, can be candied and/or glazed. Candying is done by removing objectionable parts, such as pits, seeds, bitter citrus oils and the like, and then gradually replacing the water in the fruit with sugar by soaking the fruit in successive baths of increasing concentration. After the final step the fruit is cooled and the sugar solidifies. Invert sugar can be added to prevent graininess. Pieces can be dipped in colored granulated sugar to reduce stickiness, or they can be glazed by dipping in a highly concentrated sugar syrup, draining and cooling. Whole candied cherries are often encased in fondant and dipped in chocolate. Brandy or liqueurs can be added.

Packaging of candied fruits is similar to hard candies. Excess moisture will make them sticky. Cellophane overwraps are needed. If there are no inner liners in the package, the overwraps should be a good moisture barrier cellophane. Polypropylenes will also prove superior, particularly if sharp sugar crystals are present.

JELLIES, JAMS AND PRESERVES

Jellies, jams and preserves are sweet concoctions by which traditionally the flavors of fruits can be captured for winter consumption. Solid fruit components are strained out of jellies and the clear mixture of fruit juice, boiled sugar and fruit pectins congeals on cooling to a clear gel. Jams are most frequently prepared from berries, however, peach jams and apricot jams are found. The whole fruit is included and cooked until its texture and form have been substantially broken down. Preserves contain whole small fruits or cut-up pieces of larger fruit. They are cooked only enough too avoid spoilage; the original form of the fruit is preserved as much as possible. A conserve is about the same as a preserve but frequently will contain nuts or vegetables as well. Marmalades are something between a jelly and a jam in consistency and are made from orange or grapefruit with thin slices of the peel added for additional zest. English marmalades are also made from limes, lemons and ginger.

Because fruit jellies and jams are acid, they are heated to 104.7°C, which is sufficient to kill all vegetative microorganisms, and then are packaged hot. The package serves to prevent loss of moisture. Traditionally, they have been packed in glass jars and small glazed ceramic crocks. Tops must be sealed to prevent reentry of mold spores. Since the contents are rarely consumed completely, a reclosure feature is desirable.

Courtesy of British Cellophane, Ltd.

FIG. 11.8. JAM JARS TAKE ON MODERN LOOK
The old and the new are brought together in these multipacks of jams in glass jars. The product is low-sugar dietetic and the multipack features a shrinkable polyethylene overwrap and a thin white plastic base with recesses for the jars.

Many jars have screw caps which can be replaced, rather than the older pry-off lids.

Plastic jars are already making in-roads on this market because of their lighter weight and resistance to breakage.

High density polyethylene, polyvinyl chloride and some of the newer acrylics have been used for this purpose.

Unit servings of jellies and jams are made from molded plastic cups or formed aluminum containerettes with smooth walls and flanges. Both types are closed with peelable membrane lids. Plastic caps are molded from polystyrene, polyethylene, polypropylene or polyvinyl chloride. Closures can be foil laminates, clear plastic film or metallized plastic film. The type of plastic used must be capable of heat-sealing or the membrane must have a suitable peelable adhesive or heat-seal coating, the selection of which will be governed by the material used to form the cup. In the selection of a suitable heat-seal formulation, note must be taken of the temperature of jelly fill. Jellies are filled hot (82.2°C or higher) and the actual degree of peelable seal does not become evident until it has "set" for 24 hr. Problems revolving around membrane formation, i.e., heat-seal coating separating from the inner foil ply, are a frequent plague to the material supplier. The foil used must be clean and free from oily residues to prevent separation of the laminate plies.

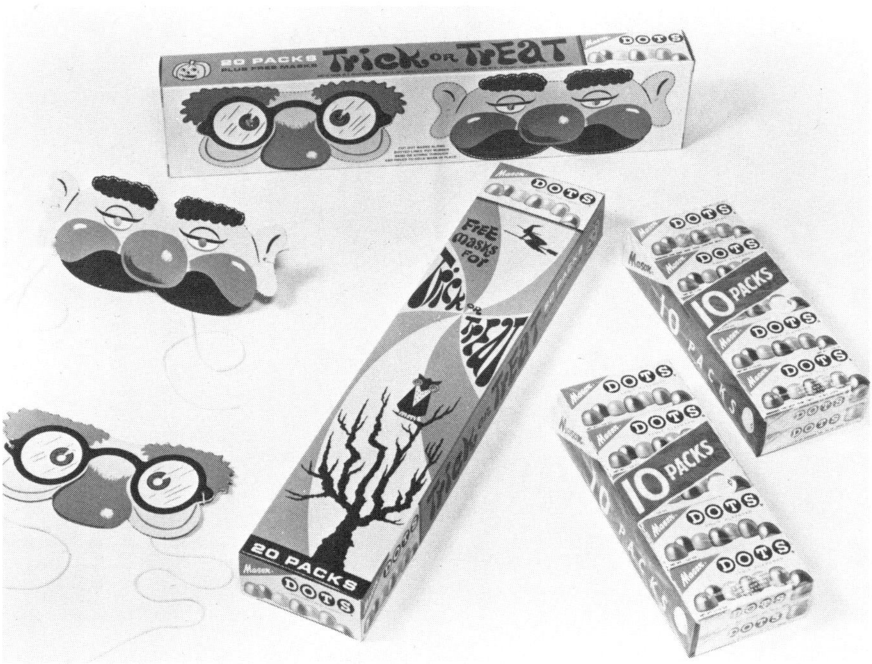

Courtesy of Mason, Au, and Magenheimer Confectionery Mfg. Co., Inc.

FIG. 11.9. HALLOWEEN MOTIF FOR CANDY CARTON
This candy carton not only packages a familiar candy product but also provides cutout masks to delight the children.

PEANUT BUTTER

The United States produces over 1.1 MMT peanuts annually, of which 63% of the edible grades go into peanut butter. Peanuts were known as early as 950 B.C. and probably originated in the Brazil or Peru regions of

Courtesy of The Metal Box Co., Ltd.

FIG. 11.10. CHRISTMAS PACKAGE COMPOSITE CONTAINER
This spiral-wound composite container utilizes a specially printed label to convert the simple container into an attractive children's game.

South America. From there they traveled to Africa and were imported back to North America with the slave trade.

Peanut butter was invented sometime around 1890. It was mostly made in the home at first, but later commercial processes were developed.

Peanut butter is a fine grind of roasted mature peanuts plus some salt, some hydrogenated fat, some dextrose and a few other additives such as glycerin, lecithin or antioxidants. Emulsifying agents are added to prevent oil separation, and the antioxidants to prevent rancidity.

Color and flavor are principally controlled by the selection of peanuts and the temperature used in roasting. Grinds vary from fine or smooth, to chunky (contains grainy particles). Exclusion of air during grinding and packaging helps prevent rancidity. The fatty portion of the oil solidifies when chilled. For best consistency refrigeration should be avoided except for initial processing.

Because of the high oil content, peanut butter is relatively stable. If humidity does not get too high, even mold is unlikely. The main problem is rancidity and staleness, which will gradually ensue during storage. Protection from oxygen and light is helpful, particularly the former. Most peanut butter is now packaged in glass jars. However, tin and aluminum cans, composite containers, plastic tubs and small flexible packages (for individual servings) have been marketed. Bulk shipment to candymakers and bakers may be in large tins or lined fiber drums.

Another product is a combination of peanut butter and jelly in a swirl design. The product is packed in glass jars with screw tops.

CHEWING GUM

Chewing seems to satisfy an inner craving in man. One will chew on a straw, a blade of grass or a twig. Various aromatic plants such as the sassafras yield twigs or roots that are pleasantly flavored.

The North American Indian chewed spruce gum, a highly aromatic resinous exudate. South and Central American Indians chewed the coagulated latex of the sapodilla tree, which they called *chicl*. The chewing of paraffin wax was common in this country in the early 1800s; traveling peddlers carried balls of the wax to sell as treats for the children on their routes. Just before the Civil War, Antonio Lopez de Santa Anna, former general and president of Mexico (better known in the United States as the commander of Mexican forces at the Battle of the Alamo), came to New York to try to raise money. Fortunes were being made in the infant rubber industry. General Santa Anna hoped to promote chicl as a commercial rubber product. In New York he met Mr. Thomas Adams, an inventor. Adams tried unsuccessfully to utilize the chicl rubber, but later he remembered seeing the general break off a piece and chew it. After some experimentation he produced a palatable mixture which he called "Adams New York Gum." Originally he sold it in chunks but later found it easier to sell if pressed into long, thin strips. After considerable success with the unflavored product, Adams brought out a licorice-anise flavored "Adams Black Jack Gum"—the oldest product still on the market. Other entrepreneurs soon jumped into the picture. Dr. Edward Beeman added pepsin and advertised "Beeman's Pepsin Gum" as an aid to indigestion. William White added corn syrup to gum formulations. This made a wider variety of flavors possible. He introduced the first peppermint-flavored gum and founded American Chicle Company. In 1892 Wrigley was selling "Spearmint" and "Juicy Fruit" and was pioneering new techniques in merchandising and advertising. Candy-coated "chiclets" were marketed by Fleer in 1897; Fleer also invented "Dubble Bubble" gum in 1928. Penny candy-coated gumballs were the specialty of Leaf Brands Company, founded in 1921. About $275 million worth of gum (regular and sugarless) were sold in 1977. The newly developed liquid-center gums such as "Freshen-Up" and "Sprout" amount to a $100 million market.

Product Characteristics

Although the first gums were chicle-based, other materials were added for various reasons—such as other gums, waxes, sugars, glycerin, water, corn syrup and flavorings. Today, synthetic formulations are being used in which polyvinyl acetate largely replaces the more expensive natural ingredients.

The most popular flavors were traditionally peppermint, spearmint and wintergreen, but many others have been tried including cinnamon, cloves and, more recently, citrus flavors.

The most popular shape has been the stick form which usually is sold in a five-stick pack. Candy-coated gum, a rectangular lozenge, is second in popularity. Ball gum is also a candy-coated product but it is exclusively sold as a penny item, usually from automatic vending machines. Bubble gum comes in many shapes, sizes and colors, nearly always in novelty packages with premiums to attract children.

The sales and consumption of bubble gum has more than quintupled in the last 10 years. More than $140 million worth of bubble gum is sold each year. A large amount of United States bubble gum is also being exported to foreign lands such as Kenya, India and Japan. One of the major problems facing the bubble gum industry is the British reticence toward gum chewing. The English consider the habit ungentlemanly and unladylike.

Packaging Requirements

Because the final operation in candy-covered gums is to give them a wax coating, the candy is well-protected against moisture. Packaging is principally used as a convenience in dispensing and to protect against soilage and physical damage. Stick gums require much more careful packaging as their large surface area permits rapid loss of moisture, flavor and aroma.Too little moisture results in a hard and brittle texture. Too much moisture produces a soggy, tough product. Because of this critical requirement, manufacture of gum is conducted in air conditioned rooms from the first batching to the final packaging.

Package Forms and Materials

Penny ball gums are usually shipped in bulk containers for vending machines. The vending machines traditionally had glass globes, but today the glass has been replaced by less breakable polycarbonate plastic.

Bubble gum packages give a minimum of protection because the product is low priced and has a fast turnover. The clientele is less fussy about flavor and texture than about the size of the bubble that can be blown and the type of premium received with the purchase. Simple printed waxed paper wrappers are used to enclose both the slab of gum and the premium card.

Candy-coated lozenges are packaged in folding paperboard cartons with cellophane windows. Cellophane overwraps are optional and used not so much for moisture barrier as for a means to prevent pilfering.

Stick gums are individually wrapped in aluminum foil-wax-tissue paper and inserted in a printed paper sleeve. Foil is usually 0.00076 cm thick or less and the paper about 6.8 kg per ream machine glazed tissue. The five-stick pack takes five of the individual sticks in a stack and overwraps them on a foil laminate wrapper. At one time the wrapper was a reverse printed cellophane-wax-aluminum foil-wax 11.34 kg per ream paper. The ends were package folded and sealed with a wax stencil. A cellophane tear tape was used to help open the outer wrapper. Today, the cellophane has largely been replaced by a printed coated litho paper wax laminated to foil. The paper is usually about 13.6 kg per ream. The printing is overcoated with a high gloss coating to simulate the former appearance of the cellophane lamination. Tear tape opening devices are still used.

Recent trends in stick packages include novelty Christmas packs, where a number of five-stick packs are put in a printed box with a cellophane sleeve overwrap to allow full visibility. There also have been larger packs introduced such as a 17-stick pack. These differ from the five-stick pack primarily in dimension. It is difficult to introduce drastic changes in the chewing gum package, as the industry has been struggling to hold down costs and maintain profits. Nearly all operations are automated and production rates of 1200 sticks per hour are common. Still, the fantastic popularity of liquid-center gums has inspired the gummakers to new heights of popularity. When Warner-Lambert marketed their "Freshen-up" gum, they were hard-pressed to find production capacity capable of meeting the demand for the product.

REFERENCES

ALLEN, R.P. 1970. Varied methods are available to test packaging materials for candy. Candy Ind. *134* (11) 93–97.

ANON. 1967A. Versatile new candy packager. Mod. Packag. *40* (6) 56–58.

ANON. 1967B. A combination of aluminum foil, paper, and polythene. Paper Packag. *June*, 17.

ANON. 1967C. Sugar turns to plastic. Mod. Packag. *41* (3) 101.

ANON. 1968A. Tube of sugar. Mod. Packag. *41* (3) 101.

ANON. 1968B. Confectionery wrapping. Confectionery Prod. *34* (4) 238–240.

ANON. 1969. Metallism on film replaces film and foil on sweets. Packag. News *16* (11) 22.

ANON. 1970A. Film pouch ousts window carton as confectionery Xmas pack. Packag. News *17* (1) 15.

ANON. 1970B. Metallized cello reflects bright candy appearance. Candy Ind. *134* (7) 23 (March).

ANON. 1970C. PVC Jar breaks glass monopoly on sweets. Packag. News *17* (12) 23.

ANON. 1978. Roasted peanuts stay fresh one year. Food Eng. *50* (12) 95.

ANON. 1979A. Tests bag-in-box for peanuts. Food Drug Packag. *40* (2) 6 (January).

ANON. 1979B. Increased automation is speeding up food packaging Can. Packag. *32* (3) 31–2.

BENCHLEY, B. 1976. The sweet taste of success. Food Drug Packag. *34* (9) 14–15 (May).

CAVALETTO, C.G. and YAMAMOTO, H.Y. 1968. Criteria for selection of packaging materials for roasted macadamia kernels. Food Technol. *22* (1) 97–99.

GRAHAM, A. 1969. Shelf life. Mfg. Confect. *49* (6) 76–77.

GUISE, B. 1978. Packaging gums. Confect. Mfg. & Mrktg. *July/Aug.*, 19–21.

JOSLYN, M.A. and HEID, J.L. 1963. Food Processing Operations, Vol. 2. AVI Publishing Co., Westport, Conn.

KLEMERT, J. 1969. Shelf life and storage of chocolates. Int. Choc. Rev. *24* (9) 366–376.

PERRY, F. 1970. Packing of assorted chocolate lines. Review of the Loesch system. Confect. Prod. *36* (2) 101–111.

PRESTON, L.N. 1967. Flexible packaging. VI. Sugar Confectionery. Food Packag. Design Technol. *April,* 20–25.

RIEDER, R. 1972. A wrap-up of materials used for candy bar. Candy Snack Ind. *137* (2) 38–40.

SACHAROW, S. 1967. Packaging confectionery in flexible packages. Alimenta *6* (20) 60–64.

SACHAROW, S. 1968. Protective packaging inhibits nut meat deterioration. Candy Ind. *130* (5) 17–18.

SACHAROW, S. 1970A. Developments in film offer greater packaging choice to fit our requirements. Candy Ind. *134* (1) 12–14.

SACHAROW, S. 1970B. From film to confections. Mod. Converter *14* (3) 38–39.

SACHAROW, S. 1970C. Moisture factor is key to package selection for fruit & nut candies. Candy Ind. *134* (5) 19–21 (March).

SACHAROW, S. 1971. Packaging confections and nut meats. Food Prod. Dev. *5* (1) 78, 80, 82.

TURNER, B. and PILLINGER, T. 1977. A confectionery manufacturer's approach to the use of aluminium foil in packaging. Presented Symp. on Alum. Foil in Packag., Nov. 16–17. London

WOODRUFF, J.G. 1970. Established ground rules for refrigeration of candies. Candy Ind. *134* (11) 35–36.

WOODRUFF, J.G. 1971A. The chilling facts of frozen candy. Part I. Candy Snack Ind. *136* (12) 24, 28, 29.

WOODRUFF, J.G. 1971B. The chilling facts of frozen candy. Part II. Candy Snack Ind. *136* (13) 30, 32.

12

Cereal Grains, Bread and Baked Foods

Man doth not live by bread alone.

Deuteronomy 8:4

Give us this day our daily bread.

Matthew 6:11

Where and when grains were first cultivated for food is unknown. Records indicate that wheat was used as food 7000 years before Christ. To people of Biblical times, grain meant life and vitality. To "sow tares among the wheat" was the ultimate revenge. Cereals were probably the first food crop to be cultivated. The name is derived from Roman ceremonies known as "Cerealia" which were celebrated in honor of Ceres in mid-April.

Wheat was unknown in the western hemisphere. Columbus found corn (Indian maize) in Haiti and introduced it to Europe. A very primitive form of corn has been found in caves in Yucatan, indicating that many thousands of years of cultivation were required to bring it to its present stage of evolution. Oats are of mysterious origin, believed by some to have been a troublesome weed that grew in cultivated barley fields. A form of oats was grown by Slavic tribes of Europe during the Bronze and Iron Ages, yet it was not regarded as an important crop by preChristian writers.

Barley was cultivated by New Stone Age men in Egypt. It was widely used as a porridge, to make bread and to brew beer. Wild barley grew in Abyssinia and in Sikkim and Tibet; these may well be the ancestors of today's cultivated varieties.

Rye is a popular bread grain, but its origin, like oats, is unknown. It probably also was a weed crop. Even as late as 1925, farmers in Afghanistan and Persia tried to kill the rye that grew in their wheat fields and were given tax incentives to do so.

Sorghum is a grain believed to have originated in Africa at least as far back as 700 B.C. It is believed to have traveled to the United States in slave ships.

Rice had its beginnings in Southeast Asia, where it was grown at least as far back as 2800 B.C. Sophocles mentioned it in his writings in 495 B.C., and it became an important European crop in the medieval period when the Saracen invaders brought it with them.

Millet is a relatively minor crop, probably originating in Egypt and Arabia before history began. It was popular in medieval times, but when other grains became available it dwindled in popularity.

Other cereal grains include the wild rice grown by the American Indians and adlay, a grain of excellent food value grown in Brazil and the Far East. Some other plants produce a seed which is used in making a flour useful in baking. These include buckwheat and soybean.

Almost 70% of the world's harvested land is used to grow over 900 MMT of grain annually. The United States ranks first among the wheat exporting nations. The per capita consumption of grain products varies with family income, age levels and geographical region. Low income families tend to use more grain products than high income families. Cereals are used largely by children and older persons rather than middle-aged adults. Consumption of corn meal and hominy grits is greater in the south than any other geographical region.

Sales of bakery foods in supermarkets continued to rise sharply in 1977. The market amounted to $4,599 million with a 10% increase over 1976 figures. In-store bakeries are now a popular supermarket addition.

Three basic classifications of grain products exist. Whole grain products, where only the husk of the grain is removed, include brown rice, whole grain meals, popcorn, rolled oats, cracked wheat, shredded and puffed grains, and most hot breakfast cereals. Milled grain products are made by removing the bran and usually the germ of the seed and then crushing the kernel into various sized pieces. They include wheat flour, bulgar, seminola flour, hominy, corn grits and meal, rice and rye flour, parched barley, some hot and cold breakfast cereals, and thickeners used for soups, gravies and other prepared dishes. Soy flour is used as a grain substitute in these products. Buckwheat groats are also included in this category.

Since breakfast cereals take many forms, are made from both "whole" grains (partly milled) and milled grains, are served hot and cold, and may be previously cooked or uncooked, they will be discussed together as a separate classification.

Baked goods form the third category. They are made mainly from flour or meal and include yeast breads, rolls, quick breads, biscuits, doughnuts, crackers, cakes, cookies and pastries.

The fantastic diversity of products stemming from grains requires an almost limitless variety of packages.

WHOLE GRAIN PRODUCTS

Brown Rice

Brown rice consists of rice where not more than 50% of the hulls are removed. Wild rice is an entirely different plant and its seeds are narrow, long and purple in color.

Most of the rice consumed in this country consists of parboiled and converted types; however, packages of brown rice do appear on supermarket shelves in paperboard cartons with cellophane windows. There appears to be a trend in rice packaging toward the use of polyethylene bags. Protection from excessive moisture and insect infestation is an important requirement of a rice package, but most packaging ignores the latter. At present, paperboard boxes represent about 50% of rice packaging; however, form-and-fill bags of polyethylene are becoming more and more popular. There is a trend toward gourmet meals where several packets made from paper-polyethylene-foil-polyethylene or similar laminates are put in a printed paperboard carton. One packet contains rice, another a sauce mix, and the third a topping or spices. The final recipe has an exotic name like "Rice Teriyaki," "Rice Provence" or "Rice Milanese."

Whole Grain Meals

Whole grain meals such as wheat, corn and sorghum are not usually found on supermarket shelves. They are more often packed in large jute, paper or plastic bags for commercial use and export distribution. But there are several retail packages on the market. One of the most popular is Quaker Oats "Masa Harina." This is a kind of fine cornmeal used in making tamales and thickening chili con carne. In many foreign lands, jute packages are used since they are economical and strong. The most critical packaging requirement is protection from insects and rodents. Jute bags have been treated with insecticides; however, the risk of grain contamination has retarded this development.

In export, whole grains such as wheat are delivered in bulk; rice is received in bags. At certain Indian ports, wheat is unloaded from the vessels with slings and then bagged. At most other ports, bagging is done within the holds of the ship. Storage usually consists of stacking the bags of grain in damp-proof, well-ventilated warehouses. Methods of disinfestation currently used consist of bag dusting with various chemicals and other methods of fumigation. In India, it is not permitted to mix certain

Courtesy of E.I. du Pont de Nemours and Co., Inc.

FIG. 12.1. WHOLE RICE GRAIN PACKAGES
Polished white rice is packaged in printed polyethylene bags.

insecticides with grain. Recent trends indicate that irradiation will reduce insect attack to a minimum. Since 3–5% of food grains are destroyed by insects and rodents in India, gamma ray irradiation is being considered. It is economically equal to chemical fumigation.

Popcorn

Popcorn is a small-grained variety of maize which pops when roasted. It is graded on the basis of popping expansion, uniformity and degree of maturity. The caramelized variety should pop into a smooth, mushroom-shaped grain, while the "butterfly" type is used for buttering. Yellow popcorn has become more popular than white.

Popcorn requires protection from moisture and oxygen, especially if it is caramelized. Raw kernels for home popping are sold in tinplate cans and paperboard boxes. The popped variety is sold in duplex cellophane bags and other snack food films (see Chapter 13). In one convenience concept eight luncheon units are packed in a printed polyethylene bag. A formed aluminum foil tray containing the raw kernels is used with an expandable foil cover. As the tray is heated, the corn pops, the foil cover expands and no cooking vessel is needed. A printed paperboard cover is used for product identification.

Hot Breakfast Cereals

Hot cereals are made from whole grains and must be cooked before eating. The farina cereals are made from middlings that are milled from the whole wheat grain. A middling is what is left of the endosperm after the bran and germ have been removed. Wheat should be of a hard variety, and particle size should be carefully controlled to be finer than 20-mesh. Vitamins and minerals often are added. Some treatments can be added to speed its cooking time. Varieties of farina can be made by including whole wheat, cracked wheat, flaked wheat, bran or wheat germ.

Cracked wheat is the whole grain that has been partially milled to remove some but not all of the bran.

Rolled oats are made by heating the grain at 100°C for 1 hr to reduce moisture to about 6% and to dextrinize some of the starch. This also makes the hulls brittle. The grain is cooled and graded for size. Hulls are removed by passing them between two mill stones that are spaced apart enough to remove the hulls, but not crush the grain excessively. This rolling action probably accounts for the name. A mixture of hulls, broken grains and whole dehulled grains (groats) is removed and screened. The whole groats are then flaked by cutting into small pieces. The smaller the piece, the faster the cooking. A whole groat may take 15 min of boiling to prepare and will maintain texture in a heated steam table for as long as 6 hr. Quick oats are made from particles about ¼ the size of a whole groat.

Courtesy of Reynolds Metals Co.

FIG. 12.2. POPCORN AND FOIL WERE MADE FOR EACH OTHER
The wire handle makes this foil package a convenient skillet in which the whole grain popping corn can be heated. The printed paperboard cover is crimped in place to protect the product and is removed before cooking.

They will cook in 5 min but won't stand up as long in a steam table.

When corn is ground in a mill, a series of products of different particle sizes results. In the order of increasing fineness these are: grits, coarse and medium meal, cones, and sharp and soft flour.

The grits can be screened into different levels of coarseness also. Pearl hominy are grits that will pass through a No. 4 wire screen, but not through anything finer. They are popular in the south as a cooked breakfast dish. The coarser cornmeals are used to make cooked cornmeal mush and also certain baked goods (johnny cake, etc.).

Whole-milled rice can be cooked as a breakfast cereal but is not too popular.

Cold Breakfast Cereals

Around the turn of the century, various people devised ideas for changing cereals into new forms. Henry D. Perky invented a method of forcing boiled whole wheat through shredding rolls to form filaments of dough and pressing them into pillow-shaped biscuits. The product was "shredded wheat." Today, other shredded grains in bite sizes are now sold as breakfast foods.

In 1902, Alexander Pierce Anderson discovered that kernels of rice which were heated to 260°C in a testtube would expand to 10 times their original size. This process was tried on other grains, and the famous "shot from guns" puffed wheat and rice breakfast cereals were born.

Flaked cereals also were first made around the turn of the century when the Kelloggs made whole wheat flakes in a barn in Battle Creek, Mich. Whole kernels are moistened and partially flattened, cooked with water, sugar, salt, malt and caramel, then further dried, flaked and roasted. During this processing, starches gelatinize, proteins and sugars turn brown, enzymes are inactivated and the flake is made crisp due to very low moisture content.

Corn flakes start with a coarse grit which is pressure cooked together with sugar, malt, salt and water for about 2 hr at about 121.1°to 126.6°C. The individual particles are separated and dried to about 20% moisture, then flaked under great pressure and roasted. Roasting browns them, blisters them and makes them crisp due to low (less than 3%) moisture.

Granulated cereals are made by baking a loaf of bread made from wheat, malted barley flour, salt, yeast and water. The loaves are then fragmented and rebaked. A final crushing and sifting finishes the job. Fines are put back into the process as part of the bread loaf formula.

Oven-puffed cereals are made today, in addition to the gun-puffed cereals already mentioned. Rice is parboiled and pearled, then cooked with sugar, salt, and sometimes malt. After breaking up the lumps, the kernels are dried to 25–35% moisture, tempered, broken up again and dried to 18–20% moisture. After heating to 82.2°C they are flattened, cooled, then puffed by passing through ovens at 232.2°–301.6°C.

Composite dough products are made from corn cones, oat flour, sugar, coloring, flavoring (some with cocoa, others with fruit flavors) and water, and forced through an extruder into a strand. The strand is cut into pellets. Pellets may be flattened or left spherical. They are then gun-puffed.

Many types of breakfast cereals are coated with sugar by tumbling while a hot sugar syrup is dripped on. Sometimes the syrup is flavored.

Roasted brown buckwheat groats are sold whole and as coarse, medium and fine grits on the kosher shelf. A recipe on the package recommends their preparation as a side dish by heating with raw egg, then adding to boiling water. Extra seasoning is optional.

Packaging Hot Breakfast Cereals.—Hot breakfast cereals require protection from moisture, insects and dust.

In the late 19th century, oatmeal was a widely used breakfast food. The product was sold from barrels, and often it was infested with vermin and mice droppings. The American Cereal Company introduced "Quaker" brand oatmeal in round paperboard containers. The use of a consumer unit eliminated most of the problems associated with the unsanitary bulk packaging. The "Quaker" brand package today is essentially the same as it was 100 years ago.

The most common retail package for farina-type hot cereals in the past was the folded carton, sometimes with a paper liner (to prevent losses due to sifting) and always with a printed waxed paper overwrap. Coarser grained hot cereals sometimes were sold in cloth sacks, paper bags (like flour) or in the round paperboard container. Modern hot cooked cereal packages are much the same. The paperboard carton is still used. The overwrap is still printed paper, but the coating is no longer straight wax. Polyvinylidene chloride hot-melt coatings are now being used. They not only provide a means of heat-sealing the wrap but they offer good barrier protection as well. Latest trends are instant cereals in unit servings. These are packaged in paper-polethylene-foil-polyethylene pouches. The user rips open the package, adds the contents to boiling water, stirs and eats. Freeze-dried fruits may be included for added zest.

Packaging Cold Cereals.—The shredded wheat products must be packaged in cartons that can breathe, else rancid odors tend to accumulate. No liners or overwraps are used. High humidities will tend to destroy crispness.

In general, all cold cereals must be protected against moisture absorption. This is doubly true for sugar-glazed cereals. Much experimentation has been conducted on carton liners. These have ranged from waxed glassines to foil-wax-tissue laminations. The latest of the latter is a buried foil specification. Wax-polyethylene-vinyl acetate coatings on glassines and the PVDC-coated papers are giving higher barrier performances and challenging foil for those cereals where moisture protection isn't quite so critical.

Some breakfast cereals, particularly the puffed variety, have been marketed in polyethylene bags. Multipaks of 8, 12 and 16 boxes of assorted unit-serving cold cereals are bundled together using heat sealable cellophane, polyethylene or heat shrink polyvinyl chloride films.

MILLED GRAIN PRODUCTS

Flours

Most flours, wheat, bulgar, seminola, corn, rye, etc., are packed in paper bags for retail sale. They are available in various sizes and should be protected from infestation and moisture. Specialty flours are sold in paperboard boxes or in novelty concepts such as flour sacks. At home, it

FIG. 12.3. GLASSINE LINER FOR READY-TO-EAT CEREAL This oat cereal gets adequate moisture protection from a waxed laminated glassine liner.

Courtesy of Glassine and Greaseproof Manufacturers Assoc.

is advisable to store flour only in small amounts. Refrigerated storage is sometimes recommended.

PREPARED MIXES

Prepared mixes are combinations of ingredients preblended so that the user need only add a few more ingredients (usually wet), mix and cook. The first prepared mix was a self-rising flour patented in 1849. It contained flour, tartaric acid and sodium bicarbonate. This was followed by pancake mixes which were introduced after the Civil War. Before long a whole family of mixes was available for making doughnuts, piecrust, cakes, biscuits and the like. Some mixes require addition of water only; others require milk, eggs or both. Mixes may contain leavening agents depending on the product. Their advantages are a balance of ingredients, uniform mixing and the savings incurred from bulk purchasing of ingredients. Even more of an advantage is the savings in time for the user. Mixes were one of the first truly convenience foods.

Courtesy of Reynolds Metals Co.

FIG. 12.1. SUGARED CEREAL REQUIRES FOIL LINER
This sugared cereal incorporates a banana flavor and requires aluminum foil to protect against flavor loss and moisture pickup.

Packaging Requirements

Since most mixes contain fats and many contain leavening, they must be protected against oxygen and moisture. Some mixes are made from specially dried flour. Over a period of storage the mixes tend to deteriorate; that is, the quality of the product baked from them gets poorer and poorer. Date coding is highly desirable.

Packages

Lined and overwrapped cartons have long been used to package consumer size prepared mixes. Composition of the liners is varied depending upon product requirements. Subportioning of the mixes into sealed pouches using vacuum- or gas-flushed foil laminates can be expected, as this will tend to minimize product deterioration.

PASTA

Seminola wheat, which flourishes in the northern part of Italy, produces a hard flour which, when mixed with water, can be kneaded into firm dough with such a consistency that it can be extruded or molded into a wide variety of shapes. Such products were manufactured in the days of the Etruscans, who predate the Roman era. By the Middle Ages various forms of pasta were being consumed all over southern Europe. A 17th century colored engraving shows Neapolitans standing at a prototypical fast-food stand daintily eating macaroni with their fingers. Meanwhile, the text notes in a recognizably English tone "that the lower classes are wont to take out pasta, suitably seasoned with cheese and salt, in their hats." It also mentions that Neapolitan macaroni is very different from Genoese, being straight or bent at one end instead of twisted. It is understandable that Italy remains the greatest producer and consumer of pasta products. They produce over 1.49 billion kg of pasta annually, and per capita consumption is 29.9 to 34.9 kg. The United States produces about 635 million kg but per capita consumption is only about 3.6 kg.

Classification

There are two broad categories based on composition and many other classifications based on shape and size. Macaroni products are made of flour and water. A number of optional ingredients may be added in small quantities, such as spices, gluten, egg white or milk solids. Flour substitutions can be made using whole wheat or soy flour. Occasionally other vegetables are added (dried and powdered) such as tomatoes or spinach.

Products in the second broad category are called egg macaroni and include noodle products. These differ from ordinary macaroni in that they contain at least 5.5% by weight of whole egg or egg yolk solids. Modern macaroni or egg-macaroni products are nearly always enriched by additions of B-vitamins and iron. Since there are hundreds of possible shapes and sizes only a few typical examples will be listed.

Extruded Pasta.—Dough is forced through dies to emerge in the desired shape. In early times this was a batch process. Today continuous process machinery is available.

When hollow tubes are extruded, they are called macaroni; when solid round rods are extruded, they are called spaghetti (other names are used for finer diameters, i.e., vermicelli and cappeli d'angeli). When flat strips or squared shapes are extruded they are called noodles. Both hollow and solid products can be extruded with smooth or corrugated outer surfaces, and both can be cut to long, medium or short lengths. By means of special dies, elbows can be formed. Some noodles are twisted into spirals. Other shapes are cut thin, resulting in stars, alphabets, wheels, weed-like shapes or even spacemen or automobiles. The substitution of Teflon (a product of duPont) in extrusion dies has reduced frictions and greatly benefited production. This has also made it possible to extrude a wide flat sheet with excellent smooth surface which could then be cut into noodles.

Rolled and Cut Pasta.—Although noodles are extruded as mentioned above, many manufacturers roll a sheet of dough to the desired thickness and then cut it into strips and lengths.

Other shapes have traditionally been made from rolled sheets. These include bows, shells and ravioli.

Final Processing

After being formed into the desired shapes, the pasta dough is dried. This is a very critical operation, as it must be done fast enough to avoid mold or souring and slow enough to avoid warping or cracking. The drying operation determines the final color and texture of the product and influences the ultimate cooking time. In olden times macaroni was dried out-of-doors, then later in large rooms. Rooms became smaller, circulating air was added, then still smaller circulating air cabinets were used. Modern manufacturers use continuous driers.

Product Characteristics and Packaging Requirements

Properly manufactured macaroni products are dried to a moisture content of less than 13% and will keep without spoilage for years. Packaging is aimed at allowing the product to breathe so that no excess moisture will be absorbed or, worse, condensed on the product surface.

Packages

Short cuts and small-shaped macaroni products are packaged in cartons and in bags. Although the bags are paper in some countries, cellophanes, polyethylenes and similar films are now being used. In the United States the usual unit is 1 lb (0.45 kg), but other quantities are available. In Europe packages run from 250, 500 and 1000 g up to 5 and 10 kg. Most bags used today are gusseted and made on form and fill machinery. Cartons are made from glued paperboard and frequently have a window to allow product visibility.

Cartons are preferred in the north and middle Atlantic; plastic bags are used in the midwest, south and Far West.

Older types were made from cheap newsboard with glassine paperliners and were overwrapped with printed paper. Modern cartons are made from a higher-grade lined newsboard which can be printed. This eliminates both the overwrap and the inner liner and permits the use of the window. Windows may be made from cellophane, cellulose acetate, polystyrene or polypropylene films.

As long as pasta products are stored properly there is no reason why they should accumulate excessive moisture or be infested with insects or other vermin. For extremely moist climates, however, additional protection can be afforded through use of foil laminates or closed tins.

COOKED PASTA PRODUCTS

A number of cooked products have been made from macaroni or noodles and then canned or frozen. These include such items as spaghetti and meatballs, spaghetti and tomato sauce, a number of Italian dishes such as lasagna, macaroni and cheese, tuna fish and noodles, and others. In such products the formula for the pasta is adjusted to make it less susceptible to softening during the cooking and canning procedures.

Another new concept is a gourmet line in which a printed paperboard carton is sold containing three packets. One packet contains macaroni or noodles, another contains a sauce (sometimes with meat) and the third contains spices, bread crumbs or cheese toppings. Exotic names are given such as "Noodles Romanoff," "Noodles Italiani," "Noodles Almondine," "Noodles Cantony," "Turkey Primavera," "Macaroni Monte Bello," and "Beef Stroganoff."

Still another product contains three envelopes: one containing spaghetti, one a spaghetti sauce mix and the third a grated Parmesan cheese. The recipe requires the consumer to cook the spaghetti and to make the sauce by adding a can of tomato paste, water and oil to the sauce mix and simmering. This package, too, is a printed paperboard carton. Selection of the components in the laminate from which the packets are made depends upon what is to be put inside them. The pasta products need very little protection. A simple plastic film pouch is sufficient. The sauce mixes are dry and can be packed in paper-polyethylene-foil-polyethylene or something similar. Cheese toppings may require more protection against loss of oils. Inner components may be polypropylene, PVDC coatings or polyamide films.

OTHER PRODUCTS USING FLOUR

Gravies

Gravies made from flour, water, cooked-meat drippings and fat with added flavorings and colorings were invented in France about 1650 by the chef of Louis XIV. The right proportion of each ingredient is needed for the best consistency. Some prepared gravies are canned. A few "gravy mixes" consisting of the necessary dry ingredients are packaged in paper-polyethylene-foil-polyethylene packets. Gravies are used as components of other prepared foods that are sold canned or frozen.

BREAD

Bread is manufactured by a variety of techniques depending on the specific type desired. The most common variety, white bread, is produced by baking an expanded dough, composed mainly of flour and water, together with an aerating agent. The outside crust becomes dry and the inner portion retains a high moisture content. Many types of bread exist. Corn bread is made of corn meal. Brown bread is made of rye flour and corn meal. Black bread is a rye bread sometimes called pumpernickel. No matter what its form or origin, bread has long been regarded as the "Staff of Life."[1]

The manufacture of bread by large bakeries is not an old custom. Only since the turn of the 20th century has commercially prepared bread become available to most consumers. Although records exist of bread making since the earliest times, most bread was produced in the home by laborious techniques. The first commercial enterprises were small family-type establishments. Even today, there are more than 20,000 bakeries in the United States, ranging from small family companies to huge commercial producers. Because of the short shelf-life (2 to 3 days), most large firms are made up of a number of subsidiaries producing and selling locally. Distribution is almost always limited to a radius of 100–200 miles. Bread and rolls comprise over 75% of the total poundage produced by the entire baking industry.

The dominant trend in 1977 was the increased consumption of variety breads. White bread still accounted for 70% of all sales, but variety breads are steadily increasing in volume.

Manufacture

More than 70% of all bread produced in the United States is made of white flour milled from the inner part of the wheat kernel. The remaining types are made from rye, whole wheat or corn flours (variety types). Vitamins are usually added to enrich the bread. Most states require bread enrichment, and more than 80% of all commercially baked bread is vitamin enriched.

[1]The idea that bread is "the staff of life" originated in 1638. In recent times, the chief money earner in a family has been called the "breadwinner" and in modern terminology "bread" is a synonym for money.

Courtesy E.I. du Pont de Nemours and Co., Inc.

FIG. 12.5. BREAD WRAPPING HAS MANY FACES
DuPont MSD cellophane is used for fresh wrapping of Jewish-style rye bread (center). Frozen products are packaged in polyethylene. Small unsliced loaves use opaque white premade bags with twist tops (left background). Sliced loaves include a cellophane wrapper placed inside a clear polyethylene bag (extreme left and right).

Flour is put into a blender and mixed with water, milk, shortening, salt, yeast and sugar. It is then allowed to rise in a fermentation room. In an additional operation, more of the above ingredients are added. The dough is then put into a machine which divides it into specific weights. weights. The "rounding" machine takes the pieces of dough and dusts them with flour. From here they go to a molder where they are pressed, flattened, molded into loaf form and placed in baking pans. They are then allowed to rise for an additional hour and placed in the oven. A slow moving conveyor carries them through controlled heat, and after being cooled, they are ready for packaging.

Packaging Principles

The object of a suitable package for fresh bread is to conserve the moisture content, prevent staling and keep the bread in a fresh condition for as long as possible.

Moisture Protection.—A moderately effective moisture barrier material is required for optimum bread wrapping. The inner portion of bread has an equilibrium humidity in the range of 90%. It tends to dry out rapidly and harden. On the other hand, the crust has a low equilibrium humidity and tends to become soggy under moist conditions. Too good a moisture barrier has the effect of promoting mold growth on the bread and allowing the crust to become soft. Yet, if a truly poor barrier film is used, the bread will tend to dry out and spoil. The most preferred material should also be economical.

Staling.—Breads tend to become stale within 4 to 7 days of manufacture. This is an inherent property of the type of flour, method of baking and storage conditions. It is caused by the migration of water from the starch to the protein portion of the interior; the starch then becomes dry and loses texture. Since this activity is independent of the moisture content of the inner portion of bread, an effective wrapper must protect the bread until staling occurs.

The ideal bread packaging material must:

(1) be attractive,
(2) maintain adequate shelf-life,
(3) run on automatic machinery,
(4) be strong,
(5) be inexpensive,
(6) be an adequate moisture barrier, and
(7) protect the shape of the product.

Flexible Packaging Materials

The earliest bread was unwrapped. Even today, about 25% of all bread sold in the United Kingdom is unwrapped. French bakers now wrap their loaves in paper bags. The trend toward larger distribution, preslicing and hygienic measures has catalyzed prepackaging. In the United States, most bread sold in supermarkets is wrapped in the factory. Only breads sold via bakery counters and delicatessen counters are unwrapped prior to sale.

Waxed Paper.—Up to about 1910, no bread was wrapped in the United States. During the era preceding World War I, wax paper began to appear as a bread wrapper. Machines were developed at the time to automatically wrap and seal wax paper around a loaf of bread. Present day wax paper for bread packaging has been greatly improved from the 1910 version. It is normally a glazed imitation parchment paper, printed and wax-coated. Waxed sulphite paper and several other varieties have also been used. Wax paper is the most economical barrier to use and is rigid enough to run on automatic bread wrapping machines. It seals readily and retards moisture loss adequately. Disadvantages are its opacity and its increase in permeability when creased. Although most early wax paper overwraps were identified by printed paper bands, present-day wrappers are printed and incorporate end-labels. Due to the modern trend toward consumer appeal and aesthetics, wax paper for bread wrappers has decreased drastically. At one time the major material, the market for printed waxed bread wraps has declined from $60 million in 1958 to $40 million in 1964 and is continuing to drop. It is still used on inner wraps for permium and specialty loaves.

Cellophane.—In the late 1920s, cellophane was introduced as a bread overwrap. The initial problems focused around machine limitations, high material cost and handling difficulties. Since it was the first transparent overwrap, many bakers considered it a fad and novelty item. Heat sealing problems necessitated the development of special lines for cellophane, but by 1956 wax paper and cellophane were being wrapped on the same line without any significant modifications. The development of end-labels in the late 1930s coupled with improved grades of cellophanes promised a long and secure future for cellophane overwrapped bread.

Coated cellophane films of the semimoisture-proof type were used. Moisture protection was afforded to the product, and from 1930–1955 the major bread overwrap was cellophane. In the United Kingdom, cellophane never gained a foothold since its price differential from wax paper was too severe. At present, many specialty loaves are still wrapped in cellophane. These breads command a higher price and generally require premium aesthetics. Moisture-proof heat sealable cellophane is used for Jewish rye bread. These breads are often baked in three-pound loaves and sold in 1 lb (0.45 kg) portions. The exposed end must be maintained at the proper texture, i.e., firm and yet not hard. For crusty type breads, i.e., French bread, Vienna bread and Italian bread, a less moisture-proof cellophane is required. Crust sogginess is prevented and a hygienic wrapper is obtained. Many of these breads are also packed in uncoated paper bags.

Polyolefins.—In 1958, polyethylene was introduced to the bread packaging industry. Bakers were concerned with high costs, and polyethylene promised to reduce material costs. At the same time, rubber-based films appeared, but due to odor and higher cost they never really penetrated the bread market.

Polyethylene film was more economical than cellophane and offered the baker a more than 30% savings. Its major disadvantage was a machine problem in the thin gauges utilized. Static also presented early problems. At its introduction, polyethylene was made by blown extrusion; caliper control and optics suffered. Later, cast films were tried and were significantly better. These latter did infiltrate the overwrap market. Before the entire wrapping industry converted to polyethylene, however, polypropylene was introduced.

Polypropylene.—Polypropylene offered better transparency than polyethylene and greater rigidity and machinability. It tended to shatter at low temperature ranges which caused distribution problems. The early films broke at −17.7°C and below. In an attempt to improve weather characteristics, oriented polypropylene was introduced. This variety resisted low temperature breaking but was difficult to heat-seal on automatic machinery. The end seals appeared puckered and distorted.

The trend toward synthetic plastic films characterized by polyethylene and polypropylene led to the development of a coextruded variety of polyolefins. In 1962, a copolymer based on a PE-PP-PE sandwich was introduced. The copolymer was extruded from a single head and blended the feel and weatherability of polyethylene with the optics and rigidity of polypropylene. Although a bright future was predicted for these copolymers, the successful marketing of polyethylene film bags revolutionized the overwrapping industry. Whereas wax paper, cellophane, polyethylene and polypropylene had been used as a sealed bottom lap and end fold package, consumers in the mid 1960s were immediately attracted to the ordinary polyethylene bag.

Polyethylene Bags.—At present, more than 90% of all bread is packaged in low density polyethylene bags. The same package is also being used for the packaging of rolls and partially baked products. Even though polyolefin film wraps had distinct advantages as bread wrappers, they never really ran satisfactorily on automatic overwrap machinery. Preformed bags eliminated production problems, were more costly, but immediately succeeded as a convenience package in the consumer's view. The upsurge in sales forced many bakers into using preformed bags and the entire industry was swept into submission.

The major advantages of bags for the consumer revolve around reusability, ease of removal, better shelf-life and the elimination of belly bands. Most experts feel that in some respects these advantages are overridden by even greater disadvantages.

Wicketing of bag packages necessitates the use of holes which tend to leave ragged edges on the tied end. The bag must be oversized to permit the introduction of the loaf. Since bottom gusset, side-weld bags are generally used, a loose package results, with evident slice separation. In addition, bags are hard to handle, transport and stack.

Most bags are composed of 0.003175−0.00381 cm polyethylene secured by plastic-covered wire twist ties. As the bags are filled, ties are applied automatically from reels. There are two basic types of bread bagging equipment available. One utilizes preformed bags while the other utilizes rollstock. Economic considerations favor the use of rollstock machinery for long production runs. The American Machine and Foundry's Mark 50 is used extensively as a preformed bag machine. Bags are pulled over the bread and various bag sizes can be used. Rollstock machines such as FMC's work on the principle of pushing the bread into the bag.

Some bags for bread are made of polypropylene with twist tie. For warm-in and fancy bread items, foil-paper laminates are used. Other overwraps are composed of foil-PE laminates. Opaque wrappers provide better protection from thiamin loss than transparent wrappers.

Recent Trends

During the last 2 years, the bread industry has entered a period of stabilization. Presently, the market is aiming toward a combination of bags and overwraps in the same package. Equipment has been developed which overwraps a bread package in the conventional manner and gathers and ties the material on the other end. This package uses a pressure sensitive label on the end which is not tied. The bag disadvantages of ragged edges, looseness and floppy gusset ears are eliminated and a neater looking package results.

A still newer trend on the horizon for bread packages is shrink films. Soft bread tends to be squashed when the films shrink; however, shrink films do offer a neat package. Recent developments enable bread to be packaged and shrunk without serious distortion of shape by directing the hot air to the bottom of the package only.

About ½ of all bread and buns are machine-bagged. Square-end bags have recently been introduced. A new automatic bun packager for large volume and smaller bakeries incorporates devices such as the double wicket, which holds 500 bags and eliminates changeover interruptions. Another machine has higher speeds and continuously forward-moving arms to feed the bags. Another makes it own bags, which saves between $2.00 and $3.00 per 1000 bags.

The bulk bin bagger makes and seals it own bag out of 0.00254 cm polyethylene film at rates of 15 packages per minute. Automatic coders and price markets are available.

Some specialty breads have been packaged in foil laminations. Usually a foil-wax-paper or foil-wax-paper-wax-paper is used. Heat seals are achieved by bleeding wax through registered perforations.

Blister packs and PVC films are still newer concepts for bread packages. Blister packages are expensive; however, they also are attractive. PVC shrink films make good tight packages for hard crust breads.

A more recent trend in bread bags has been the use of advanced printing methods creating novel "woodsy" scenes on the bags. High quality, multicolor reproductions are now printed on bread bags. Striking examples are Kingo Bakery bags with a Hawaiian background and Arnold Bakers' deli rolls. There is virtually no limit to the type of bakery items which can use this type of packaging to lure the baked goods customer to a particular product.

Canned Bread

Bread is packed in tinplate cans for use by campers, hunters, sportsmen and the military. About a 1 year shelf-life is obtained. If fillings are added, the chance of mold occurrence becomes accentuated.

Frozen Bread

Bread can be quick-frozen and sold in frozen food cabinets. This concept provides for ease of supply and convenience. After the bread is baked, it is wrapped in cellophane or polyethylene and quick-frozen. The wrapper should conform as closely as possible to the bread to prevent desiccation. It is sold in printed cartons or as plain overwraps. Shelf-life of 3 months is obtainable. Many varieties of rolls are marketed in twist-tie PE bags as frozen varieties.

Bake-In-Wrap

If bread can be baked in its wrapper, the product is completely resistant to mold spores and the wrapper provides a barrier against further contamination. Weight loss during baking is reduced and the loaves can be baked close together, insuring a more economical use of oven space.

Nylon 6 film and two specially developed heat resistant grades of cellophane have been used for this purpose.

Sandwiches

Sandwiches contain two thin slices of bread with a wide variety of fillings. They must be wrapped for display and convenience in retailing. Because of their relatively high surface area, they tend to to dry out and spoil before natural staling occurs. The proper film must maintain a sufficient moisture barrier and yet not too good, or bacterial spoilage of the filling might be encouraged. Cellophane is often used in the form of bags or direct wraps. Shrink PVC, polypropylene and PE are also used for tray overwraps. The shelf-life of wrapped sandwiches rarely exceeds 2 days. Sandwiches have also appeared as frozen units wrapped in film overwraps or packed in coated cartons. One of the latest adjuncts to this is the marketing of a frozen sandwich filling. The product is thin and sandwich size and comes individually wrapped in a film, similar to the individually-wrapped processed cheese slice. The consumer keeps the product frozen, removes the wrap and puts the filling in the sandwich. Within a few hours at room temperature it has thawed, or the sandwich can be toasted to thaw the filling in minutes.

Courtesy of Reynolds Metals Co.

FIG. 12.6. SANDWICHES ARE WRAPPED WITH SHRINK FILM
Reynolon heat-shrink PVC film overwraps are used to wrap bun type sandwiches in trays for a catering service. The Reycon Case-Redi wrapping system shown here is faster than other hand-operated systems and saves up to 19% in amount of film used.

Refrigerated Dough

Premixed and kneaded bread, roll and biscuit doughs are sold in spiral-wound paper canisters with metal ends and easy-open devices such as tear threads. Wet strength kraft papers are used to prevent weakening of the package in the moist refrigerator. Foil, glassine, nylon or polyvinylidene chloride inner liners are used as barriers against moisture loss and fat migration. When refrigerated, the product has a good shelf-life, but packages are date-coded to prevent possible loss in quality due to deterioration of the leavening agent. The consumer removes the dough and, if necessary, shapes and allows it to rise before baking.

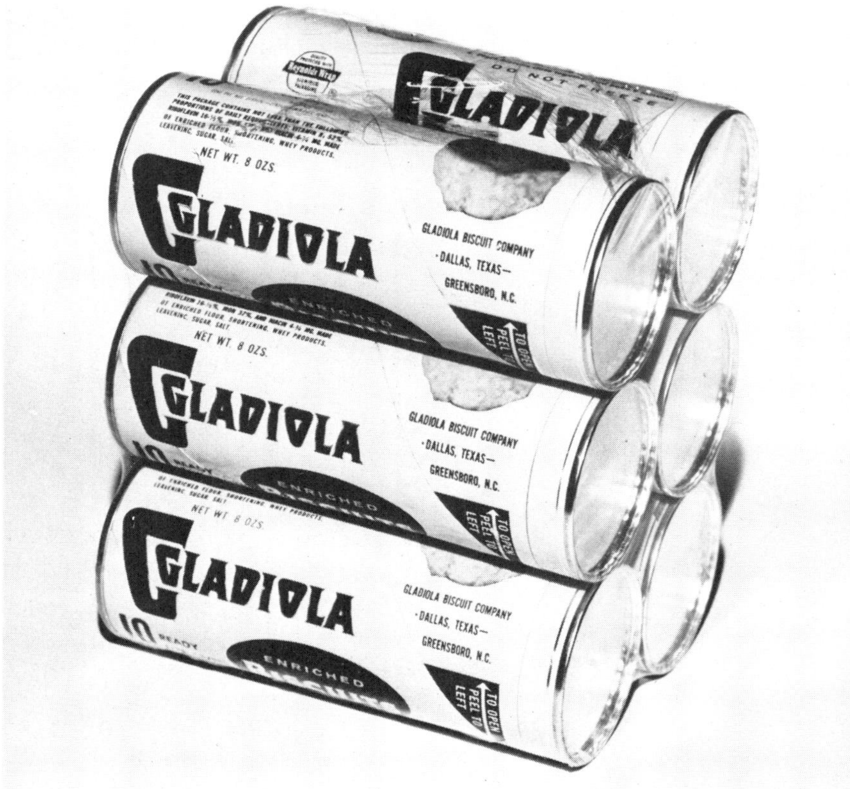

Courtesy of Reynolds Metals Co.

FIG. 12.7. REFRIGERATED DOUGH IN SPIRAL-WOUND CANISTERS
Biscuit dough packed in canisters is another convenience food popular with consumers. Shrink film such as Reynolon PVC film can be used to bundle several containers for multiple unit sales.

Frozen Bread Dough

Frozen premixed and kneaded bread dough is sold in aluminum foil containers with polyethylene bag overwraps. The consumer removes the batter, allows the bread dough to thaw and rise, then bakes it in the foil container. Frozen products have a much longer shelf-life than the refrigerated doughs.

Frozen, Refrigerated and Nonrefrigerated Prebaked Brown and Serve Breads

A variety of breads and rolls are prepared and partially prebaked. All the consumer needs to do is give them an additional browning in an oven.

Depending upon the product and the degree of prebake, these items can be found in the freezer or refrigerator display or the open shelf. Most packages are simply polyethylene bags. Some are tied shut; others are closed with saddle labels. Fancy prebaked rolls are placed in shaped foil containers and overwrapped or bagged. The consumer finishes the baking job right in the compact metal foil container.

Miscellaneous Speciality Bread Items

Bread crumbs and cubes of bread prepared from leftover ("day old") bread are sold for use in cooking. Crumbs are used as toppings for a variety of dishes; the cubes are used in poultry stuffing. Both of these products are sold in polyethylene bags. Flavored buttered toast is packaged in paperboard stiffener trays with printed cellophane overwrap. Tins of Melba toast and Zwieback are found in gourmet departments.

OTHER BAKED GOODS

Cakes, pies, pastries, sweet rolls, cookies and doughnuts comprise about 25% of all baked goods production. Many of these items are sold frozen, but most are sold fresh and unrefrigerated.

Throughout history, pastries have been an important part of the human diet. Among the ornate engravings of the 17th century are those by Abraham Bosse showing a pastry cook at work and a family celebrating Shrove Tuesday by making sweet fritters in an open fireplace.

Cakes

The hundreds of different sizes, shapes and forms of cakes sold in the prepackaged form makes this product more complex than bread packaging. Most stores aim at a shelf-life for cake between 5 and 7 days. Cakes differ as to moisture content and tendency for mold growth depending on their composition. In the vast majority of cake packages, a transparent film is required for consumer appeal. In cake baking, a mixture of flour, fat, sugar, water, eggs, spices and leavening is used.

Sponge Cake and Swiss Rolls.—Sponge cakes and Swiss rolls are essentially plain types of cake which have a low ratio of sugar to moisture, that is, they are the most moist of all cake with equilibrium humidities between 80 and 95%. In order to prevent mold growth, some drying of the surface of the cake is essential.

The most commonly used film is a semimoisture-proof grade of cellophane. If the cake is at the lower end of the humidity scale, dehydration can be prevented by the use of a more moisture-proof grade of cellophane. The trick is to utilize the proper film in order to extend shelf-life and still not encourage mold growth. During cold weather, a fully moisture-proof grade of cellophane can be used, since mold growth does not flourish. It is absolutely essential to evaluate the specific film required with its product compatibility prior to use.

There are several other factors which affect the overall shelf-life. If the cake is left in the pan too long prior to cooling, moisture may be driven out from the product. The humidity in the plant also affects the finished item. All these factors have a bearing on the moisture content of the final product.

Polyvinyl chloride films are suitable for sponge cakes and Swiss rolls because they have a moderately good moisture barrier property.

Fruit Cakes and Liquor Cakes.—Cakes containing fruit additives or liquor flavorants are usually baked to a lower equilibrium moisture in the range of 75–85%. They are less susceptible to mold growth; a moisture-proof cellophane film can be used satisfactorily. Other usable films include coated oriented polypropylene, polyethylene-polypropylene-polyethylene combinations, cast polyethylene and wax-laminated glassines. The choice is dictated by machinability and economics required. Fruit cakes have the longest shelf-life of most baked goods. Laminates of polyester-polyethylene are used as well as metal cans. Generally, the package appearance is better with polyolefin films at the end of the shelf-life period than with cellophanes.

Iced Cakes and Fancy Items.—There are hundreds of varieties of fancy cakes containing different icings and flavorants. If the cake contains fruit and no icing, its equilibrium humidity will be low enough to permit the use of a moisture-proof barrier. Moisture-proof cellophane, coated oriented polypropylene, polyethylene or foil-paper laminations can be used. If the cake is iced, a moisture-proof barrier presents problems. The atmosphere within the package is humid and the icing absorbs water and becomes sticky. A chocolate cake with chocolate icing has a different hygroscopic factor than white cake with white icing. Too good a barrier will cause the icing to run.

Icings and glazings complicate cake packaging. Most contain at least 80% sugar. Sugar icings are extremely hygroscopic, and if a high moisture barrier film is used, the icings absorb moisture from the product and atmosphere. They tend to lose stability or trap the moisture between the icing and the product. The net result is that the icing becomes soft, sticks to the film and separates from the cake. This problem area is under attack by researchers who are seeking stabilizers. In addition, the time before packaging and icing laydown must be properly controlled. If too long a time elapses, the icing may tend to crack and chip during packaging. Polypropylene is better than cellophane in icing release by about 50%.

Courtesy of E.I. du Pont de Nemours and Co., Inc.

FIG. 12.8. FRUIT CAKES IN METAL TINS
Film-wrapped fruit cakes are packaged in metal tins for long-term protection.

Doughnuts.—Doughnuts are very similar to cakes except they are fried in deep fat. The greasiness of the product demands grease-proof liners in packaging. Some doughnuts are dusted with powdered sugar; others are glazed with sugar or honey; still others are iced. Sugary types must be protected against excess moisture getting to the sugar either from the interior of the doughnut or from the atmosphere. A moderate moisture barrier is needed.

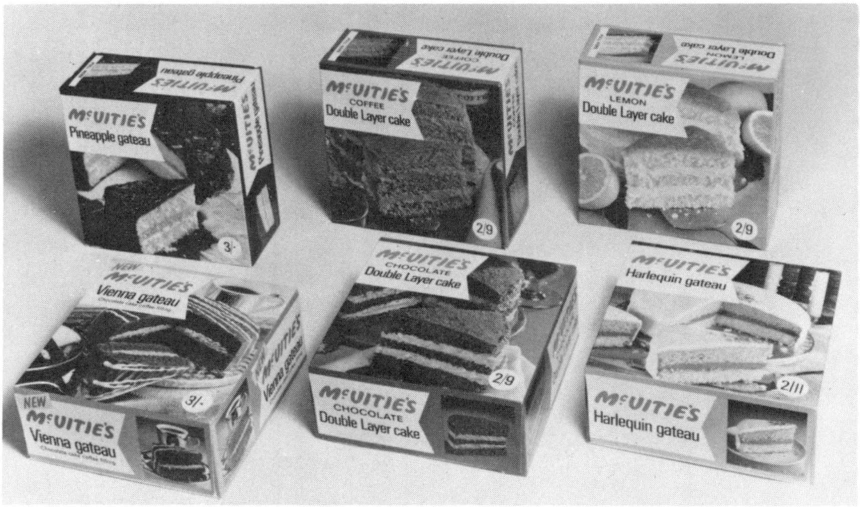

Courtesy of The Metal Box Co., Ltd.

FIG. 12.9. ICED CAKES PACKED IN CARTONS
The graphics on these cartons present appetizing sales messages. Cakes are put in boxes automatically.

Fancy Baked Pastries (Cake-like)

Sweet rolls, sometimes called Danish pastries, with fruit fillings and iced toppings, are generally baked to a low equilibrium humidity. They tend to absorb moisture and become soggy under normal conditions. When jam is added, it contributes a high humidity. Since both of these opposing effects are present, these products are very difficult to package properly in order to prevent sogginess, staleness or mold. Shelf-life is about 2 to 3 days.

Pies, Tarts and Puff Pastries

Unlike the fluffy Danish pastries, pies, tarts and puff pastries contain no leavening (yeast or baking powder). Pastries are made from flour, water and shortening with small amounts of sugar and salt for flavor. Sometimes milk solids are added. Flaky crusts are obtained by not thoroughly mixing the flour and fat. Mealy crusts are thoroughly mixed. Ingredients are kept chilled to keep the fat in discrete particles. Puff pastries are obtained by rolling the dough into thin layers and spreading shortening between the layers. Some of the best pastry doughs contain a high percentage of lard. Other fats such as butter, margarine, vegetable

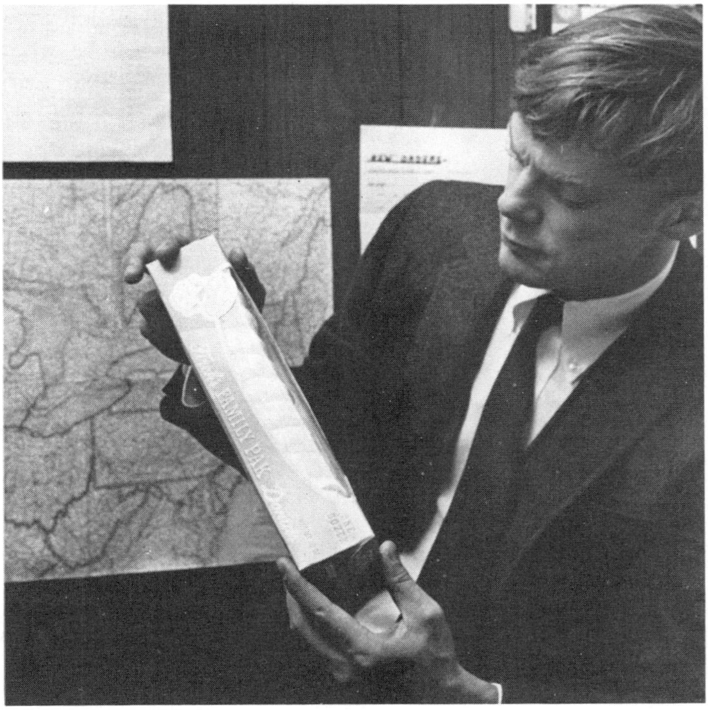

Courtesy of E.I. du Pont de Nemours and Co., Inc.

FIG. 12.10. DOUGHNUTS IN WINDOW CARTONS
Consumers like to see doughnuts and appreciate the transparent film window carton illustrated here.

shortening, chicken fat or oils may be substituted in part. Pie crusts are, therefore, susceptible to rancidification and will stain absorbent materials. The amount of moisture protection required depends principally upon the fillings. Some will dry out faster than others, some are more susceptible to mold than others. Since the pie crust is low in moisture and the fillings are usually high in moisture, the two tend to spoil one another.

Packaging of Cakes, Rolls, Pies, etc.

A direct wrap is used for most sliced cakes, with and without paperboard inserts for rigidity. The film is heat-sealed at the bottom by hand or by machine. Transparent bags are used for many pastries. If the cake is iced, contact with the film is to be avoided. Paperboard boxes with cellophane windows are used, and icing damage is prevented by making sure the box is taller than the cake.

Courtesy of British Cellophane, Ltd.

FIG. 12.11. FROZEN PUFF PASTRY PACKAGE
This coated "Cellophane" package for frozen puff pastry is reverse-printed and vacuum-metallized, then polyethylene-coated.

Pies used to be sold in tinplate pie pans. Now they are sold in paperboard or aluminum foil pans. Window boxes or overwrapped paperboard trays are used for added protection, as with cakes. The amount of protection provided by the package is governed by the type of filling used. A variation of an actual paperboard box consists of an open-top carton with a transparent film overwrap. It is particularly useful for iced cakes and meringue pies, since the overwrap does not contact the product. Doughnuts use glassine liners or overwraps to protect from grease stains.

The coming trend in baked goods packaging seems to be blister units. Thermoformed blisters made of polyvinyl chloride or polystyrene are the top of the package, and the bottom consists of either a foil tray or paperboard insert. The packages provide both aesthetics and protection.

Many sweet goods are also found in polyethylene bags. This is an offshoot from the revolution in bread bagging.

Prolonging of Shelf-life for Cake-type Products

Most cakes have a shelf-life of about 14 days. Several methods are used for obtaining longer keeping quality. These include chemical preservatives, post-bake sterilization, post-bake pasteurization and freezing.

Preservatives.—In many cakes and rolls, shelf-life extension is provided by the incorporation of a mold-inhibiting agent. Sorbic acid or its salts are used. When added to the mix, 0.03–0.125% of the weight of batter is recommended depending on pH, or the sorbate may be incorporated in a wrapping material, 2.5–5.0 g per 160 sq cm being adequate.

Post-bake Sterilization.—Cakes may be placed, wrapped, in an oven at 135°C for 20 min or longer. The material must be able to provide a hermetic seal and to withstand the high temperatures used for sterilization. Drying out must also be prevented. Films used include polyethylene, polypropylene and certain polyamides. Shelf-life of up to 1 year is possible.

Post-bake Pasteurization.—Another fairly new technique developed in France consists of infrared pasteurization in order to produce cakes with more than seven months shelf-life. Cakes and pastries are cooked in aluminum foil containers and then packed in polyamide film bags. After sealing, the cakes pass into an infrared tunnel for exposure to radiation which destroys surface mold spores. The stabilized product is then packed into printed boxes.

A test program in the United Kingdom uses a heat sterilizable nylon-PP laminate to package raisin buns, white and brown rolls and French bread sticks for 6–8 month shelf-life.

Post Bake Quick-freezing.—Although the practice is only about 20 years old, more than 60% of the output of wholesale bakeries is being frozen. This amounted to approximate volume of $903 million in 1977. Over $100 million of bread is frozen, $250 million of pies and about $550 million of other bakery products. These bakery products vary as to packaging prior to or immediately after freezing. Most baked goods are frozen after packaging. The unwrapped product would lose too much moisture in the freezer and become dehydrated. This is especially true for most yeast doughs, which require more time to reach subzero conditions.

Although freezing will preserve a cake, there are some undesirable features. Chocolate cakes lose volume during freezing and storage. Devil's food cakes become a little more crumbly. Although flat or fondant type icings are not affected by freezing, condensation during thawing can cause undesirable wetness. Few problems are encountered with frozen pies as they are harder than cakes and easily wrapped with automatic equipment. Frozen pies are principally sold unbaked. The consumer cooks the product. Tarts (small pies) are sold frozen and unbaked. A popular item is a thin rectangular tart in a pouch. The tart is cooked in a household toaster.

Muffins, waffles, pancakes, Danish and French pastries, cheese cake, and some doughnuts can be frozen successfully.

Freezing Conditions.—The actual freezing time for cakes depends primarily on the type of cake and its moisture content. The minimization of ice crystals is extremely important. With pies, rapid freezing is essential to prevent breakdown of the fruit.

Packages and Package Materials.—A proper package for frozen baked goods should be:

(1) attractive,
(2) protective against freezer burn,
(3) grease proof,
(4) free from odors, and
(5) nonabsorbent.

Six basic package types are used for frozen baked goods:

(1) waxed paperboard,
(2) foil containers,
(3) waxed paper,
(4) cellophane,
(5) polyethylene, and
(6) laminated products,
  (a) paperboard laminates,
  (b) film laminates,
  (c) foil laminates (rigid and flexible).

If paperboard cartons are used, it is important to select materials capable of resisting subzero temperatures. Most cartons are coated with polyvinylidene chloride, wax, polyethylene or wax-resin blends. If there are sticky icings on the product, silicones are used for release properties. The window material on a carton, dependent upon the product and storage temperature, may be either polyester, oriented polypropylene or cellophane.

Aluminum foil containers are very popular since they may be reheated in the oven directly from the package. For example, cake batters are placed in the foil-pan, baked and iced, then blast-frozen. Cartoning occurs as the last step prior to palletizing.

Muffins are sold loose in polyethylene bags, although some specialty items are set in deep cavity aluminum foil trays. Partially cooled waffles and pancakes are sold in individual polyethylene bags. The consumer removes them from the bag and places them in the household toaster to thaw and finish cooking. English muffins are sold frozen (already split in half). They are packaged in a polyethylene bag with or without a paperboard tray insert. The toaster tart is packaged in a paper-polyethylene-foil-polyethylene heat-sealed pouch. Multiple packets are overpackaged with a printed paperboard carton.

A number of combination packaged products are now being marketed which in a sense could be called a cooking kit. One example is a printed paperboard carton containing about 14 oz (316.9 g) of a fancy "streusel." The carton has hot melt glued end flaps with a tear strip opening device. Inside are four other packages. One is a spiral wound paperboard canister with a foil liner, metal ends and a printed foil-kraft paper label. The canister contains pastry dough rolled up into a cylinder. The consumer unrolls the sheet of dough onto a greased cookie sheet. A second package contains a fruit filling which is spread over the dough. The fruit filling is packaged in a plastic film such as oriented polypropylene, which has been formed into a tube by means of a fin-seal and closed at each end by gathering and applying a metal clip as is done for a chub pack. The third package is a form and fill pillow pack made from an opaque imitation glassine with a PVDC coating; a hot melt has been used in the seal area. This contains the streusel topping which is highly flavored. The final package is a small "tetrapak" made from a plastic film and containing a creamy icing. This is a prime example of how modern packaging technology has been used to put together a convenience package for the consumer that otherwise would be quite impossible.

COOKIES AND CRACKERS

Cookies (called biscuits in England) are prepared from doughs containing flour, fat, sugar, water and flavoring or coloring agents. The dough is rolled and cut into desired shapes. It is then baked and packaged. The composition of the dough and the type of flour used determines whether the cookie will be soft or hard. Sugar wafers are made by pouring a very thin batter onto a heated mold.

Many varieties of cookies and biscuits exist. Some have high sugar contents, others high fat contents. Some have chocolate or flavored fondant fillings, others are enrobed. The primary characteristic of all cookies and biscuits is their very low relative humidity. In order to prevent the absorption of moisture from the air, a high moisture barrier is needed. Since many cookies contain fat, a grease-proof material is also demanded. Light may also be detrimental to colors or cause fat oxidation. A suitable packaging material must be able to run on automatic machinery and protect the brittle product from physical damage. Cookies containing fruits may be susceptible to mold. Those containing nuts are susceptible to rancidification.

Crackers are similar to cookies except that they contain little or no sugar. They are flavored with salt, cheese or other nonsweet materials.

Courtesy of British Cellophane Ltd.

FIG. 12.12. CREAM FILLED COOKIES IN CARTONS
These cookies are packaged in "Cellophane" overwrapped printed cartons with a tear-strip opening device.

Packaging Materials and Package Types

Moisture-proof cellophane, coated oriented polypropylene aluminum foil laminates, polystyrenes, waxed glassines and corrugated glassines are used in cookie packaging. Polyvinylidene chloride-coated cellophane offers an excellent moisture barrier, heat-sealability and gloss. It is a fairly tough film, capable of resisting the sharp corners of many cookies. Oriented coated polypropylene offers increased economies; however, it is a little more difficult to run on automatic machinery. The introduction of polyvinylidene chloride papers have led to their use in cookie packaging. Paper is easily machined and easily coated.

The shelf-life for most cookie items is about 3 months. Laminates of aluminum foil provide for a longer shelf-life under moist conditions.

Cookie Packages.—A common cookie package consists of a printed paper bag with a film liner. The bag is secured by twist ties at the top. Many cookies are also packed in paperboard boxes with printed paper or film overwraps. A fairly new package is composed of thermoformed plastic trays of polystyrene and a cellophane overwrap. The product rests nicely in compartments, and the ends of the package are crimped for easy opening.

Courtesy of Amer. Viscose Div., FMC Corp.

FIG. 12.13. FIG BARS IN TRANSPARENT BAG
Fig bars are held in place by a corrugated divider tray and the laminated cellophane bag.

Other packages consist of decorated tins for gift packages, window paperboard boxes, direct film overwraps and foil-paperboard overwraps. Corrugated glassine paper is often used as separations to prevent damage. Even the plain brown bag has been used for cookies. The Famous Amos Cookie Company has recently replaced this with a two-ply structure consisting of a PP liner, specially treated for stiffness, and a "Scotch Ban" treated nonwicking white paper bag. The brown color of the present bag is achieved via printing.

Refrigerated cookie dough has been sold in plastic film chub packs for a number of years. A new package consists of a window carton containing 36 individual slices of chocolate chip cookie dough. This package provides added convenience in that the dough is already sliced and the package is easier to open and reclose.

Cracker Packages.—The bakers of saltine crackers have pioneered many concepts in cracker packaging. They were one of the first to use waxed glassine liners and waxed paper overwraps on cracker packages.

They introduced inner packs which subdivided the carton contents into individually wrapped units. By not opening all of the package at once, crispness of the product in the unopened section was protected for a longer period. Their latest innovation is to subdivide the package further into small packets of a few crackers each.

Cracker packages are either lined printed cartons or plain cartons with polyvinylidene chloride coated paper overwraps. Flavored crackers have made great inroads in the snack field (Chapter 13).

PRETZELS

The name pretzel is derived from the Latin *pretiola*, meaning a small reward. The first pretzels were made about 1800 years ago as gifts to children who learned their catechisms well. The shape was symbolic of a child's arms and hands folded in prayer. As the story goes, originally they were not so hard and crisp, but one day a batch was accidentally overcooked and the baker dipped them in salt to mask the scorched flavor. The result was a popular new baked item which remains popular to this day. In 1977 United States sales of pretzels amounted to $160 million, of which 75–85% was in supermarkets.

Before 1953, pretzels were twisted completely by hand. With the arrival of mechanical twisters in 1953, the industry was introduced to modern production methods and grew accordingly.

The first commercial plant in America was the Sturgis Bakery in Lititz, Pa., which opened its doors in 1861. Today there are 81 pretzel bakeries throughout the United States.

Product Characteristics

A pretzel is either soft or hard. If it is soft, it will generally be large, thick and doughy. This type of pretzel is baked only for local consumption because it stales quickly and has a very short shelf-life.

The hard or "brittle" pretzel is composed of combinations of hard and soft wheat, butter, shortening, salt, malt and yeast. These ingredients are uniquely formulated to give the pretzel a distinctive flavoring and supply the product with strength and endurance properties. The formula for a durable pretzel will have about 25% soft wheat and 75% hard wheat. Stronger gluten properties in the hard wheat are known to reduce product breakage.

Hard pretzels are found in at least 20 different varieties. Some of the more common are Thins, Rods, Sticks, Nuggets, Dutch, Beer, Big Boys, Butter, Teething and Cocktail Pretzels.

Since the pretzel is considered to be stronger than the potato chip and will travel extensive distances without breakage, it is distributed by some bakers on a national basis. This means that although the concentration of pretzel bakers is in the Pennsylvania area, it is still possible for them to compete in markets throughout the country.

Packaging Requirements

The two biggest factors that produce undesirable product are breakage and moisture. Although returns are generally less than 1%, it is evident that both these areas are troublesome and costly.

It is important that the pretzel retain a moisture content in the vicinity of 3–5%. If the moisture content becomes too high, the pretzel has a tendency to stale at a more rapid rate. Some moisture is necessary to give the product strength and make it less subject to breakage.

It should be noted that unlike the potato chip, pretzels have a low shortening content and are not generally affected by oxidative rancidity problems. Still, the flavor of the pretzel is delicate and should be protected with a packaging material that provides an adequate odor barrier.

With the exception of the oxygen barrier properties needed for potato chip packaging, a film for pretzels would be basically the same.

A film for packaging pretzels should have the following characteristics:

(1) grease resistance,
(2) durability for a packaging shelf-life of 8 weeks,
(3) coating (one or two sides depending on the application),
(4) machinability (should handle well on existing equipment),
(5) MVTR (not more than 0.4 g),
(6) appearance (low shrinkage when sealed),
(7) resistance to puncture and tear by the pretzel's salt crystals, and
(8) stiffness (similar to cellophane to support package).

Winter breakage is a problem in pretzel packaging. Current indications are that the manufacturer is over-packaging to offset breakage.

Humidity can work against the pretzel, and for this reason it is not uncommon for the baker to use single wall bag constructions during the winter months and double wall constructions in the summer periods of high humidity.

Packages and Materials

Pretzels are packaged in cans, cartons and film bags. Most cartons are wrapped with an overprinted cellophane film. A more recent trend has been the use of printed cartons with a more protective inner wrap, i.e., waxed glassine.

Segments of this industry believe that the bagging of pretzels offers better protection than a box, because the bag will give with the product where a box will not. In box packaging, pretzels are forced against rigid

Courtesy of Amer. Viscose Div., FMC Corp.

FIG. 12.14. PRETZELS
Pretzels sometimes are jumble-packed. This package uses a molded polystyrene tray to position the pretzels in the cellophane bag to reduce breakage.

sides of the carton which can contribute to breakage. Other industry leaders, however, note that the box is a stronger package. There is, therefore, less chance for the product to be crushed in transit to the retail outlet and finally to the customer's home.

Both sides seem to have valid arguments, but bagging is believed to be winning preference because of product visibility and its relation to impulse sales. With substantial ink coverage on a pretzel bag, the broken pretzels can be adequately hidden, while, with a box, the consumer will sometimes shake the contents to determine if any pretzels are broken.

The sharp salt crystals on pretzels can puncture and cause packaging films to tear. These sharp crystals are one of the reasons the pretzel bakers use the heavier gauge films for packaging.

Glassine.—Because of their low shortening content, a packaging material that protects pretzels from sunlight is not necessary. Glassine bags alone are, therefore, not thought to be a factor in this market. Glassine is, however, used extensively as an inner wrap or fractional package in pretzel boxing.

Polyethylene.—Prefabricated polyethylene bags are now being used to package pretzel rods by some bakers, and at least one is using polyethylene for large size bags which are made on a Wright form-fill machine.

Cellophane.—Transparent films are used in pretzel packaging as bags, carton overwraps, tray overwraps and fractional packages within another package. Polymer-coated cellophanes are used extensively for these applications because these films have proven to be stronger and more durable than the nitrocellulose-coated films.

New Trends

A company in New Jersey started what could develop into a major trend in box packaging of pretzels. By replacing a waxed-liner box with a cellophane twin-pak box, sales were doubled 6 months after introduction.

Another area of pretzel packaging that is cited for growth is the use of the six-pack. Until very recently, this method of packaging had been used for potato chips and other snacks, but not pretzels. If this trend continues, it will open new opportunities for additional film overwrap usage in the industry.

Many pretzel bags are now being made from laminations of cellophane and oriented polypropylene film. The composite takes advantage of the properties of both films. It is resistant to salt abrasion and provides an excellent moisture barrier and improved flexibility at low temperatures.

REFERENCES

ANON. 1967A. Breakaway for Sara Lee. Mod. Packag. *40* (8) 172.

ANON. 1967B. Longer life for crusty bread. Packag. Rev. *87* (7) 32.

ANON. 1967C. Heat-sealed bags choice for flour shipments. Food Process. Market *28* (9) 81.

ANON. 1967D. Cartons for cake baking. Packag. Rev. *87* (10) 48.

ANON. 1967E. Close up: Cookies, crackers and cakes. Packag. Eng. *12* (10) 108–111.

ANON. 1967F. Flour packing line. Austr. Packag. *15* (10) 51.

ANON. 1967G. Swedish biscuit makers employ cartons instead of wrappers. Packag. News *14* (10) 22.

ANON. 1967H. Nitrogen filled soft pretzel container overcomes short storage life problem. Quick Frozen Foods *29* (12) 76–78.

ANON. 1967I. Freezing of bakery products. Amer. Inst. Baking Bull. (July) *8*.

ANON. 1967J. Bread is baked in film bag. Packag. News *14* (12) 1–49.

ANON. 1968A. Reusable expanded polystyrene trays for Swiss rolls. Food Trade Rev. *38* (1) 83.

ANON. 1968B. Swiss rolls stay fresh longer in polypropylene film. Packag. News *15* (6) 46.

ANON. 1969A. Bread stays fresh and moist in cheap polypropylene film. Packag. News *16* (10) 16.

ANON. 1969B. Which frozen food package probes herald future breakthroughs? Quick Froz. Foods *31* (12) 35–37.

ANON. 1969C. Keeps bread fresh. Food Eng. *41* (12) 161.

ANON. 1970A. Composite bags for crusty bread. Neve Varpack *23* (4) 441 (German).

ANON. 1970B. Stuckey's pecan log roll shipped in polypropylene film to retain freshness. Candy Ind. *134* (5) 22–23.

ANON. 1970C. Crusty bread takes a 'breather'. Baker & Bakery Mgmt. *1* (6) 13–14.

ANON. 1971A. Pecan twirls protected by styrene cups. Food Drug Packag. *24* (2) 1, 19 (Jan.).

ANON. 1971B. New technique packs spaghetti automatically. Food Eng. *43* (4) 75–77.

ANON. 1975. Keeping fragility intact. Mod. Packag. *48* (9) 30–31.

ANON. 1977. Edible pouches aid easy mixing for bakery plants. Packag. Eng. *22* (9) 44–46.

BENCHLEY, B. 1978. Packs crumbs in PP can, saves cash. Food Drug Packag. *39* (7) 3 (Oct.).

BRIDGES, C.H. 1976. Protective packaging for cereals. Cereals Foods Wld. *21* (10) 542, 545.

DAY, F. 1978. Cartons and cakes. Baking Ind. J. *11* (3) 11–12, 15.

DICKHAUSER, K. 1970. PVC packages for pastries and sweets. Verpack,-Rdsch. *21* (5) 702–708 (German).

DRODGE, A.F.J. 1967. Cellulose films for bakers. Bread Baker *155* , 40-43.

HUMNEL, C. 1966. Macaroni Products, 2nd Edition. Food Trade Press, London.

MARKINDE, M.A. *et al.* 1976. Nutritional implications of packaging systems. Food Prod. Dev. *10,* 112–19 (September).

PETERS, J.W. 1977. Glossy domes make pies stand out. Food Drug Packag. *36* (9) 3, 19 (May).

PRESTON, L.N. 1967. Flexible packaging. IV. Bread. Food Packag. Design Technol. Feb. 32–35. Ibid. V. Cakes and pies. Mar. 7–13.

SEILER, D.A.L. 1968. Prolonging the shelf life of cake. Bread Baker *156* (7) 25–26.

WILLIAMS, K.S. 1967. Advances in bread packaging. Food Mfg. *42* (9) 42–44.

13

Snack Foods

> Looks may not be everything in snack foods. But almost. Because snack foods are impulse items that have to catch the consumer's eye and whet the appetite at point-of-purchase. And for repeat sales, the snack has to be as crisp when it's consumed as it looked on the shelf.
>
> *St. Regis Paper Co.*

The meteoric rise in development of snack products has been one of the most impressive trends in recent food history. Increased leisure time coupled with consumer affluency has sponsored a total snack food market estimated at $7.6 billion annually. The term "snack food" embraces an enormous range of products from potato chips and pretzels to chocolate-covered ants. Growth of the market is estimated at 6% annually.

A snack food is a product not normally used as an essential part of a meal. It is something eaten between meals and usually consumed under conditions of recreation, sports or relaxation. By definition, the term includes potato chips, pretzels, nuts, snack crackers, cheese spreads, many cookies, biscuits, puffed snacks, cracker sandwiches and additional products. The packaging of these many different snack foods involves differing principles, but considerable overlapping exists. This chapter will be devoted to those snack foods cooked in oil as well as the new range of formed chips currently being sold in composite cans. Pretzels, crackers and nuts will be found in those chapters involving baked goods and confectionery products.

POTATO CHIPS, CORN CHIPS, ETC.

The retail sales of potato chips in 1975 amounted to over $2.5 billion. And while dollar sales of chips showed a moderate increase in 1977 over 1976, this was primarily due to the effect of higher retail prices. The "junk food" label is seriously affecting snack food sales.

Since all chip type items represent similar packaging problems, they will be treated in one category. More development has been reported on potato chips than any other chip item; thus, the following data is representative of all chip-like foods.

Product Characteristics

Potato chips are prepared from thinly sliced potatoes which are cooked in hot oil. Nutritionally they consist of 2.0% water, 6.7% protein, 37.1% fat, 51.1% carbohydrates and 3.0% minerals and other substances. The best potato for chip processing is round with a 5 to 7.6 cm diameter. It should have a high solids content since this reduces tendency to absorb cooking oil and yields more chips per kilogram. A low content of reducing sugars gives a lighter and more flavorful product.

In the processing of potato chips, the best results are obtained from second crop potatoes grown in a dry, sandy soil. Useful varieties include Idaho Russet and Colorado.

Preparing the potato for frying involves peeling by machine, cutting out the eyes and slicing by machine into a tank of cool water. The amount of soaking time is quite important and varies depending upon the particular type of potato. Times used are between ½ and 2 hr. After removal from the water, the potato slices are drained for several minutes and fried.

The chips are lowered into the frying oil by means of wire baskets. Oil temperature is between 93.3° and 110°C. For every kilogram of sliced chips, there should be about 25 kg of fat. Frying time is between 3 and 5 min depending upon the degree of browning desired.

After frying, the chips are thoroughly drained either by hanging the wire basket above the oil or by using a metal draining shelf. The chips must be kept warm during draining. After the chips are drained of excess fat, salt may be applied. The amount of salt depends on taste. Chips are then cooled prior to packaging. This prevents sweating and sogginess.

Packaging is generally conducted via automatic form-fill-seal machinery of the vertical type. The cooled chips are loaded into a hopper and filled into the bag via an inner tube. As one bag is filled with chips, another is being automatically formed around the mandrel. The filled bag is sealed, severed and conveyed to the packing room.

Package Requirements

Important functions of a suitable potato chip package are to protect against rancidity, moisture, loss of odor or entry of foreign odors, and product crushing. Additional factors imposed by marketing requirements include attractiveness, no oil wicking or staining, easy opening and machinability. Potato chip packages exist in many sizes ranging from the small 15¢ bag to the 1 lb (0.45 kg) family size unit. Each package size influences the selection of materials and package structure and the price at which the product is offerred for sale. Due to the ratio of surface area to product volume smaller bags require the highest barrier protection but are least able to justify the cost. Many chippers compromise on the properties of small bags depending upon quick turnover of the product for assurance of freshness.

Rancidity.—Rancidity may occur because of the large quantity of oils used in processing potato chips. The oil content of potato chips ranges between 35 and 40%, and in some cases even higher. Since the overall shape of a chip involves a large surface area, rancidification becomes a critical problem. BHA and BHT antioxidants may be used up to 0.02% by weight of the residual oil to reduce tendency for rancidification.

Oxygen and light produce oxidative rancidity in the oils. This action is accelerated further by heat, free circulation of air, presence of trace quantities of copper and blue or ultraviolet light. Red and yellow light in the visible spectrum have minor effect. Ozone is particularly a bad actor as are any other peroxidic compounds.

The most suitable transparent chip package should have an oxygen barrier less than 1 cc per 1.6 sq cm per 24 hr at 1 atm and 23.8°C. Since opaque materials block off light, the oxygen barrier properties of these materials would be suitable in the 20–25 cc range. The relatively poor light barrier properties of transparent film packages for chips is a major reason for the increasing use of aluminum foil laminates. In addition, glassine provides a fairly good light barrier and requires less oxygen barrier to provide adequate protection.

Moisture.—Moisture protection is required to prevent the potato chip from becoming soggy and tough. Potato chips containing moisture in excess of 3.5% are tough and unpalatable. The preferred MVTR of a packaging material for chips should be below 0.4 g per 1.6 sq cm per 24 hr at 37.7°C and 95% rh. The development of moisture pickup by potato chips is a greater problem to the chip industry than oxidative rancidity.

Fragility.—In successful merchandising, the extremely fragile nature of potato chips makes package stiffness essential. The use of vertical racks and display cabinets also requires a stiff package. If the material is too flexible, a poor package results.

Odor Permeation.—A good barrier to odor permeation is required since the product is oily and odor permeation causes aroma or flavor alteration.

Wicking and Staining.—Oils and grease tend to stain a material having an inadequate grease barrier. Delamination and/or ink deterioration may result from such staining, in addition to its unattractive appearance.

Other Properties.—Since the package may be shipped to various temperature zones, cold weather durability is essential. High gloss is needed for sales appeal and easy-opening features are mandatory. In all cases, the material must be capable of being handled economically on the packaging machinery at high production speeds without material deterioration.

Package Making and Filling

There are two basic methods used to fabricate packages of potato chips. Either premade bags or automatic form-and-fill machinery may be used.

The success and rapid growth of the snack food industry during the late 1940s led to an increased use of automatic form-and-fill machinery. Since waxed glassine and other materials for premade bags did not function on these newly developed machines due to technical and economic reasons, the need for better packaging materials became apparent.

Development of materials suited to automatic machinery started with cellophanes which graduated from the first small 5¢ bags to the present day 1 lb (0.45 kg) sizes. Now a wide range of laminated transparent materials are available for use on automatic machinery. Opaque materials have developed at a slower rate. At present, glassine aluminum foil laminates and metallized structures are used.

Premade Bags Versus Form-and-fill Machinery.—Economics offered in the use of rollstock often is the deciding factor in the selection of automatic form-and-fill machinery. For large volume processors, automatic machinery is an economic necessity. Small volume chip producers or large volume producers for small volume products will still order premade bags but over 95% of the chip industry will use automatic machinery in future years.

Transparent Versus Opaque Packaging.—Chip manufacturers have been divided into two different philosophies with respect to transparency. Some favor transparent packages and others favor opaque packages. Based on the preferences of certain geographical locations, chippers are often reluctant to change a successful package. The history of the potato chip industry is based on regional manufacturers serving selected geographical areas. For various reasons, the first processor in a region often sets marketing policy. Consumers are slow to change and processors do not argue with success. In the United States, transparent packages are widely used in the east, southeast and northeast. Opaques are widely used in the southwestern, western and midwestern sections of the United States and Canada. Total sales are about evenly divided between opaque and transparent packages.

FIG. 13.1. OPAQUE GLASSINE POTATO CHIP BAGS ARE POPULAR IN CANADA AND PARTS OF THE UNITED STATES
This bag is a plasticized grade of opaque white glassine. Special coatings are applied to the glassine to improve moisture barrier and facilitate heat sealing.

Courtesy of Glassine and Greaseproof Manufacturers Assoc.

Technically, opaque packages are superior to transparent packages. When exposed to light, potato chips are unsaleable in 6 days. The same chips shielded from light remain fresh much longer. Since oil rancidification is a major source of off-flavor in potato chips, the ultraviolet light barrier present in opaque materials provides superior flavor protection. Other advantages in the use of opaque materials are the nonvisibility of brown and broken chips. If the potato used in chip making contains too high a content of reducing sugars, brown chips result. Poor handling, cooking and consumer misuse cause broken chips. Opaque packages mask these problems and add to the overall saleability of the product.

Specific advantages in using opaque aluminum foil laminates are their positive barrier properties and aesthetic appeal. In addition, foil packages may be heated for a different taste treat in snack foods.

Transparent packages offer product visibility which may enhance merchandising appeal. Transparent packages also are "sparkly" or "glossy." Newer opaque laminates now are available which match the gloss of transparent materials. Reverse-printed films or high gloss heat resistant lacquers provide satisfactory aesthetics. High quality printing techniques are able to print vignettes on opaque packages which depict the product inside.

Courtesy of Glassine and Greaseproof Manufacturers Assoc.

FIG. 13.2. "SNACK PACK" FEATURES EIGHT SMALL POUCHES OF CHIPS IN A LARGE OUTER BAG
Coated glassine/form-and-fill bags are used for individual packs of potato chips. Eight small bags are then packaged in a printed, coated paper bag which can be reclosed.

The growing trend toward material standardization among chippers seems to favor opaque materials. Regional processors will still market packages aimed at their specific markets. Larger processors will probably standardize, and opaque materials do offer greater product shelf-life.

Natural Potato Chips[1]

The rise of interest in "natural" potato chips has caused many chippers to reexamine both aluminum foil and metallized film laminates. Natural

[1]New regulations proposed by the Federal Trade Commission state that the phrase "health food" would be prohibited as being too difficult to define. Advertisers would not be allowed to claim that a food is "natural" if it or any ingredient has undergone more than minimal processing, or if it contains artificial ingredients. The word "organic" could not be used if the product were grown with certain types of fertilizers or if synthetic chemicals such as pesticides were used.

chips tend to use safflower oil as opposed to conventional cottonseed oil. The vast majority of manufacturers of natural potato chips, products that do not use chemical preservatives, feel that foil offers the best protection against rancidification. It is required to retard spoilage and give the products a reasonably long shelf-life. It is also much more expensive than an all-film unit. Yet, some metallized film converters argue that their product is just as good. Rediscovered in 1973, 40 year old metallized film surged to a 600% gain in 1976 in the snack food area.

TABLE 13.1. ORIENTED POLYPROPYLENE FILM VARIETIES

Type	Gauges Available	Description
Oriented polypropylene (homopolymers)	50–125 ga. 12.7–31.7 microns	Used in laminations such as OPP/LDPE.
Additive modified overwrap film	75–100 ga. 19–25.4 microns	Used for carton overwraps and lightweight snack packages.
White opaque oriented polypropylene	150 ga. 38.1 microns	Recently developed, outstanding potential.
Polymer-coated oriented polypropylene	70, 75 and 100 ga. 17.8–19 and 25.4 microns	Versatile film.
Acrylic-coated oriented polypropylene	73–130 ga. 18.6–33.0 microns	Attempting to replace single film cellophanes.
Coextruded oriented polypropylene	70, 90 and 125 ga. 17.8, 22.9 and 31.7 microns	Used now in snack food bags. Outstanding potential for future.

Metallized films used for potato chip packages include metallized polyester (MPE) and polypropylene (MPP). "Alure" is St. Regis Paper Company's trademark for its MPE and has been used by many chippers such as Cabana Foods Processing Company and Frito-Lay for their natural type potato chips. "Alure CX" is St. Regis' laminate of MPE/high density coextrudeal film. It has been used by several chippers and does effectively extend the product's shelf-life.

MPP is used as a single ply material as well as in a MPP/PVDC-Glassine laminate. MPP achieves a fairly high gloss on a 19 micron web as compared with 12.7 micron PET film.

Packaging Materials

Premade Opaque Bags.—The earliest premade opaque bags used by the potato chip industry consisted of waxed glassine and aluminum foil-paper laminates. Duplex waxed glassine bags, wax-laminated glassine bags and aluminum foil-paper laminates were used in the 1940s and 1950s. Improvements in material machinability and economics are the major reasons for the continued sale of opaque premade bags.

The largest selling premade bags were made from waxed glassine. At first they were filled manually and stapled shut. Later the "thermotop" bag was developed which provided a heat-sealed closure and eliminated the staples. Protection was related to the quantity of wax used; ultimate protection was obtained by the use of double-walled, waxed glassine bags.

Other structures used were coated glassines and duplex combinations of waxed sulfite paper. Premade aluminum foil laminate bags have been tried by several chippers. This trend is growing.

For packages containing "twin packs," the inner pouches are two glassine bags. The outer bag may be sulfite paper or glassine. New developments in premade opaque bags include the introduction of PVDC-coated glassine, PVDC-coated glassine laminated to oriented polypropylene and sulfite paper laminated to oriented polypropylene.

Although the overall use of premade opaque bags has dramatically declined, the economics of a waxed material is an important factor in their continued use.

Premade Transparent Bags.—The most popular type of material used in premade transparent bags are duplex cellophane combinations. When less rigidity and protection will suffice, single wall cellophane bags are used. Barrier protection is improved by means of polyvinylidene chloride-coated varieties; however, earlier duplex cellophane bags were nitrocellulose-coated.

Duplex-coated cellophane bags offer excellent protection to the potato chips but suffer from low flex endurance, poor cold weather flexibility and marginal stiffness. In addition, many manufacturers prefer a single film.

Nabisco's new line of snack items use PE bags. These are intended to compete directly with Frito-Lay's line of snack food items.

Duplex cellophane bags are being displaced by cellophane laminates on form-and-fill machinery.

Opaque Form-and-fill.—The development of opaque materials for form-and-fill machines has proceeded at a slower rate than transparent materials. The early success of aluminum foil laminates in the chip industry was hindered by the short supply of aluminum in World War II. Only within the last 10 years has there been a concerted effort by both foil and glassine manufacturers to capture this market. Even more recently, coextruded high density polyethylene (HDPE) has been used as a snack food film.

Opaque bags for form-and-fill machinery may be classified according to size. Small size packages (below \$0.40) usually utilize a single web. Larger size (above \$0.40) employ a composite lamination as the packaging material.

Two of the most used materials for small size chip packages are: (1) heat seal coating-glassine-PVDC coating, and (2) wax-EVA coating-glassine. Most small size chip packages consist of material No. 1. The use of a PVDC coating provides good barrier properties and heat-sealable coating provides good gloss and back seam seals. Material No. 2, used in significantly less volume, provides fair barrier protection. The material works on small 15¢ bags if oxygen barrier and flex endurance are sacrificed. The low hot-tack of the coating often causes sealing problems. The major problem in running glassine materials on automatic machinery is that they tend to fracture on formers and during shipping.

TABLE 13.2. SOME RECENT SNACK FOOD PACKAGING MATERIALS

Type	Trade Name	Firm
Metallized polyester film	"Alure"	St. Regis Paper Co.
Oriented PP films (uncoated, coated, coextruded)	"Bicor"	Mobil Chemical Co.
Coextruded HDPE	"Co/Ex"	St. Regis Paper Co.
Nylon coextrusion	"Crown-Gard"	Crown-Zellerbach Co.
LDPE/EVA Film	"Tuf-fine"	American Can Co.
HDPE coextrusion	"Zeelon"	Crown-Zellerbach Co.
Reinforced cellophanes	"R C"	Olin Mathieson Co.

Larger size chip packages normally are made from laminates of glassine or sulfite paper with transparent films. Materials such as barrier coated glassine-adhesive-oriented polypropylene, PVDC-coated cellophane-polyethylene-PVDC-coated glassine, oriented polypropylene-adhesive-hot melt coated glassine are used. The stiffness required is provided by the glassine or sulfite papers while the use of either cellophane or polypropylene as the outer ply provides package gloss.

Coextruded HDPE is also being used as a natural opaque, chocolate and cream-colored film. Its high haze gives the film a glassine or paper-like appearance. Speeds of up to 50 pks./min are possible with this material.

A new market is now available in aluminum foil laminates for form-and-fill machines. For many years, foil specifications which were economical enough to use would not run on automatic machinery. The major problem had been the successful running of a foil material around the severe turns in the machine forming head. The introduction of foil-paper-Elvax materials was the first economical lightweight specification that could be run successfully on these machines. Various other foil materials may be used for chip packaging ranging from cellophane-polyethylene-foil-polyethylene to foil-polyethylene-PVDC-coated glassine.

Olin has also marketed their type of "reinforced cellophane" to the snack food market. It is composed of a foil core with cellophane. Several major processors are currently using paperfoil laminates for potato and corn chip packages.

Transparent Form-and-Fill.—The earliest form-and-fill pouches for potato chips were made from duplex cellophane bags. The printed web was obtained from a converter, while the other web was obtained from the raw material manufacturer.

In 1961, a relatively simple process was introduced which combined two rolls of polymer-coated cellophane by the application of heat and pressure. Although thermal laminates may also be prepared by adhesive bonding, the former are cheaper and easily made by small converters. Laminated cellophane-cellophane and polypropylene-cellophane materials offer greater stiffness and superior aesthetics to duplex bags.

Polymer coated cellophane-polymer coated cellophane laminations may be made by several different methods. In dry adhesive lamination, the two webs are joined after the adhesive is applied to one web and all residual solvents removed. Wet lamination is rarely used since the two webs would be combined prior to solvent removal making the chance of solvent retention very high. Thermal bonding combines both webs by heat and pressure with steam injection as an optional technique. The use of steam allows for the restoration of moisture lost in printing and lamination.

A polymer coated cellophane-polymer coated cellophane laminate offers excellent moisture and oxygen protection. It runs on automatic machinery and yields a stiff package. Problem areas have been the lack of cold weather durability and resistance to hot and dry climatic conditions.

Oriented polypropylene-polymer coated cellophane laminations are available in many varieties. By the addition of oriented polypropylene, cold weather durability is increased and an economical, low cost material results. A major disadvantage in the use of oriented polypropylene is its fairly narrow sealing range. In recent years, several coated varieties have been introduced to remedy this effect. Since polymer-coated cellophane will not seal to uncoated polypropylene, several different methods have been used to secure a back seam seal.

The "Arbond" lamination consists of uncoated bioriented polypropylene-adhesive-polymer coated cellophane. The principal feature of this material is the use of the "cut back" edge. The oriented polypropylene sheet does not extend the full width of the lamination. It terminates between 1.27 and 1.58 cm from one edge of the lamination while being flush on the other side. The cellophane web is coated with the adhesive used across the entire width of the lamination. By the use of the "cut back" edge, the end user is able to produce a bag with a seemingly conventional top seam. In essence, the seal is really a cellophane-to-cellophane created by folding the extended edge back on itself. The cellophane seals to itself and the adhesive coating on the cellophane activates under heat to tack down the uncoated oriented polypropylene. Advantages include the elimination of the cost of using coated polypropylene. Disadvantages are the fin back seam, soft roll edge and poor tracking on the machine.

A thin strip of heat-sealable coating (thermal strip) can be applied to the polypropylene and it forms a back seam upon contact with cellophane. The major problem in this approach is that a strong back seam is not always obtained with a thermal strip. In recent years, the thermal strip approach has been preferred by most chippers.

The introduction of coated oriented polypropylenes has opened up new areas of development. The overlap back seam, desired by potato chip manufacturers, can readily be made using coated polypropylene. A heat-sealable coating on polypropylene offers many machine advantages to the chip manufacturer.

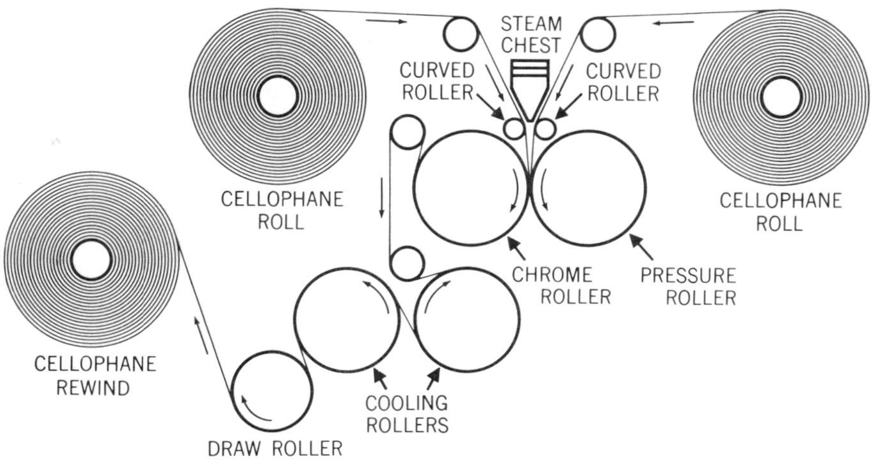

Courtesy of Amer. Viscose Div., FMC Corp.

FIG. 13.3. FMC THERMAL LAMINATOR COMBINES TWO FILMS BY HEAT AND PRESSURE
Two webs of polymer-coated cellophane can be combined on this machine by heat and pressure to form a high barrier duplex laminate for snack foods.

Packaging Machinery

A large variety of automatic form-fill-seal machines can be used for filling potato chips.

Wright Machinery Company offers a full range of form-and-fill machinery for opaque and transparent materials. Other firms supplying machinery include the Woodman Company and Mira-Pak, Inc. Three factors of great significance in proper machine production are "former" design, electric eye positioning and design of the sealing jaws. Each of these must be carefully evaluated prior to selecting a flexible material.

TABLE 13.3. TYPES OF SNACKS

Type	Market Value (1977)	Package
Cookies	$2.8 million	(see Chapter 12)
Potato Chips	nearly $3 billion	bags, boxes
Fabricated Chips ex: "Pringles"	$105 million	composite can, pouch
Natural Chips	10% of total market	bags (mostly foil)
Corn Chips ex: "Doritos"	$428 million	bags
Extruded Snacks ex: "Whistles," "Bugles," etc.	$175 million	boxes and bags (mostly bags now)

Recent Developments

The use of inert gas packaging represents a tool capable of extending the current shelf-life of chip-like products. Current distribution methods used for most chip products revolve around direct store delivery from the regional processor. The shelf-life required for these packages is 5–7 weeks. New snack foods which require boxed cartons also require warehouse distribution and utilize entirely different package concepts. The shelf-life requirements of these products is a 6 month minimum.

Many chippers are examining the use of packaging under a nitrogen atmosphere. The significant advantage to the regional chipper is his ability to extend his market to a greater area. If inert gas packaging is to become a successful process, a shelf-life of 3 months would be mandatory.

Any packaging system to be used in inert gas packaging must provide an acceptable moisture barrier coupled with an oxygen barrier below 1 cu cm. If the amount of moisture present in potato chips rises above 3½% they become soggy. Oxygen contents greater than 1 cu cm tend to cause oil rancidification.

Several chippers have experimented with inert-gas packaging concepts. The lack of specific equipment together with an economical material has retarded progress. In addition, expensive and specialized machinery would have to be purchased by the chipper.

Possible material candidates include foil laminates-polyvinyl alcohol laminates, PVDC-coated cellophane-polypropylene-polyethylene, and amylose film laminates. In all these cases, a sufficiently high oxygen barrier is required. If a material is to be PVDC-coated, the amount of laydown must be sufficient to retard oxygen permeation.

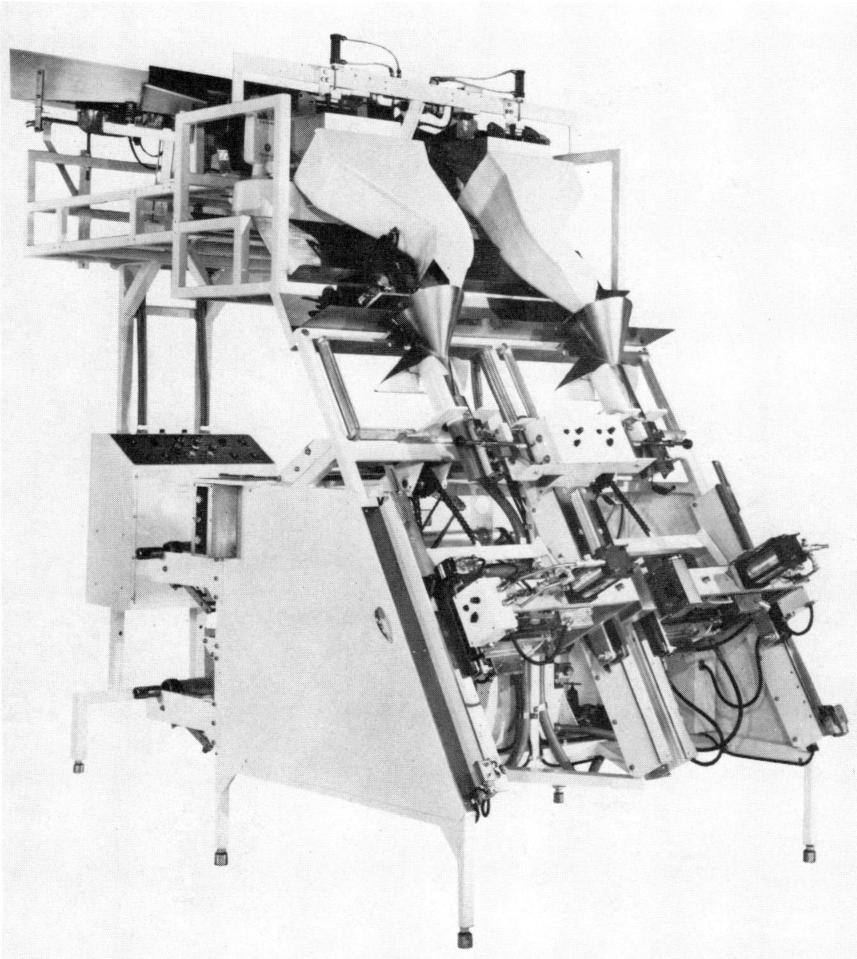

Courtesy of The Woodman Co., Inc.

FIG. 13.4. WOODMAN COMPANY FORM AND FILL MACHINE
This model DF-8 "Fleet-Weigh Dual-Pac" is a large bag automatic, net weight, form, fill and seal machine which can handle potato chips and other snack foods.

Molded Chips vs Conventional Chips.—In 1968, Procter and Gamble (P & G) introduced the original molded chip in a composite can with a PE reclosable lid. After scoring excellent market penetration, molded chips ran into strong competition from traditional type chips which stressed that they were made from sliced whole potatoes. At present, molded chips account for about 8% of the market. In an effort to withstand the natural type advertising of the conventional chip, P & G reformulated "Pringles" to eliminate preservatives. Two types of "Pringles" are now available—"Pringles Light" and "Pringles Extra Crunchy."

Courtesy of Reynolds Metals Co.

FIG. 13.5. BOXED AND BAGGED SNACKS
A wide variety of snack foods are packaged in bag in box concepts which provide needed moisture, grease and light barrier protection. Aluminum foil laminates are necessary for inner bags. Small portion pouches of the same products also feature the superior protection of aluminum foil packaging.

The use of composite cans for snack foods was also pioneered by Frito-Lay ("Crunch Barrel" and "Sun Chip"). Although these packages are not now on the market, Sealrite's "Ultrakan" is used for fried potato rings. It is a round container whose sidewall and bottom are made of PE-coated foil laminated to paper. An aluminum pull-tab top and re-closeable plastic overcover complete the container. The cans are made from die-cut blanks on a Sealrite forming machine. Savings come from the elimination of space-consuming palletized loads of composite cans, and from the use of less expensive materials: coated paperboard/foil laminate compared to more complex multiply structures.

REFERENCES

ANON. 1968A. Notes toward a definition of snack food. Snack Food *57* (1) 24–27.

ANON. 1968B. Close-up: Snack foods. Packag. Eng. *13* (3) 84–86.

ANON. 1969. Challenge for flexible materials in the 70's. Snack Food *58* (10) 53–54.

ANON. 1979. Snack foods. Packaging *50* (588) 18–20, 29 (March).

BROWN, M.L. 1969. Gas packaging for chips, snacks. Snack Food *58* (3) 44–45.

DALY, C. and JENSEN, L.C. 1969. Increases in-store exposure, doubles shelf-life. Food Process. Mktg. *29* (3) 14–15.

GIBLIN, J.P. 1968. USA Snack food packaging. Food Process. Mktg. (England) *37*, 429–432.

PERINO, D.A. 1965. Snack food sweepstakes. Paper, Film, Foil Converter *39* (10) 104–106.

PERINO, D.A. 1968. Snack foods—frustration and opportunity. Paper, Film, Foil Converter *42* (3) 70–72.

PORNER, F.E. 1971. Everything you always wanted to know about snack food packaging (almost). Snack Food *Aug.*, 64–70.

RUSSO, J.R. 1969A. Better system for bag-in-box packaging. Food Eng. *41* (10) 84–86.

RUSSO, J.R. 1969B. Bags crackers gently in stacks. Food Eng. *41* (12) 86–87.

SACHAROW, S. 1969A. Update on packaging snacks. Food Eng. *41* (3) 82–84.

SACHAROW, S. 1969B. Foil and snack food—a happy marriage. Snack Food *58* (3) 39–41.

SACHAROW, S. 1970. Converting for the snack food industry. Flexography *15* (3) 16–59.

SACHAROW, S. 1975A. Snack packs: Where the action is. Part 1, Package Promotor Die cut. *21* (10) 6–8.

SACHAROW, S. 1975B. Snack packs: Where the action is. Part 2, Package Promoter Die cut. *21* (11) 6–7, 60.

SCHREEP, C.H. 1969. White cello—A new tool for converters. Paper Film Foil Converter *43* (10) 57–58.

TALBERT, W.F. and SMITH, O. 1959. Potato Processing. AVI Publishing Co., Westport, Conn.

TAYLOR, C.C. 1971A. Flexible packaging for snacks. Part I. History of films and laminations. Snack Food *60* (1) 61–62.

TAYLOR, C.C. 1971B. Flexible packaging for snacks. Part II. Survey of Snack packaging material. Snack Food *60* (2) 46–47.

TAYLOR, C.C. 1971C. Flexible packaging for snacks. Part III. Laminating today's snack package. Snack Food *60* (3) 65–67.

WOODHOUSE, S.F. 1969. Packaging needs of the snack processor. Snack Food *58* (10) 41–44.

ZREMBA, J.V. 1970. Better-way sandwich packaging. Food Eng. *42* (2) 116.

14

Statutory and Religious Regulations Affecting Food Packaging

Do not pervert justice when you measure length, weight, or quantity. You must have honest scales, honest weights, honest dry and liquid measures. I, the Lord, am your God who led you out of the land of Egypt. You shall observe all my statutes and precepts, and do them; I am the Lord.

Leviticus 19:37

Adulteration of food and drink without regard for the health of the consumer has been a commonplace undertaking for centuries by those who sought to exploit the unwary. Milk and wine were watered; chalk or clay was added to sugar, salt or flour. Copper salts were added to green vegetables to make them greener. Perhaps the worst extreme took place in England in the 1750s when gin mills operated to extract the last penny from the poor by serving them "gin" made from almond oil, turpentine oil, alcohol, sugar, lime water, rose water, alum, tartrate of soda and sulfuric acid.

There is little doubt that this made the consumer blind drunk. The political satirist painter James Hogarth railed against the practice with little effect. "Let the buyer beware" was the law until very recent times.

UNITED STATES FEDERAL REGULATIONS[1]

Pure Food and Drug Laws

When the first Federal Pure Food and Drug Law was passed in 1906, the federal government of the United States was given the authority to

[1]United States regulations are by no means the most stringent ones. Recently, swordfish banned for sale in Canada because of an unacceptably high mercury contamination level (more than 0.5 parts/1 million is banned) were sold in the United States (regulations allow up to 1 part/million), more than 1000 MT of swordfish are sold annually with a retail value of up to $10 million.

Courtesy of U.S. FDA

FIG. 14.1. FDA INSPECTION
An FDA Import Inspector of the New York District is sampling a shipment of olives on the docks in New York City. The sample will then be taken back to the District Laboratory for analysis.

control the food supply of the American consumer. This law constituted a cooperative effort between the canning industry and federal officials to protect the consumer through intelligent enforcement of food law. Initially, the law was administered by the Bureau of Chemistry which was the predecessor of today's Food and Drug Administration (FDA). The law of 1906 was a beginning, but it was lacking in many respects.

In 1938 the federal Food, Drug and Cosmetic Act was passed. This law was a tremendous improvement over the law of 1906, but it did not provide for the control of food additives except where such additives were known to be poisonous or deleterious substances. In 1958, the Food Additives Amendment of the Food, Drug and Cosmetic Act was passed. This is the amendment that has caused great concern among people in the food packaging business. The rules and regulations promulgated by the FDA under this amendment are of significant concern to the food packaging industry. Other important amendments to the Food, Drug and Cosmetic Act which have been passed since 1958 include the Color Additives Amendment of 1960, which establishes the rules under which colors can be qualified for use in or on foods, drugs and cosmetics, and the New Drug Act Amendment of 1962. This drug act is responsible for checking and approving the efficacy and safety of all drugs and includes provisions for approval of packaging materials and labels for drugs. An amendment to the 1962 New Drug Act was proposed in late 1977. It contained sweeping changes which would dramatically affect drug legislation. This legislation had not become law as of early 1979.

Food Additives Amendment.—The most important amendment ever made to the original law, as far as the food packaging industry is concerned, is the Food Additives Amendment of 1958. This law deals with food supply in the United States and takes cognizance of all chemical components of this supply, whether these components enter the food by direct addition or by indirect means; for example, by virtue of migration from a food packaging material. The packaging industry is concerned with the part of this 1958 amendment that deals with indirect food additives; i.e., those substances which can become a component of the food by migration from a packaging material.

When the 1958 Food Additives Amendment was passed, the FDA was faced with a great deal of work. It had to find a way to administer the new law as well as to bring under control by scientific evaluation the many thousands of substances involved in foods and in packaging materials for foods.

Rules and Regulations.—In order to accomplish this, the FDA first setup a set of Rules and Regulations for administering the new law. These Rules and Regulations included Subparts A through G under Part 121 (Food Additives) of Title 21 (Food and Drugs). The collective evaluation of each of the Subparts totally defines the full scope of the law. The Subparts had the following definitive titles:

Subpart A—Definitions and Procedural and Interpretative Regulations.

Subpart B—Exemption of Certain Food Additives from the Requirement of Tolerances.

Subpart C—Food Additives Permitted in Feed and Drinking Water of Animals or for the Treatment of Food Producing Animals.

Subpart D—Food Additives Permitted in Food for Human Consumption.

Subpart E—Substances for Which Prior Sanctions Have Been Granted.

Subpart F—Food Additives Resulting from Contact with Containers or Equipment and Food Additives Otherwise Affecting Food.

Subpart G—Radiation and Radiation Sources Intended for Use in the Production, Processing and Handling of Food.

Courtesy of U.S. FDA

FIG. 14.2. FDA INSPECTION
An inspection team composed of an FDA Inspector (center) FDA Microbiologist (right) and the USDA Plant Inspector (left) are in the process of performing a joint plant inspection of an egg-breaking and drying plant. They are looking at egg-breaking machinery.

Subpart A.—Subpart A contains several sections and relates primarily to matters pertinent to the administration of the new law. This includes the rules and regulations for the filing of petitions, objecting to proposed regulations, public hearings, judicial review, matters of evidence, etc. This Subpart makes up the substantive law upon which the Food Additives Amendment is administered.

Subpart B.—Subpart B is quite short and includes only two sections, but nevertheless affects the food packaging industry to a great extent. Under this Subpart Section 121.101 appears and lists all those "Substances that are generally recognized as safe." These substances have now come to be known as having a "GRAS" status. The forerunner of this section was the so-called "White List." The FDA does not add new materials to this list, but often removes materials from the list when evidence comes to their attention to show that the GRAS status should not be continued. Plans now include a thorough review of all GRAS substances. No product should be considered as safe until it is fully reviewed by FDA personnel.

Subpart C.—Subpart C includes approximately 75 sections and relates to "Food Additives Permitted in Food and Drinking Water of Animals or for the Treatment of Food-Producing Animals." It is not of great relevance to packaging suppliers except for the packaging of pet foods or feed for farm animals.

Subpart D.—Subpart D relates to "Food Additives Permitted in Food for Human Consumption." This Subpart includes approximately 186 sections and lists those additives which are made directly into food. It is of most value to those concerned directly into food processing; however, packagers must know which substances are permitted in a packaging material.

Subpart E.—Subpart E relates to substances for which prior sanctions have been granted. These substances are listed in the only section under Subpart E, i.e., 121.2001, which is entitled "Substances Employed in the Manufacture of Food Packaging Materials." Prior sanctioned materials listed in this section mean materials that were sanctioned by FDA for use in food packaging prior 1958. Doctor Lehman of FDA is well known for his issuance of prior sanctions prior to 1958. No further additions have been made to the prior sanctions list since 1958.

Subpart F.—Subpart F relates to "Food Additives Resulting from Contact with Containers or Equipment and Food Additives Otherwise Affecting Food." This is the most relevant subpart for the packaging industry. It contains approximately 97 sections. Within these sections most components of food packages are regulated. Section 121.2501 relates to olefin polymers which include polyethylene, polypropylene, etc.

Section 121.2502 relates to nylon resins. Even though there is a regulation for a food packaging material such as polyethylene under Subpart F, it is not enough to refer to the regulation; it is also necessary to obtain assurance from each supplier that the particular material in question meets the provisions of the regulation because not all polyethylenes are food grade. Such assurance must be obtained in writing and state that the grade of polyethylene to be used in a food packaging application is food grade by virtue of its compliance with Section 121.2501 of regulations under Subpart F. Similar assurances must be obtained for all other packaging components, such as lubricants, plasticizers, coatings, adhesives, etc., citing the pertinent regulation complied with in each instance.

Furthermore, several FDA regulations under Subpart F require that extraction tests be run. This is required to make sure that the chemicals involved do not migrate into the food beyond the level established by FDA to be safe.

Subpart G.—Subpart G relates to radiation and radiation sources in the handling of food.

Clearance.—A substance can be cleared for use as a food packaging material in any one of four ways:

(1) It is GRAS that is generally recognized as safe (Subpart B).
(2) It is permitted to be used as a direct additive in food ((Subpart D).
(3) It is prior sanctioned (Subpart E).
(4) It meets the requirements of the pertinent regulation of Subpart F.

Subpart F is the only Subpart which contains regulations which require compliance since if a material can be added directly to food, is GRAS or has been prior sanctioned, it is not regulated as an indirect food additive. Accordingly, the food packaging industry is concerned primarily with these Subpart F regulations.

Food Additive Definition.—The definition of a food additive is of great value in fully understanding the law. The 1958 Amendment provides that a food shall be deemed to be adulterated if it contains a food additive. Once a food contains a food additive or a food packaging material is capable of giving rise to a food additive, penalty provisions of the act apply. The definition of a food additive is most important.[2] A food additive is not any substance added to food directly or indirectly.

---

[2]In 1940, a federal ruling prohibited listing the vitamin content of alcoholic beverages on the label as this would imply that drinking alcohol is healthy. Since *all* food additives *must* be listed, the ruling means that vitamins cannot be added to alcoholic beverages. For alcoholics, suffering from Wernicke-Korsakoff syyndrome (caused by lack of thiamin), the ruling is particularly harmful.

The Food Additives Amendment as passed by Congress included a special definition for a food additive. The legal definition of a food additive means "any substance the intended use of which results or may reasonably be expected to result, directly or indirectly, in it becoming a component or otherwise affecting the characteristics of any food (including any substance intended for use in producing, manufacturing, packing, processing, preparing, treating, packaging, transporting, or holding food; and including any source of radiation intended for any such use), if such substance is not generally recognized, among experts qualified by scientific training and experience to evaluate its safety, as having been adequately shown through scientific procedures (or in the case of a substance used in food prior to January 1, 1958, through either scientific procedures or experience based on common use in food) to be safe under the conditions of its intended use. This term does not include pesticide chemicals or any substance used in accordance with a sanction or approval granted before the enactment of this act."

The complexity of the definition has caused a great deal of confusion in industry.

A food additive is actually any substance that can become a component of food by direct or indirect means unless that it is known to be safe or is prior sanctioned. If the substance is known to be safe or is prior sanctioned, it is not a food additive. In other words, all substances that can become a component of food are presumed to be additives unless declared otherwise by FDA. The FDA has declared some substances nonfood additives by listing them on the GRAS list and has declared other substances nonfood additives by prior sanctioning. These substances appear in Subparts B and E, respectively.

The simplest check for a questionable material is to ascertain its acceptability with the FDA published rules. If the product is a film variation, adhesive or coating formulated by a supplier, a letter noting FDA compliance should be requested. Any formula containing all acceptable ingredients is considered suitable for clearance.

All other materials which are used in food directly or in packaging materials for foods are treated by the FDA in a different manner. If assurance that the material complies cannot be obtained from the raw material suppliers, a petition must be filed with the FDA. Extractability studies must be conducted and the data forwarded to Washington. The FDA requires the petition to be filed by the chemical company wanting a permit to use a product in food or in a food packaging material. A petition to the FDA must include the following information:

(1) details of usage,
(2) composition of the finished packaging material,

(3) analytical method to determine additive,
(4) identity and composition of the additive, and
(5) extractability data.

All extractability information must be obtained by using the final package. The suitable analytical method must be sensitive to within the safe limits of the ingredient. The sum total of the ingredients is additive as relates to FDA rules. After FDA reviews the toxicological information in the petition and finds it satisfactory, the FDA will issue a regulation for the safe use of the chemical in question. This regulation then regulates the use of the material. Regulations which cover direct additives for food appear in Subpart D and regulations which cover indirect additives appear in Subpart F. The regulations appearing in Subpart F are of most relevance.

TABLE 14.1. INFORMATION SOURCES FOR FOOD PACKAGING REGULATIONS

Publication	Publisher	Type
Code of Federal Regulations (CFR)	Office of the Federal Register, National Archives and Records Service, General Services Administration, Washington, D.C.	Annually
Code of Federal Regulations Index	as above	Annually on July 1
List of Code of Federal Regulations Affected	as above	Monthly
Federal Register (FR)	as above	Daily (Monday–Friday)
Federal Register Index	as above	Monthly, quarterly, annually
Food Chemical News Guide To The Current Status of Food Additives and Color Additives	Food Chemical News, Inc.; Washington, D.C.	Basic volume plus weekly replacement pages
Food and Color Additives Directory	Hazleton Laboratories, Inc.; P.O. Box 30; Falls Church, Va.	Published from 1961–1976
Food, Drug, Cosmetic Law Reporter	Commerce Clearing House, Inc.; 4025 W. Peterson Avenue; Chicago, Ill. 60646	Annually
Food and Agriculture Legislation	Food and Agriculture Organization of the United Nations; Via delle Terme di Caracalla; I-00100 Rome, Italy	Twice a year

By the issuance of a regulation for a particular substance, FDA has declared that a substance is safe and, therefore, not a food additive so long as the use of the substance complies with provisions of the issued regulation. Suppose a regulation states that migration of certain substances from a coating shall not exceed 0.5 mg per 6.45 sq cm of contact surface in the presence of a certain solvent under particular extraction conditions. If, in fact, the extraction level from a sample of a packaging material containing the coating is higher than the allowable amount, then the packaging material in question would be ruled capable of introducing a food additive into the food supply. At this point, violation of the 1958 Food Additives Amendment exists and all the provisions of the Food, Drug and Cosmetic Act apply.

It is known that if the FDA should become aware of a food packaging material which contains an unsafe food additive, they may seize the material immediately. They do not have to wait until the food has been packaged in the material and then prove analytically the presence of the food additive in the food itself. Under provisions of the Act, FDA may be able to seize the packaging material strictly on the basis of violations of packaging regulations as exist under Subpart F. With respect to the seizing of packaging materials by FDA which are alleged to contain an unsafe food additive, industry spokemen have questioned any basis for this claim in the 1938 Food, Drug and Cosmetic Act.

Extraction Data May Be Complex.—All extraction data must contain the amount of a single substance which migrates under standard conditions and the total contamination of the food product. In addition, toxicity data must be available for FDA evaluation.

Migration of additive is dependent on time, temperature, area and type of food.[3] While extraction studies may be conducted on the specific food, most tests utilize special extracting solutions. This is due to the wide range of variables inherent in food products which cause problems in obtaining accurate results. For particularly sensitive applications, radioactive tracer analysis is used. When tagged carbons are employed, excellent results are obtained. However, radioactive tracer tests are expensive and require extremely intricate equipment. Due to these factors, the process is not in widespread use.

In order to formulate a range of solvent solutions capable of simulating food extraction, several factors are significant. The various solutions types that should be included are acid, alkaline, and neutral solutions as well as alcoholic and oil materials. A representative list suggested by the United States Food and Drug Administration is:

(1) distilled water,
(2) 3% aqueous Na Cl,
(3) 3% aqueous $NaHCO_3$,
(4) 3% aqueous acetic acid,
(5) 3% aqueous lactic acid,
(6) 20% aqueous sucrose, and
(7) lard or vegetable oil.

[3]Dr. Seymour Gilbert of Rutgers University has reported that below a certain concentration a solvent dissolved in a plastic film will not migrate. This critical concentration is specific for each solvent/resin combination. This discovery will have profound effect on the establishment of minimum permissable concentrations of food additives.

The list of solutions recommended by the French authorities is more complex:

(1) distilled water,
(2) 10% ethyl alcohol,
(3) 50% ethyl alcohol,
(4) 95% ethyl alcohol,
(5) 10% acetic acid,
(6) 50% acetic acid,
(7) 1% NaCl solution,
(8) 5% NaCl solution,
(9) 10% sucrose solution,
(10) 2% citric acid solution (pH = 4.5),
(11) lard, and
(12) arachide oil.

It is essential that all extractants be used correctly to arrive at accurate determination of the migrant. Since various migrants often require different extracting solutions, the proper test method for the specific chemical must be used. For example, highly poisonous substances require extremely sensitive test solutions.

Migration studies should actually be conducted for the entire shelf life of a packaged food. In principle, the duration involved makes long term tests impractical. Most tests are conducted at accelerated temperatures and the results correlated with actual conditions. United States authorities state that the test conditions used must correlate with actual packaging conditions. If a film is autoclaved during a packaging operation, migration tests must be conducted at the autoclaving temperature. Other nations feel that the test conditions must be more severe than actual practice. Since it is often impossible to know the total packaging cycle, extremely stringent test conditions are usually employed.

The diffusion of migratory substances into food products increases with increasing temperatures. It has been reported that extraction is fairly uniform for the first 5 days at a temperature of 60°C. After the fifth day, the rate decreases and extraction ceases by the tenth day.

The thickness of the sample used should be uniform. A 25.4 micron sample is recommended since it encompasses most commonly used plastic films and sheets. Total migration should be expressed in parts/million, and a ratio of 0.3 ml solvent/cm^2 of surface is preferable for the test method.

After allowing the sample to be exposed to the solvents chosen at the specified time and temperature cycles, quantitative identification of migratory compounds is made by gas chromatography or conventional analytical methods.

Manufacturing Regulations.—Good manufacturing practice is carefully outlined in Federal regulations. Regulation 121.2500 "General Provisions Applicable to Subpart F" states that the use of any and all additive substances regulated in the 96 other regulations of Subpart F is predicated upon the use of good manufacturing practice.

Section 121.2500 of Subpart F "General Provisions Applicable to Subpart F," contains the definition of good manufacturing practice; Paragraph (a) (1) "The quantity of any food additive substances that may be added to food as a result of use in articles that contact food shall not exceed, where no limits are specified, that which results from use of the substance in an amount not more than reasonably required to accomplish the intended physical or technical effect in the food contact article, etc." This means that a food additive can be contained within a food packaging material, but that such additive must be contained at the absolute minimum required to accomplish the intended physical or technical effect.

There are special places where this provision would apply in packaging operations; for example, in the production of packaging materials where a lacquer coating is applied to a construction, where such coating is not properly dried, and where the construction is used as a food wrap with the coating in direct contact with food. Since the solvent retained in the coating would normally migrate into the food, the definition of a food additive applies. Any time that a packaging material gives rise to a food additive in excess of the minimal amount allowed by this provision under good manufacturing practices, industry must face application of the Food, Drug and Cosmetic Act penalty provisions.

In other Subpart F regulations, reference also is made to good manufacturing practice. In the Adhesives Regulation 121. 2520 the quantity of adhesive that contacts packaged dry food must not exceed the limits of good manufacturing practice.

The quantity of adhesive that contacts packaged fatty or aqueous foods must not exceed the trace amount at the seams and at the exposed edges that may occur within the limits of good manufacturing practice. In the case of adhesive lamination this means using only the amount of adhesive required to obtain firm bonding of the seams and laminate components.

Color Additive Amendment.—In 1906, the original Food and Drug Act of the United States provided for a voluntary system of certification of the purity and safety of dyes used in or on foods. Under the 1938 revision, limits were established on the minute amounts of lead, arsenic and other heavy metals that could be formulated in the colors. It also added a provision setting down the exact specifications of each individual color. Manufacturers submitted samples to the FDA for certification. Covered under this act were the coal tar colors but not the nonaniline colors such as iron oxide, titanium dioxide and naturally occurring colors.

The Color Additive Amendment of 1960 embraces all colors, both natural and synthetic, used in foods or in packaging materials contacting foods. It was designed to replace the outmoded provisions which governed the use of color in foods under the 1938 act. Tolerance limitations were established when required and other conditions for safe use. The system of official certification was continued.

In essence, three groups of colors are permitted for use in coloring of packaging materials for food. Certified FDA dyes and alumina lakes are acceptable. FDA approved pigments and FDA sanctioned colors are also permitted. Any colorant that has an impermeable barrier between it and the food is not subject to the Color Additive Amendment. The colorant may be nonapproved; however, the burden of proof relative to migration, offset, bleeding, etc., rests with the user.

An ink should be formulated by using certified or FDA approved colors. It should also use ingredients that are generally recognized as safe or approved materials under the Food Additives Amendment. Lists of acceptable ingredients are available in the various periodical editions of the *Federal Register*.

"Fair Packaging and Labeling" Bill.—In 1966, the Federal Packaging and Labeling Act (FPLA) was enacted. The FPLA became effective July 1, 1967. Specific extensions have been authorized by the FDA, but otherwise all food products must comply with the laws. A revised FPLA (The Magnuson Act) was introduced to Congress in late 1977.

The regulations state that the following information is required to appear on the label of food packages:

(1) The food must be identified. This must appear on the principal display panel and be in bold type. It should be in lines generally parallel to the package base when the unit is displayed. The common name of the food should be used whenever possible. If the food is offered as whole, sliced or chopped, this must be clearly visible to the consumer either by a statement or vignette.

(2) The name and address of the manufacturer, packer or distributor must be identifiable. It should be conspicuous and denote the type of firm supplying the product.

(3) The net quantity of contents must be stated. The use of "Jumbo Pound," etc., is forbidden. This must be placed within the bottom 30% area of the display panel and appear as a distinct item. If the food is liquid, fluid measure is used. If solid, weight is used. For

fresh fruits and vegetables, the bushel peck system is necessary. The net contents statement must be accurate and may not include the weight of the packaging material. For packages containing between 1 and 4 lb or between 1 pint and 1 gal., a dual declaration of net weight or net volume is required. The first statement must show total ounces or total fluid ounces. The second statement in parentheses must express the total in pounds and ounces or pounds and fractions or decimal fractions of a pound for solids or in quarts, pints, fluid ounces and fractions thereof for liquids. A separate declaration of net contents in terms of metric units may also be added. Net contents must show the total amount of food in the package. Where part of the contents is not used as food but thrown away, such as with the brine in a bottle of olives, only the amount of solid food should be stated.

(4) A statement of the net contents of a serving is required. This may appear in units of weight, liquid or dry measure or any other commonly used term, i.e., tablespoons. If the contents are noted in terms of servings, the description of a serving size is needed.

(5) All ingredients must be mentioned. This must be legible and appear on a single panel of the label. It must appear in decreasing order of ingredients. If a special ingredient is advertised as exceptionally valuable to the food, a declaration of percentage content is required.

OTHER GOVERNMENTAL REGULATIONS

In addition to the FDA several other governmental organizations regulate packages. The Meat Inspection Act of 1906 established all Meat Inspection Division laws under the Department of Agriculture (USDA). Labels and the overall visibility of all meat packages must be approved by the MID (now Technical Services Div. of Consumer and Marketing Services). Poultry and poultry products are also subject to USDA approval.

The Treasury Department supervises the packaging of alcoholic beverages and tobacco. Narcotics are supervised by the Bureau of Narcotics. Imports are subject to the Bureau of Customs and postal shipments are enforced by the Post Office Department.

Often, overlapping jurisdiction occurs and a product must meet FDA and MID regulations. In such cases, approval is obtained by each agency. It is important to note that rules and regulations differ considerably. The most stringent regulations are FDA laws, and if a product meets these regulations, additional governmental approval is fairly easy.

A package considered perfectly legal in one state may be illegal in a neighboring state. All state laws must conform to the basic provisions of the FDA rules; however, many states have adopted their own interpretations of some of the ambiguous parts of FDA regulations.

Extraction and migration data are standard in all states. Variances occur with package design and labeling. Colors are controlled and the size of lettering may become a part of state law. It is important to note that the limited financial ability of most individual states serves to retard intrastate court action. Prosecution is cumbersome and expensive for state authorities.

Even cities have their own packaging rules. Large metropolitan areas often have regulations relative to deceptive packaging and product visibility.

RELIGIOUS REGULATIONS

Kosher Rules

> Though some explain Jewish dietary laws as ancient hygiene, others suggest that the laws encourage self-discipline and sanctity. An observant Jew is denied innumerable pleasures of the palate because of their danger to the soul.

Many Jewish consumers subscribing to *Orthodox* (not Reform) Judaism examine products for Kosher approval. A recent figure stated that more than 2000 products from over 400 companies are manufactured according to the laws of *kashruth*.

In the United States, about 45 large companies manufacture kosher-labeled processed meat products such as salami, wieners and pastrami. The gross wholesale business of these firms averaged $85 million. The total retail volume of all kosher-labeled processed meats, including small plants operated by kosher butcher shops, is about $200 million. It is significant to note that a substantial portion of all kosher-labeled products are bought by the nonJewish who have developed a taste for kosher meat products.

A kosher product must be packaged in an approved material which is subject to the same regulations as the product. Package approval is not easy to obtain but it is an absolute necessity for certain markets.

Definitions.—*Kosher.* A Hebrew adjective meaning "proper," used to describe a food or related product prepared and packaged in accordance with Jewish Dietary Law. Kosher is a religious term which has nothing to do with taste. The term kosher flavor is often confused with a garlic

TABLE 14.2. KOSHER REGULATIONS CHART

Biblical Source	Statement
Genesis	Ban on consumption of flesh from a living animal and sinew of an animals' lip.
Deuteronomy	"Thou shalt not eat any abominable thing"
Leviticus	Ban on certain animals including camel, hare, swine, eagle, vulture, raven, owl, hawk, cuckoo, swan, pelican, cormorant, stork, heron, lapwing, bat and "all fowls that creep, going upon all four . . . " "Whatsoever parteth the hoof and is wholly cloven-footed and cheweth the cud among the beasts, that shall ye eat."
	Forbids eating of fish that are without fins and scales, and the eating of blood except from fish.
Exodus (stated twice)	Jews are forbidden to seethe a kid in its mother's milk. ("Seeth"-to cook by boiling.)
Deuteronomy	Same prohibition as in *Exodus*
	Eggs, young may be taken from a nest but the mother sitting on the eggs or young must not be taken.
Mishna (laws and commentary of the Talmud, often called impossible)	The "seething" ban in *Exodus* and *Deuteronomy* has become a prohibition of cooking, eating or benefiting from a mixture of meat and dairy foods; products such as fruit, grain and eggs are *pareve*, or neutral, and may be eaten with meat or milk. The very vessels in which food is prepared must sometimes undergo ritual immersion.
	The ban on taking the mother from its young or eggs in *Deuteronomy* has become "If the dam hovered over the nest and her wings touched the nest a man must let her go . . . A man may not take the dam and her young even for the sake of cleansing the leper."
Gemara (as above, commentary on the Talmud)	Similar prohibitions, complex law often called irrational.

flavor. The two are not synonymous in any way, although some kosher foods may have a strongly spiced or garlicky flavor. Many of the dietary laws are found in Chapter 11 of *Leviticus*. Others are located in *Deuteronomy*.

Kashruth. The noun of kosher. The five requirements of kashruth are the following:

(1) All pork products are inherently unable to be kosher. The law also forbids consumption of birds of prey, e.g., owls, vultures and eagles. Worms, snails, oysters, crabs, clams and shrimp are also forbidden. The use of various byproducts of nonkosher animals are not permitted.

TABLE 14.3. COMMONLY USED RABBIS[1]

Rabbi	Examples of Approvals
Rabbi Bernard Levy Brooklyn, N.Y.	*International approvals*—Japan (rennet), Taiwan (mushrooms), Spain (olive oil), England (sweets), Portugal (sardines), Belgium (chocolates). *Domestic approvals*—Maxwell House Coffee, Philadelphia Cream Cheese, Log Cabin syrup, Kraft margarine, Barricine candy, etc.
Rabbi Jacob Cohen New York, New York	All approvals including Breakstone cheese, Swee-Touch-Wee Tea, etc. (plus many others).
Rabbi Johoseph H. Ralbag New York, New York	Borden Foods, Standard Brands, Coca-Cola, General Mills, General Foods, fruit industry (California).
Rabbi Yacov Lipshutz New York, New York	(1) Ritual slaughtering of chickens, vegetable and cooking oils. (2) Pickles, etc.

[1] There are many other Rabbis used for approval. Any rabbi may decide to exercise kosher supervision.

(2) Acceptable animals include the flesh of all quadrupeds with cloven hoofs, i.e., deer, cattle, sheep and goats, but not the horse. The barnyard fowl and scaly fish with fins are permissible.
(3) Fresh vegetables and fruits in their natural state are inherently kosher.
(4) The acceptable animals must be slaughtered and prepared according to certain regulations. If this is not followed, the entire food is nonkosher.
(5) No dairy and meat product may be mixed or consumed at the same time. Separate cooking vessels and dishes must be used for these products.

The above list is only a quide to the general requirements of the laws of kashruth. Other regulations exist.

The term *Glatt Kosher* is used to describe fresh meat from cattle or sheep whose lungs were checked by *bodkim* (trained religious Jews) and were found free of any adhesions. If the adhesions can be removed from the lobes of the lungs without leaving any trace of perforations, the meat is kosher. The term "Kosher-style" on meat products is a euphemistic term for nonkosher and is very misleading. Its use is permitted in most states.

An additional term in kosher terminology is *Parvae* or *Parveh*. This is applied to foods which can be classified either as dairy or meat dishes. They may be eaten with either type and do not infringe upon the dietary laws. Care is needed to keep parvae utensils separate and not to cleanse them together with either dairy or meat utensils.

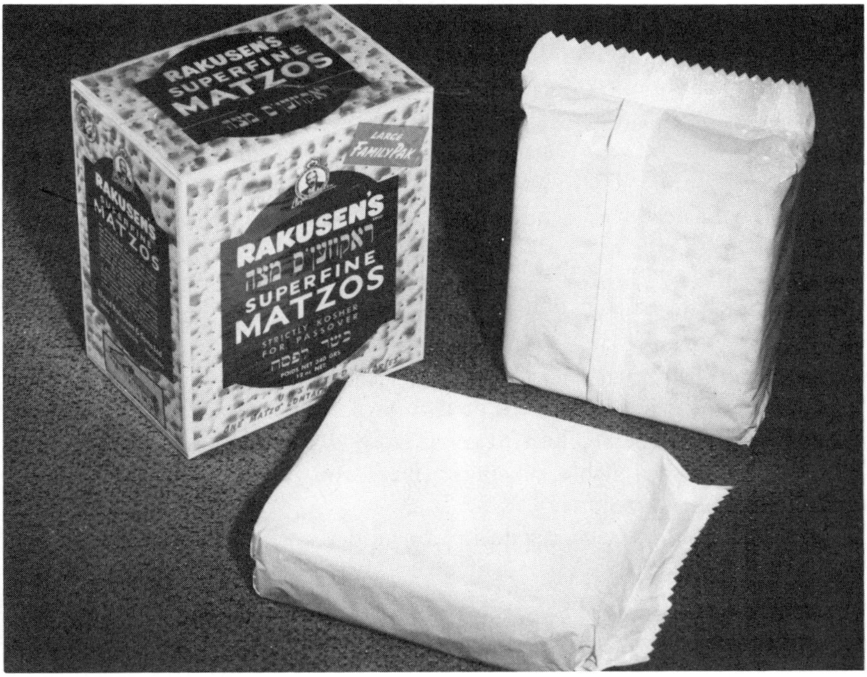

Courtesy of British Cellophane, Ltd.

FIG. 14.3. THIS MATZOS PACKAGE INDICATES PASSOVER ACCEPTABILITY OF CONTENTS
To attain this labeling extremely strict care is taken in manufacture and packaging of food. The white opaque "Cellophane" film conveys the message of purity and hygienic preparation. Two inner packs insure freshness.

The term "Kosher for Passover" means that there is no leaven in the food product. Due to religious regulations, it is strictly forbidden to eat leaven during the Passover season. Labels are attached to the package denoting Passover acceptibility. These are often pressure sensitive labels which can be removed after the Passover season.

The Interpretations of Kosher Laws.—Since there is no authoritative Jewish hierarchy in the United States, an orthodox Rabbi's interpretation of the kosher laws is binding. It is important to note that not all Rabbis are qualified to judge the acceptability of a product. Judaism is divided into three basic sects: Orthodox, Conservative and Reform. All Orthodox Rabbis subscribe to kosher laws and may interpret their tenets. In seeking approval, only Orthodox Rabbis should be consulted.

In the manufacture of food packaging materials and the packaging and processing of foods, a large number of chemicals are used. Each one must comply with kashruth. Cleaning materials must be free from any ingredients derived from animals forbidden by Jewish Law. Emulsifiers, stabilizers, defoamers and lubricants should be obtained from vegetable and/or synthetic sources. Glycerol widely used in cellophane manufacture, must be obtained from vegetables and not animals.

Although all natural fruits and vegetables are inherently kosher, processed fruits and vegetables are subject to investigation. Fresh scaly fish is inherently acceptable but not frozen fish. The preservative solution used for freezing must not have been used for freezing any nonkosher product. The freezing of poultry involves an analogous situation. Kosher poultry must not be processed with any chemicals or baths used for the nonkosher variety.

Aluminum foil must be rolled using acceptable lubricants. This applies also to the lubricants used in container forming. Waxes, "fatty acids," esters, amines, and alcohols derived from tallow are forbidden. No animal glues are acceptable. All plasticizers and flexible film additives are subject to investigation.

Acceptability is based on definition and interpretation of Rabbinical law. Many issues are decided after careful study and each product is a specific case. A layman cannot decide whether a product adheres to the Jewish law. Only a fully trained Orthodox Rabbi is capable of arriving at an answer. The steps involved are simple and the results may be significant. The difficulties in relating a modern product to ancient laws may be illustrated by an interesting example.

Most Orthodox Rabbis considered nonkosher any cheese made with rennet. Since rennet is derived from the stomach of a calf and nonkosher cattle are used, they reasoned, the product is not kosher. In order to process rennet, the stomach of a young calf is cleaned and stored for a considerable amount of time. Chemical additives are then added to yield a fine powder. Rennet is also used in pudding type desserts. In Biblical days this process was unknown. A study made by a Rabbi determined that rennet is kosher, even if derived from nonkosher cattle. His reference was a quotation from ancient law stating: "The skin of the stomach is sometimes salted and dried so that it is rendered like wood, and if then it is filled with milk it is kosher, because after the stomach has been dried, it is just like wood, and does not contain any moisture of meat."

How to Get Approval.—There are three basic organizations offering converters a method of attaining kosher approval.

The Union of Orthodox Jewish Congregations of America created a national service in 1925. At the request of the interested company, a completed application detailing specific ingredients is submitted to the

UOJCA central office. After an initial investigation of the product, the UOJCA sends rabbinical investigators into the plant. Supervision consists of observation of day-to-day practice and plant conditions. Packaging and processing equipment are carefully checked and investigated in order to determine compatibility with the kosher laws. All data submitted and observed are fully confidential. If the laws of kashruth have been fulfilled, the company is permitted to use a (U) symbol on the specific products. The entire study involves a fee based on the total cost to the UOJCA.

The Joseph Jacobs Organization in New York City offers another supervisory service. An investigation and visit is made by an Orthodox Rabbi in order to determine adherence to kashruth. Approval is then given to display a "K" on the product. All information obtained is confidential and the fee is based on the total cost.

In Massachusetts, the Vaad Harabonium (Rabbinical Council) authorizes the use of a "VH" symbol. Certification is granted to products only if their manufacture remains under the strict supervision of the Vaad Harabonium. Companies are required to submit a detailed list of their ingredients and an investigation is made. Once a product has been cleared, the manufacturing plant and equipment are cleaned and sterilized. Manufacture then starts under the supervision of an Orthodox Rabbi assigned to the plant's production and quality control departments. As in the prior cases, all information is confidential and a rated fee is paid.

In addition to the above three organizations, an agreement may be made with an individual Orthodox Rabbi for approval. The degree of supervision is wholly dependent on the Rabbi. Approval will allow for his signature on the package vouching for its kosher purity.

When a product has been granted approval for adherence to the kosher laws, the three organizations provide a variety of marketing and counseling services for the product. The potential increase in sales from obtaining approval may be substantial.

Other Rules

In contrast to the rules of Orthodox Judaism, no other religion has formulated a set of rules for food packaging in the United States. This may be due to the lack of immigrants from other than Christian lands. Until 1966, members of the Roman Catholic faith could not consume meat on Fridays, but this did not prohibit the use of packaging materials containing glycerine from animals.[4]

---

[4]Catholics in Spanish-speaking countries had centuries ago been relieved by Church dispensation from the obligation to abstain from meat on Friday and during Lent.

Kosher meat products are bought by Seventh-day Adventists and Seventh-day Baptists who observe the Sabbath day on Saturday. They do not eat any products which may contain an ingredient of pork or lard. Thus, religious precepts differ not only among faiths but also in the degree of strictness of adherence to the precepts.

Although not formalized in the United States, members of the Hindu and Islamic faith follow strict rules as to food and indirectly to packaging. Many Moslems purchase their meat from kosher butchers to ensure an acceptable source of supply. The Sepoy Mutiny in India of 1857 resulted from the introduction of the new Enfield rifle by the British. The cartridges used had to be bitten open before loading. In order to protect the metal from the hot, humid Indian climate, the cartridges were coated with grease. For a caste Hindu to bite on the fat of the sacred cow was an inconceivable sin; to a Muslim, the fat of the unclean pig was insufferable pollution. Though a deep social unease had made the Indian Mutiny inevitable, it was false rumors about the greased cartridges which set it in motion.

Regulations Affecting the Packaging of Margarine

The manufacture and sale of margarine in its early days was undertaken by so many that inevitably some very poor quality products were marketed. It was also easy for the unscrupulous to sell the cheaper substitute at the higher price of butter. To insure product identification and reasonably good quality, nearly every manufacturing country passed legislation regulating the composition and the packaging of margarine. In some instances powerful dairy industries influenced repressive legislation which prohibited the sale of margarine. Most of the prohibitive legislation has since been repealed.

It is not within the scope of this book to consider all legislation relating to the composition of margarines, but we are interested in those laws affecting the packaging of the product. Austria, parts of Australia, Argentina, Belgium, France, West Germany, Norway and Switzerland all require retail packages of margarine to be cubical in shape. Portugal says the retail package cannot exceed 1 kg, Spain says it must be more than 100 g, South Africa limits it to ½ lb. Most countries require special markings. Some require red bands on bulk packaging. Nearly all require prominent labeling with large letters showing the name margarine. In France, this must be on four sides. It will require special legislation to change these rules before newer margarine packages can be adopted.

REFERENCES

ANON. 1955. Customs and Traditions of Israel. Joseph Jacobs Organization, New York.

ANON. 1970. Food wrappings: A legislative dilemma. Food and Cosmetic Toxicol. *8* (5) 568-572 (October).

ANON. 1971A. Japanese authorities move against PE bottles. Med. Plast. Int. *1* (1) 16.

ANON. 1971B. Food packaging and health; migration and legislation. Inst. Packag. (Dec.).

ANON. 1979A. A conversation with FDA's Stuart M. Pape. Food Drug Packag. *40* (1) 22–25, 28.

ANON. 1979B. Standardisation of produce transport. Paperboard Packag. *64* (1) 104–106, 108.

ATTMORE, M. 1978. Regs. should assess risk-benefit ratio, says FDA official. Food Drug Packag. *38* (10) 1 (May).

EGAN, H. and HUBBARD, A.W. 1971. Food additives specifications. Oherm Ind. No. *42* , 1181–1186 (October).

ESTEVEZ, J.M.J. 1967. Toxicity in plastics: The problem as typified by plastic packages. Plastics Inst. Trans. J. *35*, (116) 448.

GOLDBERG, C. 1971. Trace chemicals contaminants in food: potential for harm. Food Cosmetic Toxicol. *9* (1) 65-80 (Feb.).

HIGGINS, J.C. 1968. Food packaging regulations. Flexible Packg. Conf. Ppr. No. *5*. Brit. Bus. Publ., London.

JONES, A. 1967. Toxicity, plastics, and packaging. Rubber Plastic Age *48* (4) 340.

KALAN, L. 1970. Toxicity tests and the food consumer. New Sci. *47*, 428–431 (Aug.).

KATZ, M.C. 1968. Deception and Fraud with a Kosher Front. Private publication by the author. Endicott, New York.

KORFF, S.L. 1966. The Jewish dietary code. Food Technol. *20* (7) 76–78.

MAHADEVIAH, M. 1970. The alternatives to sanitary tin can for packaging food products. J. Food Sci. Technol. 7 (1) 3–7 (March).

SACHAROW, S. 1968. Plastic processors and the U.S. Food and Drug Administration. Aust. Plastics Rubber J. *23*, 31–35.

SACHAROW, S. 1970. FDA Regulations that affect food packaging. Food Prod. Dev. *4* (7) 58, 60, 62.

SACHAROW, S. 1971. Other federal regulations affecting food. Food Prod. Dev. *4* (8) 48, 50.

Index

Other AVI Books

CONSUMER BEHAVIOR
Redman

FOOD ANALYSIS: THEORY AND PRACTICE
Pomeranz and Meloan Revised Edition

FOOD AND THE CONSUMER
Kramer

FOOD DEHYDRATION
van Arsdel et al. Second Edition Two Volumes

FOOD MICROBIOLOGY: PUBLIC HEALTH AND SPOILAGE ASPECTS
deFigueiredo and Splittstoesser

FOOD QUALITY ASSURANCE
Gould

FUNDAMENTALS OF FOOD CANNING TECHNOLOGY
Jackson and Shinn

FUNDAMENTALS OF FOOD FREEZING
Desrosier and Tressler

HANDBOOK OF PACKAGE MATERIALS
Sacharow

LABORATORY MANUAL FOR FOOD CANNERS AND PROCESSORS
Natl. Food Processors Assoc. Third Edition Two Volumes

NUTRITIONAL EVALUATION OF FOOD PROCESSING
Harris and Karmas Second Edition

PACKAGE PRODUCTION MANAGEMENT
Raphael and Olsson Second Edition

PACKAGING REGULATIONS
Sacharow

QUALITY CONTROL FOR THE FOOD INDUSTRY
Kramer and Twigg Third Edition

THE FREEZING PRESERVATION OF FOODS
Tressler et al. Fourth Edition Four Volumes

THE TECHNOLOGY OF FOOD PRESERVATION
Desrosier and Desrosier Fourth Edition